Filter Troubleshooting and Design Handbook

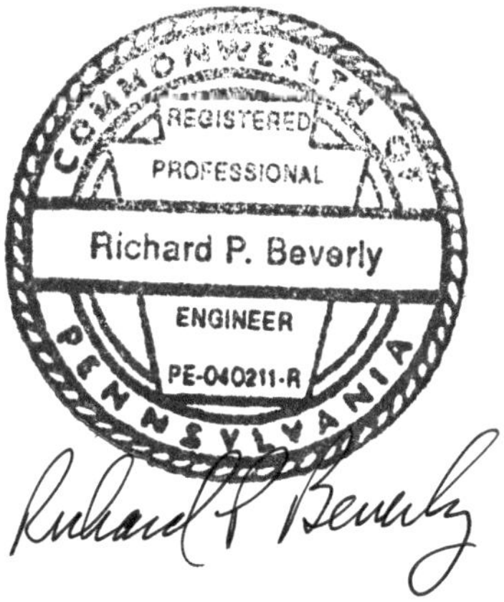

Filter Troubleshooting and Design Handbook

English Language
First Edition

Richard P. (Phil) Beverly, P.E., W.R.E., W.T. III

American Water Works Association

Science and Technology

AWWA unites the drinking water community by developing and distributing authoritative scientific and technological knowledge. Through its members, AWWA develops industry standards for products and processes that advance public health and safety. AWWA also provides quality improvement programs for water and wastewater utilities.

Printed in the United States of America.

Project Manager/Editor: Melissa Christensen
Produced by: Glacier Publishing Services, Inc.

Disclaimer

This book is provided for informational purposes only, with the understanding that the publisher and author are not thereby engaged in rendering engineering or other professional services. The author and publisher make no claim as to the accuracy of the book's contents, or their applicability to any particular circumstance. The author and publisher accept no liability to any person for the information or advice provided in this book, or for loss or damages incurred by any person as a result of reliance on its contents. The reader is urged to consult with an appropriate licensed professional before taking any action or making any interpretation that is within the realm of a licensed professional practice.

Library of Congress Cataloging-in-Publication Data has been applied for.

ISBN 1-58321-349-X

American Water Works Association

6666 West Quincy Avenue
Denver, CO 80235-3098
303.794.7711

Contents

Acknowledgments

The author wishes to thank the staff of Lee Engineering, Inc., for helping make this book possible.

Technical Editing:

The Lee Engineering Water Service Group:

- James R. Shaver, P.E.
- Michael B. Warren, P.E.
- Edward P. Hodges, P.E.
- Andrew Szatkowski, P.E.
- Kenneth Condit, P.E.
- Jeffrey G. Marchioro, P.E.

Peer Review:

Thanks also to the following peer review group:

- Kevin Batridge, Veolia Water, Wilsonville, Oregon, Water Treatment Plant
- Mark Cage, South Fork Water Board, Oregon City, Oregon
- Rob Cummings, Clackamas River Water, Clackamas, Oregon
- Chuck Kingston, Joint Water Commission Water Treatment Plant, Hillsboro, Oregon
- Henry Thomson, Lake Oswego, Oregon, Water Treatment Plant
- John Adams, Canby Utility Board, Canby, Oregon
- Elizabeth Edwards, Canby Utility Board, Canby, Oregon

Graphics:

- Mike Grose, Technician
- Daniel Lee, Assistant

Word Processing and Organization:

- Nancy Jelinek

Author

Mr. Phil Beverly has a B.S. degree in Engineering from the Virginia Polytechnic Institute. He has professional licenses in civil, mechanical, and environmental engineering and is also a certified Water Rights Examiner and a Level 3 Water Treatment Plant Operator with the filter endorsement in the State of Oregon. He has served on the AWWA B604 Standards Council Subcommittee for Granular Activated Carbon as Filter Media and is currently serving on the regional AWWA Seminars Committee. He was on the Environmental Registry Who's Who—1992.

Mr. Beverly has over 30 years of experience in troubleshooting filters in both water and wastewater applications. He has helped rehabilitate approximately 200 filter plants and has been retained as a special consultant to resolve problems in numerous filtration facilities for water and water treatment systems across the country.

His background includes working for several water/wastewater equipment suppliers and several consulting engineering companies. As a part of that experience, he operated a research lab for an equipment supplier, which resulted in his obtaining three patents on filter underdrain improvements.

He is currently Vice President of Lee Engineering, Inc., in Oregon City, Oregon. To contact the author, call 1-800-848-9731.

CHAPTER 1

Introduction

The purpose of this handbook is to provide the necessary information for troubleshooting granular media filters, including conventional gravity plants, package plants, traveling bridge filters, slow sand filters, and pressure filters. Traveling bridge and slow sand filters are considered to be specialty applications of conventional gravity plants and receive little mention in this handbook.

Although the figures in the first nine chapters primarily illustrate conventional gravity filters, most of the technical issues are applicable for any type filter. Chapter 10 discusses issues specific to pressure filters.

Typical Filter Operation

There are many filter plants operating on both potable water and wastewater. Conventional granular media filters are available in both package plants and concrete plants. The filters can also be provided with a variety of underdrains and pretreatment processes. A typical filter in filtration mode is shown in Figure 1-1. When the media becomes fouled with accumulated solids, the flow is reversed. The filter is backwashed, and the solids are flushed to waste, as shown in Figure 1-2.

The filters are normally provided with either surface wash or air scour to assist in cleaning the media during backwash. In either case, each filter will often have six valves, as shown in Figures 1-1 and 1-2. It should be noted that these figures illustrate the use of air scour.

Once the filter is clean, the media is reconditioned in filter-to-waste mode, as shown in Figure 1-3. Following this, the filter may be returned to service.

Process-related issues can result in poor filter performance. However, most of the physical upsets or "mechanical" problems that occur are related to the backwash process.

Many sources are available regarding the process issues involved in operating a filter, such as coagulation and flocculation. However, the mechanics of conventional filters

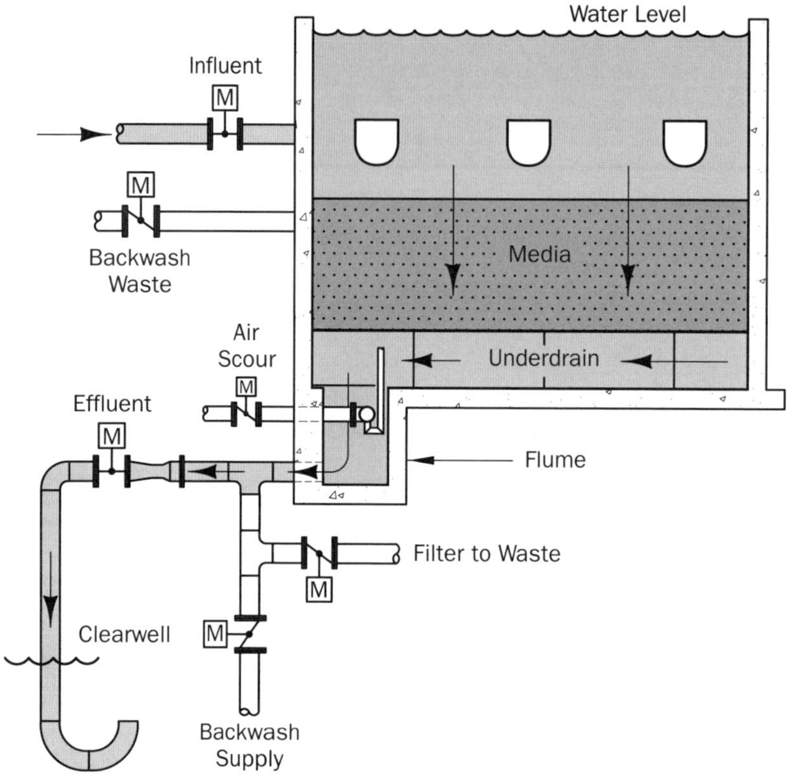

FIGURE 1-1. Typical filter in filtration mode

and the related equipment is not widely understood. The goal of this handbook is to provide background information on the mechanics of filters and tools for the operator and design consultant to use in solving and preventing filter problems.

Process Issues

Poor filter performance can be the result of process issues, including inadequate pretreatment, changing raw water conditions, excessive use of chemicals, and hydraulic overloading. Filter performance can be monitored using effluent turbidity and particle counting. It is also important to monitor historical records, such as changes in clean head loss over time, changes in the rate of head loss development, and whether or not filter runs are terminated on head loss or effluent quality. If a filter is performing poorly, an investigation should be made to optimize the process issues first.

Many sources are available regarding the optimization of filtration processes. However, optimization is only briefly mentioned in this handbook. If filter performance cannot be resolved by optimizing process issues, an investigation should be made into the mechanical components of the plant, which is the primary subject of this handbook.

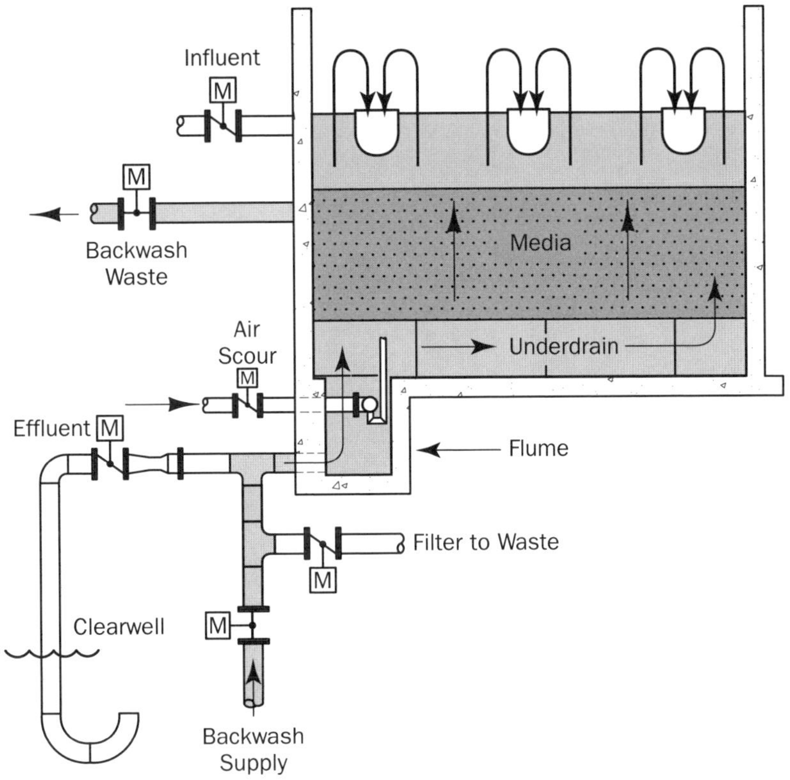

FIGURE 1-2. Typical filter in backwash mode

Mechanical and Material Issues

Most filter failures are caused by equipment and components outside the filter box, such as pretreatment efficiency; uncontrolled air, surface wash, or air scour systems not operating properly; quick opening or improperly sized valves; and control sequencing problems. The most common and disruptive problems are the result of hydraulic shock and uncontrolled air in the backwash cycle. A brief discussion of filter efficiency and pretreatment is included in this handbook. The primary focus of this handbook is on the mechanical and material components of the filter itself and its auxiliary equipment.

Troubleshooting

When troubleshooting a filter, it is important for the investigator to understand the significance of what they see and to know what caused the problem. A practical working knowledge of filter media, underdrain hydraulics, and plant operations is required to adequately troubleshoot a filter. An investigator with that knowledge will better understand what it will take to resolve the problem.

If a major filter problem occurs, the troubleshooting effort can be conducted by either the plant operators or an outside consultant. However, it is recommended that the plant operators be as knowledgeable as possible on the equipment they use. The

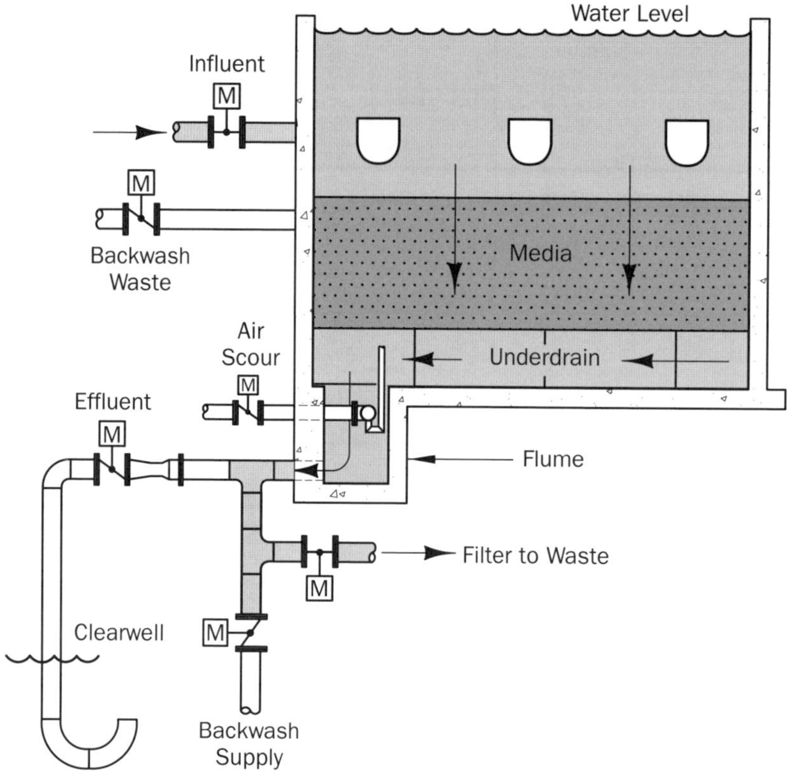

FIGURE 1-3. Typical filter in filter-to-waste mode

material contained in this handbook is intended for use by both plant operators and consultants.

Plant Operators

Plant operators do not have to be conversant with all design aspects of a filter plant to do troubleshooting. However, it is important for them to be able to recognize the issues and know what they can or cannot resolve with the equipment they have.

Consultants

Consultants with troubleshooting knowledge will be better able to identify typical problems, such as uncontrolled air and hydraulic shock, and to recognize specific issues related to filter hydraulics, selection of filter media, and underdrain design.

How to Use This Handbook

Chapters 2 through 8 contain the bulk of information needed for troubleshooting. These individual chapters are intended to be relatively independent. To resolve specific issues, the investigator may be able to refer to the pertinent chapters for the necessary information.

An experienced investigator may be able to identify the problems in a short period of time. However, for a complete investigation, use the procedures and series of checklists in Chapter 9 for gravity filters. Issues specific to pressure filters are in Chapter 10.

Finally, the maintenance task lists in Chapter 11 should be used once the filter(s) are operating properly.

Summary

The information presented in this handbook is based on the author's personal hands-on experience and research spanning 30 years. It includes a basic understanding of the design and physical aspects of filtration, including driving head, plenum hydraulics, support gravel, filter media, underdrain, optimization of the backwash process, and controls. The intent is to provide the investigator with a reasonable pathway to follow for identification and resolution of filter problems.

CHAPTER 2

Driving Head

Introduction

Filtration is the physical process by which suspended solids are removed from a stream of water (Figure 2-1). For the purposes of this discussion, filtration is assumed to be accomplished by granular media. The removal and accumulation of solids by granular media requires an amount of force, defined as driving head, capable of pushing the water through the filter bed and overcoming the resistance of the media and any accumulated solids. Driving head can be measured in inches or feet of water, or pounds per square inch (psi).

The resistance of the water passing through the media and accumulated solids is referred to as *head loss*, and is measured as the difference in pressure across the filter.

Several types of filters and the driving head relevant to each are discussed in the following sections.

Gravity Filters

In a gravity filter, the driving head is the difference between the water level over the filter media and the clearwell water level, assuming the discharge from the filter is submerged (Figure 2-2). If the discharge from the filter is not submerged (Figure 2-3), an air bubble will be generated that moves to the last horizontal run of pipeline. The effective driving head is reduced to the distance from that point to the water level over the filter.

Pressure Filters

The driving head in a pressure filter is the available upstream pressure minus the discharge pressure (Figure 2-4). A control valve is often provided to control the flow rate out of the filter. In that case, the filter head loss is the upstream pressure minus the pressure in front of the rate control valve. Refer also to Chapter 10 for a more in-depth discussion of pressure filters.

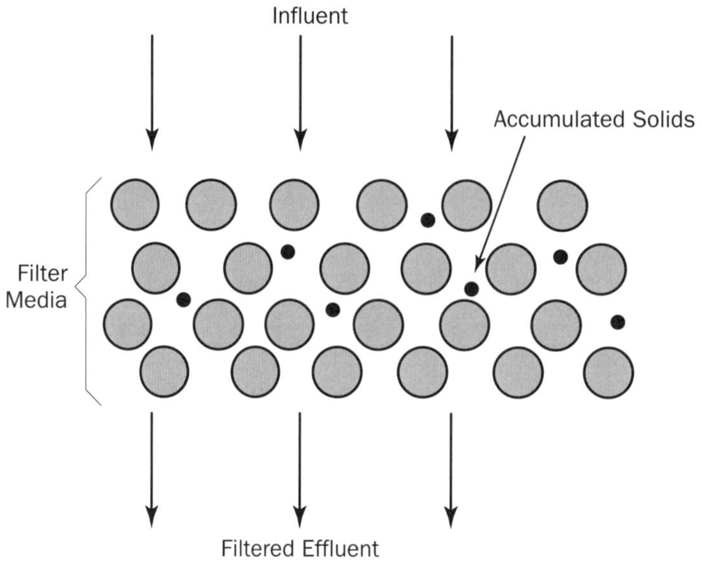

FIGURE 2-1. Granular media filtration

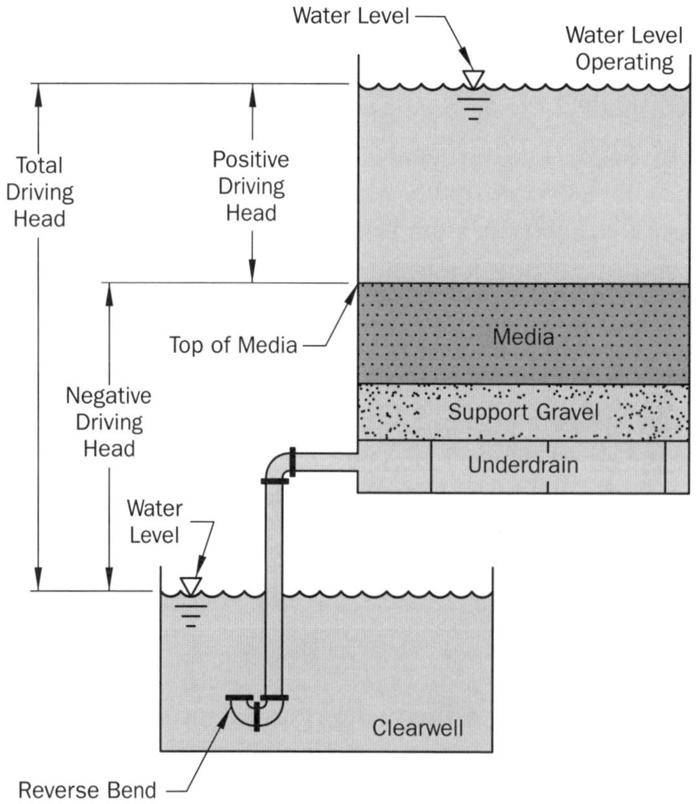

FIGURE 2-2. Total driving head

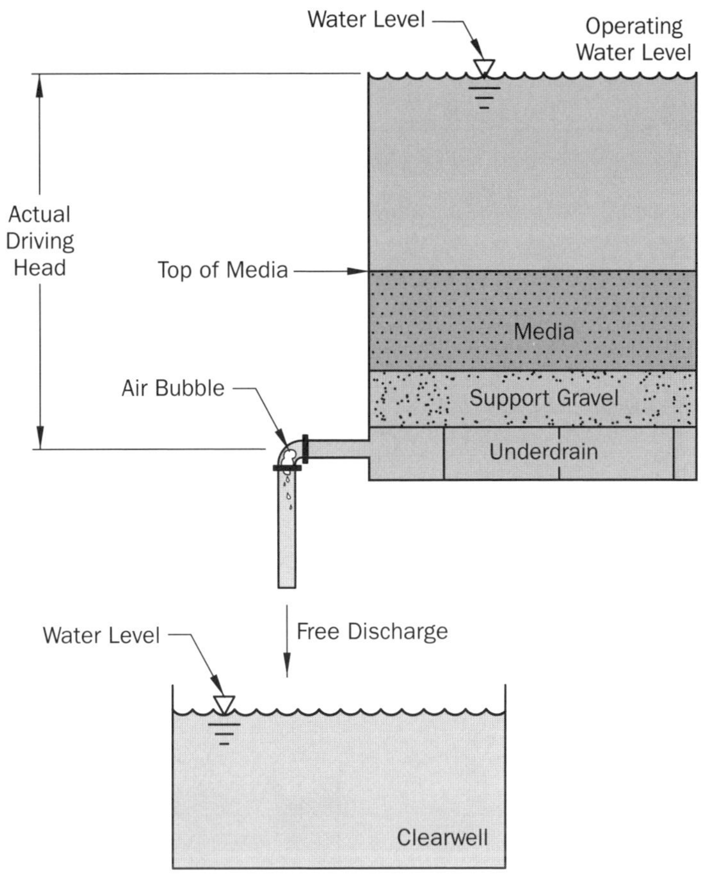

FIGURE 2-3. Free discharge

Slow Sand Filters

Driving head for slow sand filters is calculated in a similar manner as other gravity filters. It is the distance from water level over the media to water level at the point of discharge or air break.

Two different effluent arrangements are common with slow sand filters: reverse bend piping with an air vent (Figure 2-5) or effluent weirs (Figure 2-6). With either design, the intent is to maintain a minimum water level over the media.

When clean, the water level in a slow sand filter will be at the minimum level. As solids are accumulated in the media, the water level will slowly rise. An allowance of 4 ft to 6 ft of water in the filter is recommended for this purpose. When the maximum water level is reached, the filter should be taken off-line for cleaning.

Clean Bed and Operational Head Loss

In conventional filters, driving head has two components: clean bed head loss and operational head loss, each of which is discussed in the following sections.

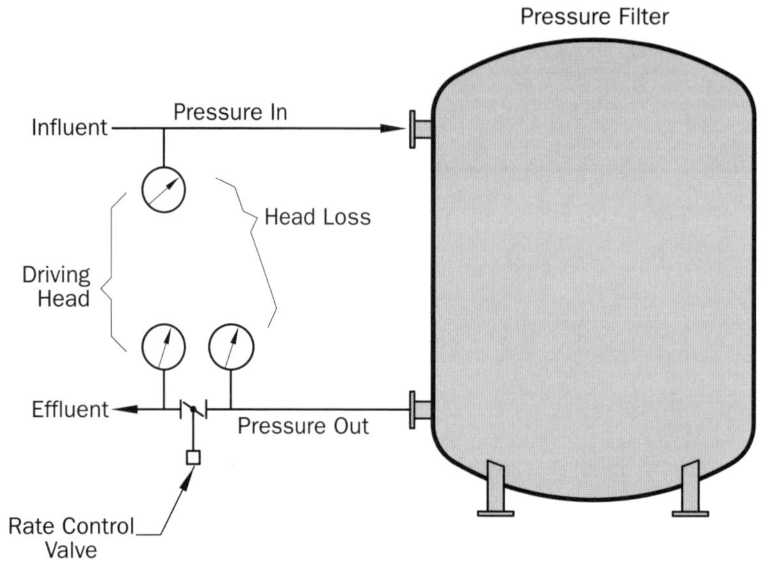

FIGURE 2-4. Pressure filter driving head

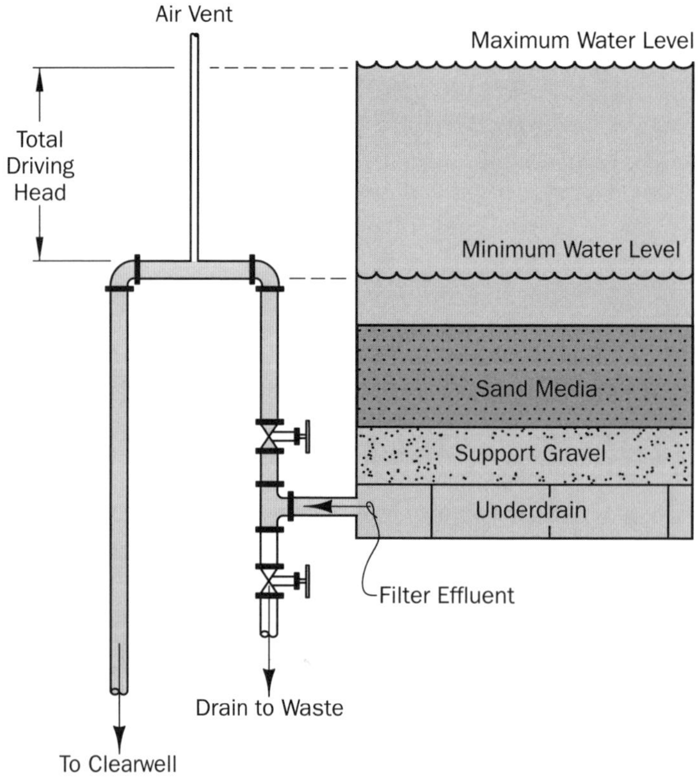

FIGURE 2-5. Slow sand filter with reverse bend piping

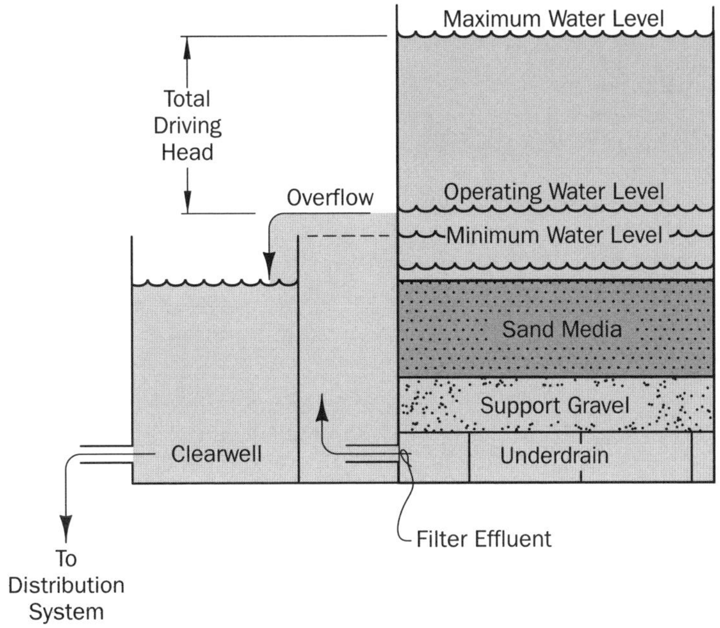

FIGURE 2-6. Slow sand filter with effluent weir

Clean Bed Head Loss

For the purposes of this discussion, the clean bed head loss is assumed to include the filter media when clean, the underdrain, and all pipe fittings and appurtenances between the filter and the clearwell, including rate control valves and flow meters. For a typical 30-in. dual-media filter, the clean bed head loss during filtration can be approximated as shown in Table 2-1.

It is important to note that the head loss during backwash will be much different (refer to Chapter 7).

Clean bed head loss will obviously vary with the type and depth of filter media, and the type and number of fittings and appurtenances in the pipeline between the filter and the clearwell. For example, deep bed monomedia or other custom designed media configurations can have a significantly higher filtration head loss. However, for many 30-in. media configurations, an estimated clean bed head loss of 4 ft is a close approximation.

Operational Head Loss

The operational head loss is the amount of force over the clean bed head loss that is required to overcome the resistance of the accumulation of solids within the filter media. The operational head loss is typically between 8 ft and 10 ft of water pressure for most treatment plants. The ability to actually develop that much head loss in the media depends on the filter configuration, the floc strength, and the condition of the other filter components.

TABLE 2-1: Clean bed head loss

Filter Component	Approximate Clean Bed Head Loss During Filtration
18 in. of anthracite	9 in. to 12 in.
12 in. of filter sand	12 in.
Support gravel	Minimal
Filter underdrain	Minimal
Filter rate control valve	12 in.
Flow meter	6 in.
Piping	6 in.
Total:	Assume 4 ft of clean bed head loss

A lower operational head loss may result in shorter filter runs, depending on the pretreatment efficiency. If very high quality water is delivered to the filters by the pretreatment processes, longer filter runs may be possible with low operational head losses.

The amount of pressure available for operational head loss is a factor of the filter configuration for a conventional gravity filter and the available line pressure for a pressure filter. In many pressure filters, higher operational pressures are occasionally used because the pressure is often available. However, filtration head losses higher than 8 ft to 10 ft are not recommended because compaction of the media and the formation of mudballs can result.

Total Desired Driving Head

The total desired driving head is the sum of the clean bed head loss plus the operational head loss.

$$\text{Clean Bed HDL} + \text{Operational HDL} = \text{Desired Driving Head (Minimum)}$$
$$4 \text{ ft} + 8 \text{ ft} = 12 \text{ ft}$$

NOTE: Head loss (HDL) is the pressure drop experienced by the flow of water through a filter.

Therefore, the distance from the operational water level of the filter to the *maximum* clearwell level should be 12 ft or greater. The maximum clearwell level will result in the lowest available driving head. As the clearwell level drops during normal operation, the available driving head will increase. However, it is recommended that the design provide the minimum driving head required at the maximum clearwell level. It is also recommended that calculations be made for the requirements of each specific application, taking into account the actual media load, pipe fittings, and other appurtenances. In some cases, pilot testing may be necessary to determine the actual head loss.

If the total driving head is less than 12 ft, it could result in short filter runs, unless the pretreatment is highly efficient. The goal for any filter should be to achieve at least 24 hr of operation before backwashing is required. A filter run of 12 hr would probably be the minimum acceptable, although not desirable. If the filter runs are less than that, an investigation should be made to evaluate the plant operation and determine if the pretreatment can be optimized, or if modifications to the filter are required. Less than 8 hr of operation usually results in the use of excessive backwash water.

Fluctuating Clearwell Level

The level in the clearwell often fluctuates as a result of system demand. Because of that, simply extending the filter discharge pipe below normal level may not be adequate. A reverse bend is normally installed on the discharge piping below water level in the clearwell. The intent is for the reverse bend to function like a p-trap to ensure a water seal on the pipeline (Figure 2-2). Air cannot then enter the pipe at any water level, and the driving head cannot be disrupted.

Another aspect of fluctuating clearwell level is that the driving head will change with the level. As the clearwell level decreases, the total driving head will increase and vice versa (Figure 2-7). Depending on the method of control used, the effluent rate on the filter could change; with some control types, the filter rate may increase. With other modes of control, the effluent rate valve on each filter may modulate to maintain a constant flow rate from each filter. Chapter 8 discusses control philosophy.

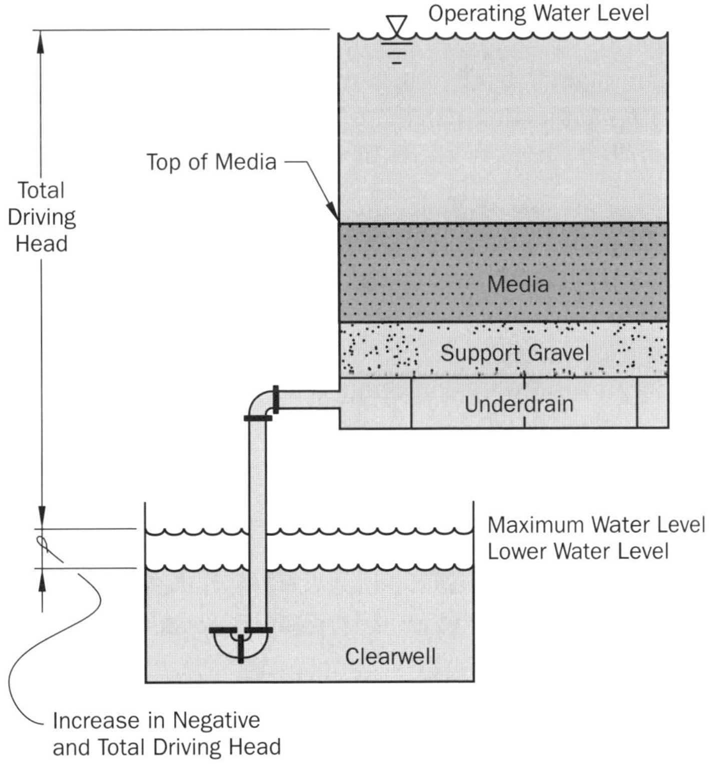

FIGURE 2-7. Increase in negative and total driving head

Positive Driving Head Versus Negative Pressure

Assuming a total driving head of 12 ft, using the previous equation, it makes a great deal of difference where the filter media is located within that span. It is common for a significant amount of operational head loss to be generated in the top few inches of the media. The bulk of the incoming solids tends to be retained on or near the surface, where the raw water first comes in contact with the media. Significant head loss can also be generated further down in the media, at the interface between layers in dual- or multimedia bed designs. However, for the purposes of this discussion, it is assumed that all the operational head loss occurs at the media surface, which is a conservative approach. For exact information, pilot studies may be required.

The positive driving head is the distance from the operational filter water level to the surface of the media. The pressure from the height of water pushes on the filter media, while the balance of driving head is negative pressure, which "pulls" the water through the media (Figure 2-2). Depending on how much positive driving head is available, the actual pressure within the filter can go negative as the filter run progresses. Figure 2-8 illustrates a typical progression of pressure changes within the media during the filter run, assuming 6 ft of positive driving head. It should be noted that actual conditions may change.

In Figure 2-8, static conditions with no flow are represented as a straight line from the water surface to the clearwell, assuming a submerged discharge. As water begins to pass through a filter, the media generates head loss. The approximate clean bed head loss is represented by a curve that bends to the left within the media before straightening out and resuming a direction parallel to the static pressure line. At normal filtration rates (3–5 gpm/ft^2), support gravel (if used) and the underdrain are assumed to have negligible head loss. When the water enters the effluent piping, additional head losses are generated by the rate control valve, flow meter, and pipe fittings.

As solids are accumulated and the head loss increases, the pressure line is gradually extended further to the left, until it reaches a zero pressure condition at approximately 6 ft of head loss. If the filter is run to an operational head loss of 8 ft above the clean bed head loss, a negative pressure condition will occur over the last 2 ft of head loss development.

Many types of water supplies that need filtration have dissolved air or gas in the water. Examples are wastewater effluent, well water, and raw water supplies from cascading pipelines or rivers. Negative pressure has the effect of pulling dissolved air or gas out of solution. The air or gas forms small bubbles that attach themselves to the filter media. The end result is the rapid buildup of artificial head loss, resulting in shorter filter runs.

Air problems from negative pressure are relatively easy to detect. When the filter goes into backwash, numerous little air bubbles will be visible uniformly across the filter. Air problems of this sort do not damage the filter gravel, because it occurs higher in the media. However, the little air bubbles will tend to carry media out of the filter in the backwash water.

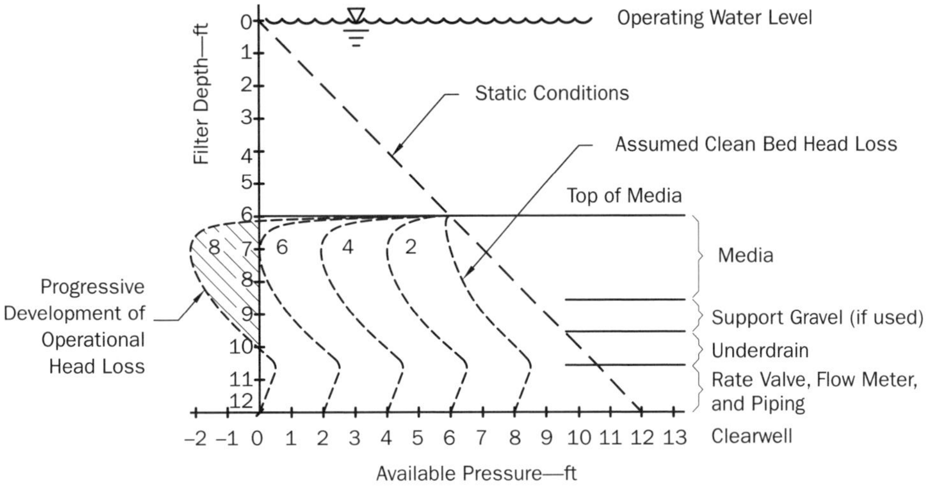

FIGURE 2-8. Negative pressure development with 6 ft of positive driving head

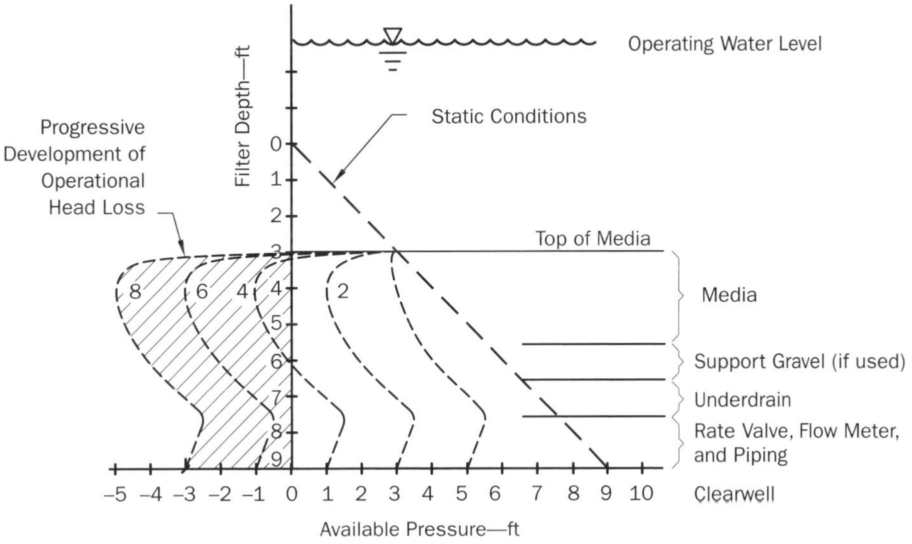

FIGURE 2-9. Negative pressure development with 3 ft of positive driving head

Figure 2-9 shows the impact of having only 3 ft of positive driving head. With that small amount of positive driving head, the negative pressure experienced is much greater and will exist for most of the filter run. A relatively low positive driving head is typical of many package plants and some older concrete filters. Negative pressure is not a problem that normally occurs in pressure filters.

Many filters do not experience problems with dissolved air or gas, and less driving head may be acceptable. However, when dissolved air or gas is a problem, higher positive driving head is necessary. For those cases, the designer should carefully review the application before the plant is constructed. If dissolved air or gas is a problem in an

Problem	Probable Cause	Proposed Solution
Short filter runs	• High clean bed head loss	• Skim media
	• Air binding (entrained air) caused by negative head	• Backwash more frequently
	• Low operational driving head	• Extend effluent pipe deeper into clearwell and add a reverse bend. Lower clearwell operating level if possible.
	• Vacuum or air release valves	• Remove air/vacuum release valves
Poor effluent quality	• Fluctuating effluent clearwell level	• Revise controls if possible to stabilize clearwell operation

FIGURE 2-10. Driving head troubleshooting guide

existing filter, it will be necessary to backwash them more frequently to avoid negative pressure conditions, with the associated loss of media.

Air/Vacuum Release Valves

Most filters experience some amount of negative pressure during part of the filter run. As a result, the effluent piping must be sealed between the filter and the clearwell. Any leaks allowing air into the piping will reduce the driving head in a manner similar to a free discharge, as shown in Figure 2-3.

In the same way, air or vacuum release valves will allow air in when the pressure in the effluent piping is low or negative. As a result, air or vacuum release valves should never be installed on filter effluent piping.

Summary

The total driving head available is an important factor in determining how well a filter will operate. The amount of positive versus negative pressure is equally as important.

Where possible, gravity filters should be designed with a positive driving head of at least 6 ft. Including freeboard and allowances for media, support gravel and underdrain will result in a filter that is at least 12 ft deep, not including distribution flumes. Deep filters will normally provide better service by reducing or eliminating problems with dissolved air or gas.

Shallow filters with less positive driving head, such as package plants, are acceptable for use on many raw waters. However, the designers should pay particular attention to the raw water quality in those cases, as the negative driving head will be greater. On the other hand, operational head loss in pressure filters should be limited to approximately 12 ft to prevent mudballing and potential damage to some underdrain types.

Figure 2-10 is a driving head troubleshooting guide.

CHAPTER 3

Plenum/Flume Hydraulics

Introduction

During filtration, the underdrain collects the filtered effluent and discharges it to a common flume or plenum, which in turn routes the water out of the filter. During backwash mode, the flow is reversed and water enters the flume or plenum and is directed to the underdrain.

The intent of this chapter is to illustrate the principles of design, and therefore the techniques used to evaluate the hydraulics of a filter for distribution efficiency. A number of different types of filter failures can be attributed to poor plenum or flume design.

The information contained in this chapter is relatively technical and is primarily intended for the design consultant. Once a filter is built, there is little an operator can do to improve backwash flow distribution in the flume or underdrain. The primary interest for operators may be in learning flume terminology as contained in Figure 3-1, and learning to recognize flume-related problems, as shown in Figures 3-25 and 3-26.

Plenum/Flume

A plenum is a large area or reservoir below the underdrain, as shown in Figure 3-2. A flume is a smaller, more confined conduit, which serves the same function (see Figure 3-3). As used in this handbook, a manifold is a type of flume consisting of a large pipe that is connected to a series of laterals.

Plenums are used with false deck underdrain types, while flumes can be on the end of the filter or below the underdrain in several locations. One type of plenum and several types of flumes will be discussed in this handbook. However, the guidelines and recommendations contained in this handbook apply equally to plenums, flumes, or pipe manifolds.

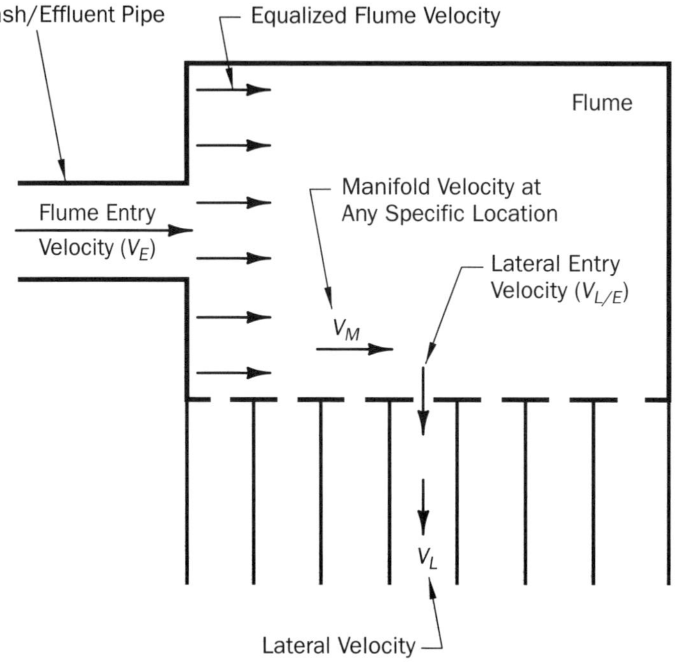

NOTE: The lateral entrance velocity ($V_{L/E}$) is usually different from the velocity in the lateral (V_L).

FIGURE 3-1. Flume terminology

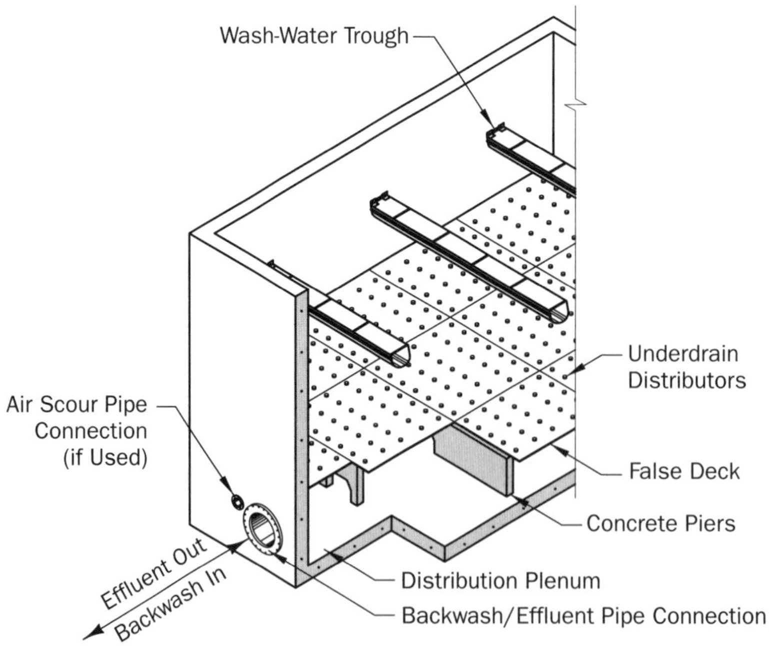

FIGURE 3-2. Typical false deck plenum arrangement

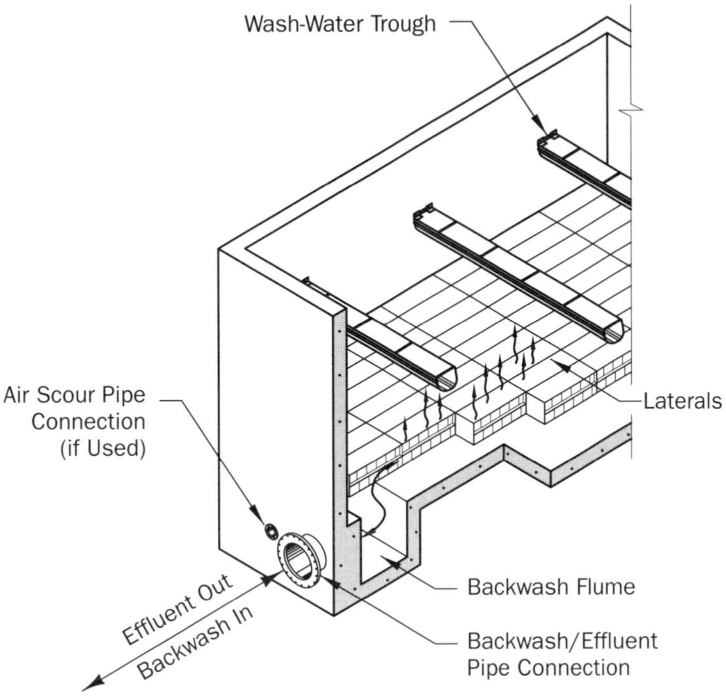

FIGURE 3-3. Typical lateral-type filter arrangement

Maldistribution

The backwash cycle usually dictates the design of the underdrain and the flume or plenum, because the flow rate is higher than that during filtration. The goal of the backwash cycle for any filter is to uniformly distribute the wash water across the entire filter area to ensure proper cleaning of the media. As the backwash flow enters the filter, the velocity of the water is reduced, depending on the size of the flume. The velocity increases as the water turns to enter the various laterals. The relationship of the entry velocity, V_E, the flume or manifold velocity, V_M, and the lateral entrance velocity, $V_{L/E}$, determine the distribution efficiency of the filter design. The distribution efficiency of the wash water across the filter is measured as percent *maldistribution*, with lower desired values.

Maldistribution is the difference in percent in the distribution of backwash water across the filter between two locations in the filter. A testing procedure, along with the supporting calculations, is given later in this chapter. Perfectly uniform distribution of backwash water would result in zero maldistribution. However, in actual practice, the goal is to obtain a maldistribution of 3 to 5 percent.

An underdrain system has a specific maldistribution efficiency. In addition, the plenum or flume contributes a maldistribution component of its own, as shown in Figure 3-4. Each underdrain lateral may produce acceptable maldistribution. However, the flume must distribute the backwash water equally to the various laterals or

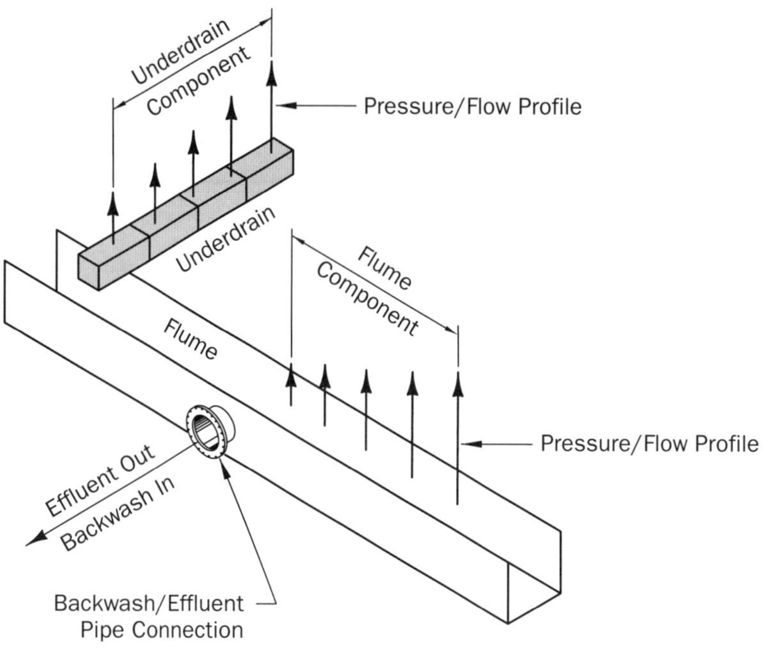

FIGURE 3-4. Components of maldistribution

nozzles. Minimizing the overall maldistribution is, therefore, very important and requires that the plenum or flume be properly designed.

Underdrain Types

Most underdrain types can be classified as either a lateral or a false deck type.

Lateral Types

With a lateral-type underdrain, a series of conduits, or laterals, underneath the filter media collect the water, route it out of the filter, and deliver it to a clearwell for final distribution (Figure 3-3). The laterals are typically close together (12 in. on center or less) and may be separated from the media by support gravel or may consist of, or include, a direct media retention system. The laterals may connect to a central manifold or flume within the filter or to a collection/distribution flume outside the filter box. Lateral underdrain types include:

- Dual parallel lateral configuration in either clay tile or plastic
- Folded plates
- Wedge wire
- Pipe laterals

A description of the various lateral underdrain types is included in Chapter 6. The dual parallel lateral underdrain is used for illustration purposes in this chapter.

False Deck Types

A *false deck* underdrain involves the construction of a structural membrane, or false deck, suspended above the filter floor by a series of piers or short walls, creating a *plenum space* (Figure 3-2). A plenum is also intended to provide an adequate volume to slow and evenly distribute the incoming backwash flow. The support gravel (if used) and the filter media sit on top of the false deck. A series of discrete underdrain fixtures, usually a type of nozzle or strainer, transmits filtered effluent down through the false deck to the plenum. During backwash, the flow is reversed and the water goes up through nozzles in the false deck, and then through the support gravel (if used) and the media. The Wheeler, Hydrocone, and porous plate underdrains are other examples of false deck types.

Maldistribution

The recommended goal for the filter is to have less than ±5 percent overall variation, or maldistribution, in backwash flow. Some underdrains individually distribute the water with less than ±3 percent maldistribution. Therefore, the contribution of the flume or plenum would have to be less than ±2 percent. If the underdrain maldistribution is greater than ±3 percent, the flume or plenum component must be reduced accordingly in order to satisfy the stated goal.

Hudson's dividing flow equation* is used to predict the distribution of water from a flume or manifold to a series of laterals. Using modifications of that equation, it can be shown that the ratio of the flume or manifold velocity, V_M, to the lateral entry velocity, $V_{L/E}$, determines how equally the flow is divided among the various laterals (Figure 3-5). In other words, the velocity ($V_{L/E}$) and therefore the head loss in the laterals should be high enough to control distribution from the flume. Hudson's dividing flow equations suggest that the *cross velocity ratio*, $V_{L/E}/V_M$, should be ≥2.3. It should be noted that the lateral entrance velocity ($V_{L/E}$) is usually different from the velocity within the lateral (V_L), because of restricted openings. One exception is for pipe manifold systems where $V_{L/E}$ and V_L are usually the same.

Higher lateral entrance velocities ($V_{L/E}$) result in higher lateral head losses and higher velocity ratios ($V_{L/E}/V_M$). Under these conditions, the laterals exert more control on distribution, which is desirable. High flume velocities (V_M) result in lower velocity ratios with greater maldistribution within the flume.

The filter size dictates, in part, the lateral velocity, which in turn determines the flume size required for low maldistribution. The only other variable affecting maldistribution is the location of the flume. The impact of the location of the flume on its size is discussed later in this chapter.

* Herbert E. Hudson, Jr., *Water Clarification Process*, Van Nostrand Reinhold Co., 1981.

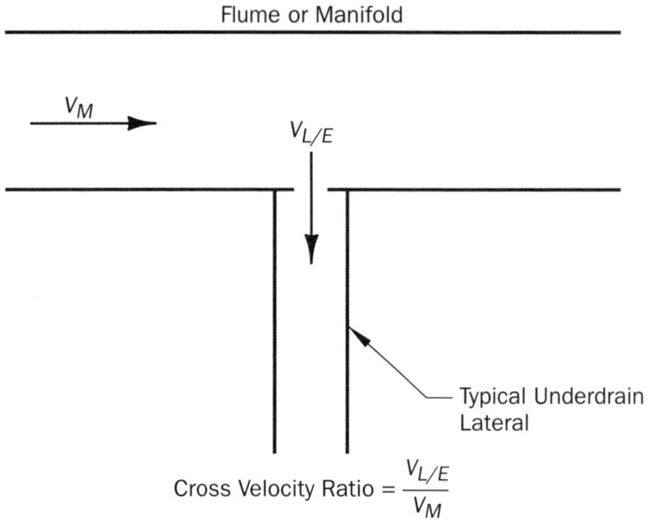

FIGURE 3-5. Flume cross velocity

Design Issues

There are two issues to be considered in order to finalize the design and configuration of a flume: the flume entry velocity head in relation to the underdrain head loss, and the cross velocity ratio ($V_{L/E}/V_M$).

Flume Entry Velocity

Effluent water from the filter exits the flume or plenum typically through a single pipe. The effluent pipe also serves to deliver backwash water into the filter. The size of the effluent/backwash pipe can impact the flume hydraulics significantly. As discussed previously, the backwash flow typically governs the design of the flume and underdrain system because of the high flow rate. Smaller effluent/backwash pipes with high velocities entering the flume (V_E) will result in high maldistribution, even when there is a large flume, as shown in Figure 3-6. The first few laterals can experience lower than desired flow if the flume entry velocity is too high. Maldistribution in the flume is the result of both the flume velocity during backwash (V_M) and the flume entry velocity (V_E).

Bernoulli's equation can be reduced to the following for a flume or manifold application as shown in Figure 3-7:

$$\frac{V_1^2}{2g} + P_1 = \cancel{\frac{V_2^2}{2g}}^{\,0} + P_2 \quad \textbf{(Eq 3-1)}$$

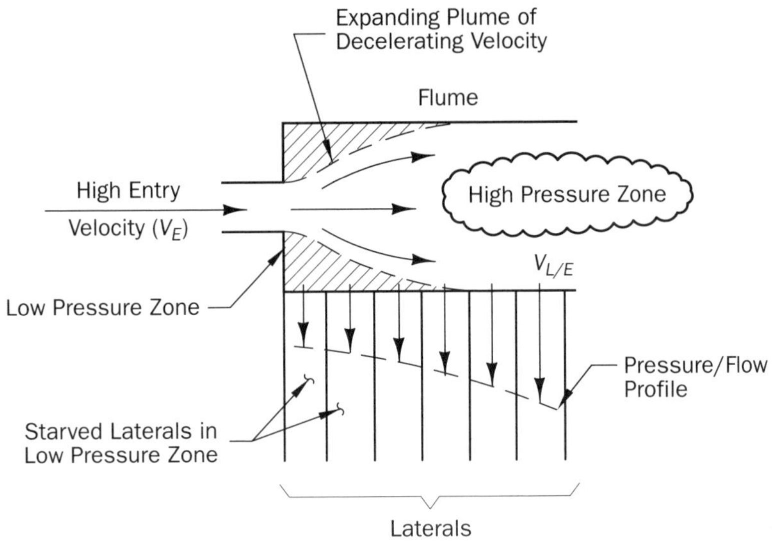

FIGURE 3-6. High flume entry velocity

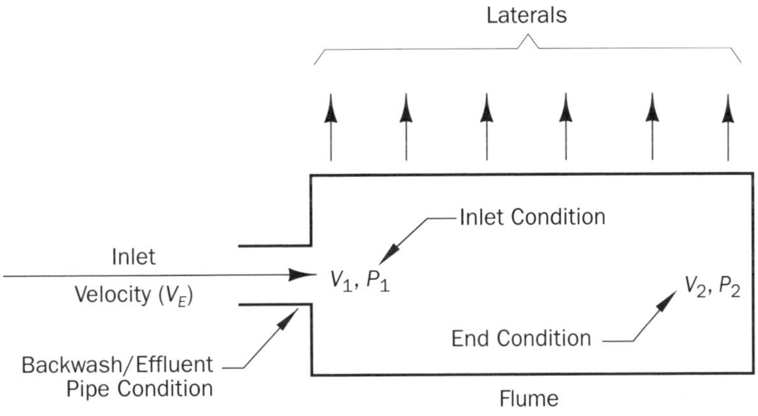

FIGURE 3-7. Flume pressure differential

where V_1 = velocity at the front of the flume

P_1 = static pressure at the front of the flume

V_2 = velocity at the end of the flume

P_2 = static pressure at the end of the flume

g = acceleration of gravity (32.2 ft/s/s)

The flume velocity at the entry point is a positive value, while at the end, after the entrance to the last lateral, all the water has been directed upward out of the flume to the underdrain. Theoretically, the velocity at the end of the flume, V_2, is zero as illustrated in Figure 3-7.

Bernoulli's equation can then be reduced to:

$$\frac{V_1^{\ 2}}{2g} = P_2 - P_1 \qquad \textbf{(Eq 3-2)}$$

In other words, the difference in pressure from one end of the flume to the other, and therefore the entry maldistribution, varies with the velocity head of the incoming water. The lower the flume entry velocity, the lower the maldistribution caused by the flume.

As a practical matter, head loss always controls distribution. In order to distribute the incoming backwash flow evenly across a filter, the underdrain should have a much higher head loss than anything else in the filter. The exception would be the use of coarse nonfluidizing monomedia that has higher head loss than the underdrain at the low backwash rates used for that design.

A recommended rule of thumb, when using fluidizing media, is for the underdrain head loss to be at least five to six times greater than the entry velocity head. The head loss of dual parallel lateral underdrains is typically in the range of 3 ft of water when using backwash rates of 18 to 20 gpm/ft^2. If the underdrain head loss is assumed to be in that range, the entry velocity head and flume velocity head should be 0.5 ft to 0.6 ft or less.

A flume entry velocity of 8 ft per second (ft/sec) would result in a velocity head of 0.99 ft, which is larger than recommended.

$$\frac{V_E^2}{2g} = \frac{(8\ \text{ft/sec})^2}{2 \times 32.2\ \text{ft/sec}^2} = 0.99\ \text{ft} \qquad \textbf{(Eq 3-3)}$$

On the other hand, an entry velocity of 4 ft/sec results in a much lower velocity head, which is acceptable (<0.5 ft or 1/6 of underdrain head loss).

$$\frac{V_E^2}{2g} = \frac{(4\ \text{ft/sec})^2}{2 \times 32.2\ \text{ft/sec}^2} = 0.25\ \text{ft} \qquad \textbf{(Eq 3-4)}$$

Therefore, it is recommended that piping be provided in an adequate size to lower the flume inlet velocity to 4 ft/sec or less (Figure 3-8). If the backwash piping cannot all be that large, a reducer should be provided at the filter wall to enlarge the piping entering the filter and, therefore, reduce the velocity of the backwash water before it enters the filter (Figure 3-8).

Equalized Velocity

When the backwash water enters the flume or plenum, there are usually underdrain laterals or false deck nozzles/orifices immediately in front or to one side, as shown in Figures 3-2 and 3-3. To ensure equal flow distribution (low maldistribution), the incoming backwash water must make a sharp turn to the underdrain. For that to occur, the flume/plenum velocity should be low. Theoretically, the lateral or underdrain entry

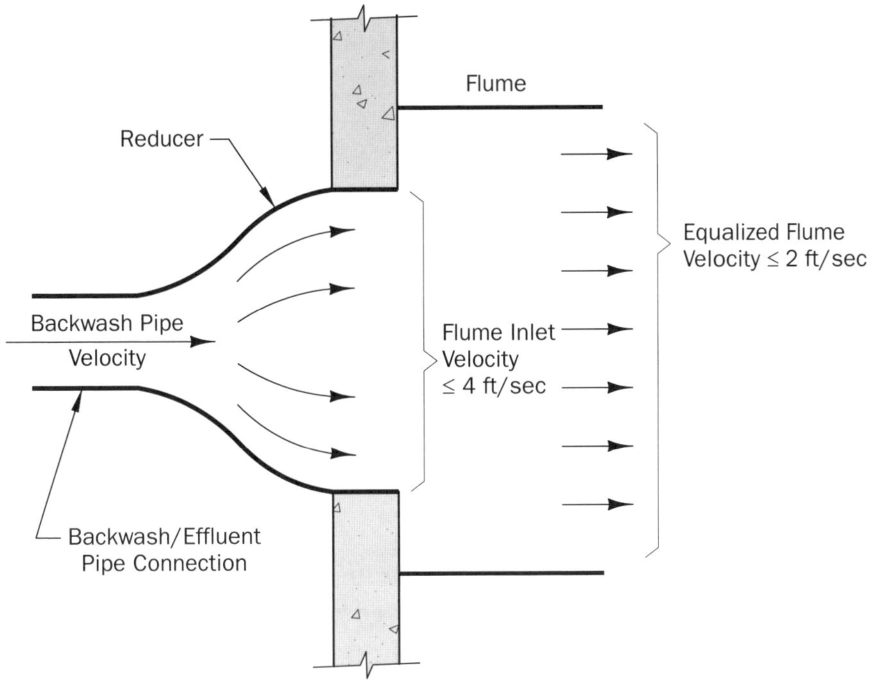

FIGURE 3-8. Equalized flume inlet velocity

velocity could be increased to maintain the desired cross velocity ratio ($V_{L/E}/V_M \geq 2.3$). However, there are practical limits to that approach. The best distribution will occur with a high cross velocity ratio and a low or negligible flume velocity head.

The incoming backwash flow does not immediately spread out equally across the flume. However, for design purposes, an equalized velocity can be assumed, as long as the flume velocity head is very low or negligible and the flume entry velocity (V_E) is below the limits listed previously. It is recommended that the equalized velocity at the front end of the flume not be greater than 2 ft/sec (Figure 3-8). The velocity head in the flume itself is calculated as follows:

$$\frac{V_{EQUAL}^2}{2g} = \frac{(2\ \text{ft/sec})^2}{2 \times 32.2\ \text{ft/sec}^2} = 0.064\ \text{ft (negligible)} \qquad \textbf{(Eq 3-5)}$$

The designer must ensure that the underdrain lateral entry velocity ($V_{L/E}$) is at least 2.3 times greater than the assumed equalized flume velocity, as discussed previously.

$$\begin{aligned} V_{L/E} &= 2.3\ V_{EQUAL} \qquad \textbf{(Eq 3-6)} \\ &= 2.3 \times 2\ \text{ft/sec} \\ V_{L/E} &= 4.6\ \text{ft/sec (minimum)} \end{aligned}$$

The lateral entrance velocity ($V_{L/E}$) may not actually be as high as 4.6 ft/sec. In that case, an orifice should be constructed at the entry to each lateral, to generate the necessary velocity and head loss to force equal distribution to the various laterals. The addition of an orifice at the lateral entry also helps maintain the relationship between the underdrain head loss and the flume entry velocity (V_E).

False Deck Underdrains

For underdrains with a false deck and plenum, the theory is similar. The impact of the entry velocity is the same, as is the cross velocity ratio. The recommended ratio between the underdrain head loss and the flume or plenum entry velocity head is also valid. In place of the lateral entry velocity ($V_{L/E}$), the velocity of the nozzle or strainer tail pipe, or the thimble velocity under a Wheeler pyramid or cone is used.

The biggest impact of analyzing the hydraulics of one underdrain versus another is their head loss. Underdrains with a relatively low head loss will require large flumes or plenums.

Air Header Pipe Blockage

For filters that require the use of an air scour underdrain with the air header in the flume, the space taken by that header must be added to the size of the flume, so as not to affect the flume velocity. The air header pipe will vary in diameter, depending on the size of the filter. The blockage caused by the air header pipe and its support brackets will probably be at least one square foot for most small- to medium-size filters. For large filters with large air header pipes, higher values will be required.

For distribution purposes, it is best that the air header pipe be installed above the top of the backwash inlet pipe. If the air header pipe is installed directly in front of the backwash inlet pipe, the resulting turbulence could adversely affect the distribution efficiency of the underdrain blocks over the flume. An allowance of 16 in. to 18 in. is typical for clearance of the air header pipe (H_{AHP}), as shown in Figure 3-9. For more accurate information, the manufacturers of the equipment should be contacted for recommended sizing on specific projects.

Constructability

For constructability purposes, the backwash entry pipe should be at least 6 in. to 8 in. above the flume floor to provide adequate clearance for a waterstop in the wall and a seep ring on the pipe (Figure 3-9 and 3-10). As an alternative, the walls and floor of the flume could be constructed in one concrete pour. In either case, a structural evaluation may be warranted to ensure that a watertight joint is obtained.

Maintenance Accessibility

Periodically in the life of a filter, it may be desirable or necessary to access the flume for maintenance. For example, if the media becomes upset and penetrates into the underdrain, access will be necessary to clean the flume. Chapter 5 should be referred to for a description of flushing an underdrain. Access may be through a separate hatch in the flume or by taking the backwash/effluent piping apart, if large enough. It is recommended that the flume be a minimum of 2 ft wide for access purposes.

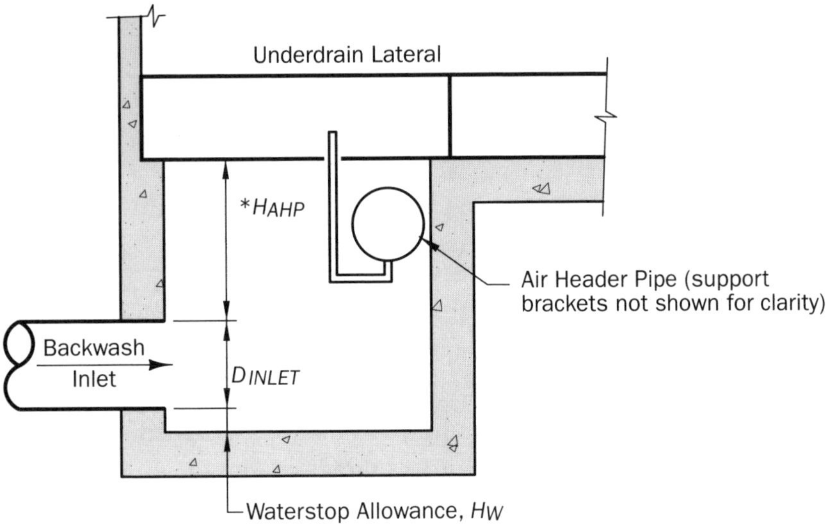

*16 in. to 18 in. clearance for AHP and risers is common.
Flume Depth = H_{AHP} + D_{INLET} + Additional Hydraulic Height (H_H if required) + H_W

FIGURE 3-9. Flume depth

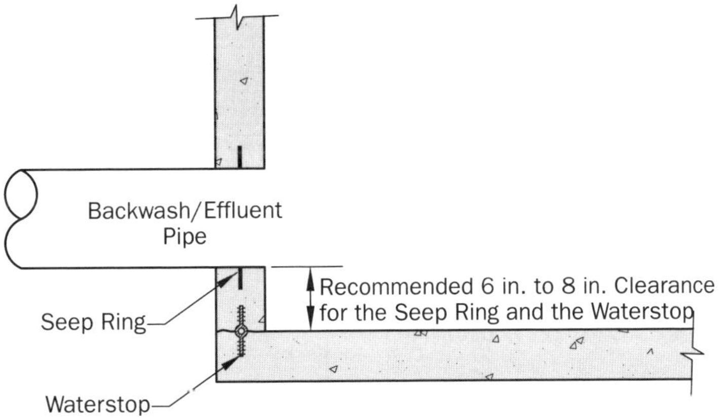

NOTE: As an alternative, the walls and floor of the flume could be constructed in one concrete pour. In either case, a structural evaluation may be warranted to ensure that a watertight joint is obtained.

FIGURE 3-10. Clearance for seep ring and waterstop

Flume Depth

The depth of an end or middle flume is a combination of the clearance required for the air header pipe (H_{AHP}), the diameter of the inlet pipe (D_{INLET}), the additional height required for hydraulic purposes (H_H) if any, and the waterstop and seep ring clearance (H_W). (Refer also to Figures 3-9 and 3-10.)

$$\text{Flume Depth} = H_{AHP} + D_{INLET} + H_H + H_W \quad \textbf{(Eq 3-7)}$$

It should be noted that for smaller filters, the required equalized flume velocity (V_{EQUAL}) may be achieved without an additional height for hydraulic purposes. In that case, the flume depth can be calculated as follows:

$$\text{Flume Depth} = H_{AHP} + D_{INLET} + H_W \qquad \textbf{(Eq 3-8)}$$

Flume Configuration

The location of the flume also affects its size.

Front Flume

A front flume with a center feed has the highest cross velocity ratio, $V_{L/E}/V_M$, for a given flume size, as shown in Figure 3-11. The incoming velocity is split in two directions, which reduces the flume or manifold velocity (V_M) accordingly, and results in smaller flumes. An example of a front flume is shown in Figure 3-12. They are typically used for small- to medium-size filters.

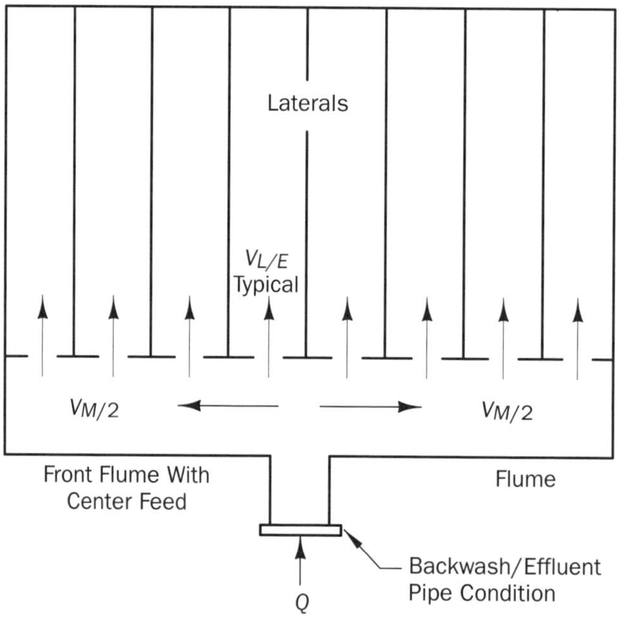

OTE: If the backwash water enters at the middle of the flume, the flow in either direction is ›duced by half. The manifold velocity is reduced by half and the cross velocity ratio is therefore oubled. The flume can be reduced in size while maintaining the desired cross velocity ratio.

$$\frac{V_{L/E}}{V_{M/2}\downarrow} = \uparrow \text{ (Doubles)}$$

FIGURE 3-11. Front/middle flume cross velocity

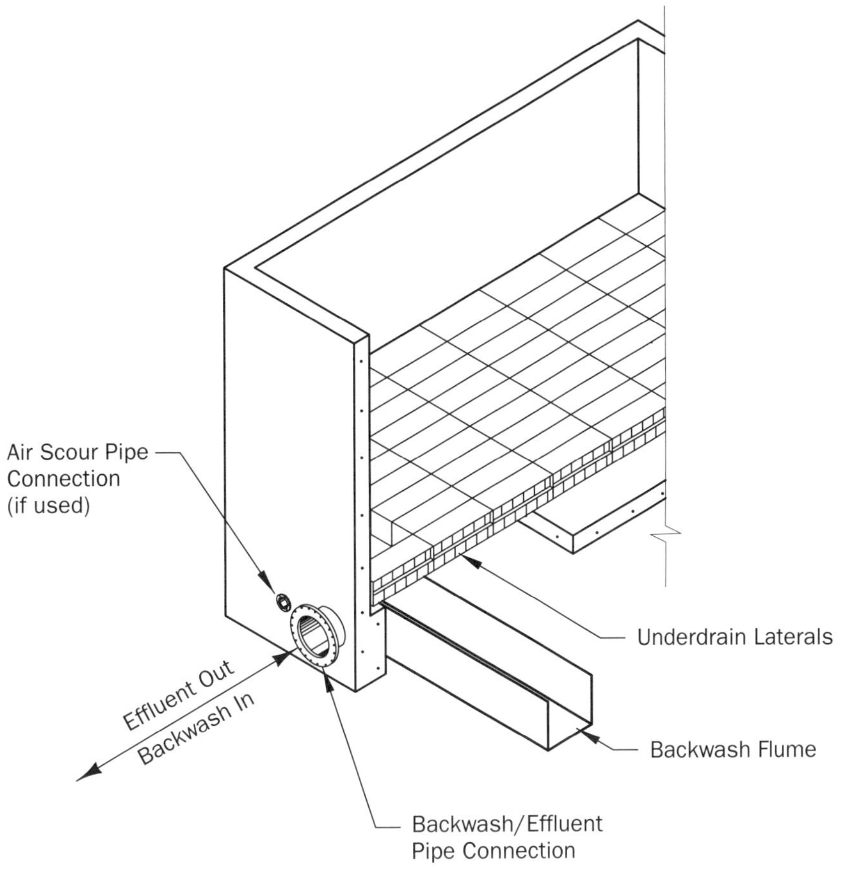

FIGURE 3-12. Typical front flume filter arrangement

Center-Fed Middle Flume

Larger filters are often designed with a center-fed middle flume as a result of the very high entrance velocities ($V_{L/E}$), which would otherwise result from the long laterals. Too high of an entrance velocity can cause maldistribution to increase within the lateral. A high air flow rate in long laterals with air scour underdrains is also an issue, which can be mitigated by using a center fed middle flume. Chapter 6 also discusses long laterals. A center-fed middle flume will, however, be larger than a front flume for the same filter configuration. Because the laterals are fed in two directions, the lateral entry velocity, $V_{L/E}$, for a middle flume will be half that of a front flume design. A larger flume will then be required to satisfy the $V_{L/E}/V_M$ requirement. A typical center-fed middle flume is illustrated in Figures 3-13 and 3-14.

Very high lateral velocities, V_L, can result in higher than desired maldistribution within the lateral. In addition, problems can occur distributing air (when used) when the laterals are longer than 36 ft to 40 ft. Using a center-fed middle flume allows both

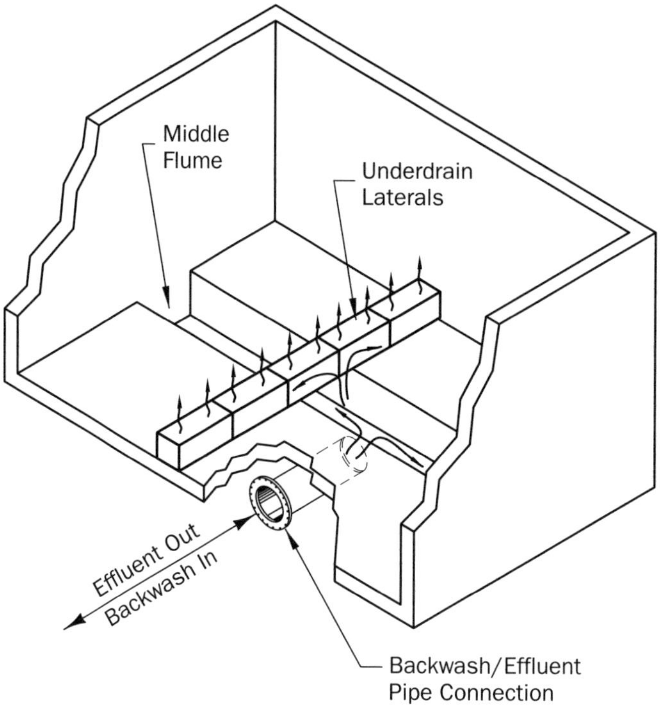

FIGURE 3-13. Center-fed middle flume arrangement

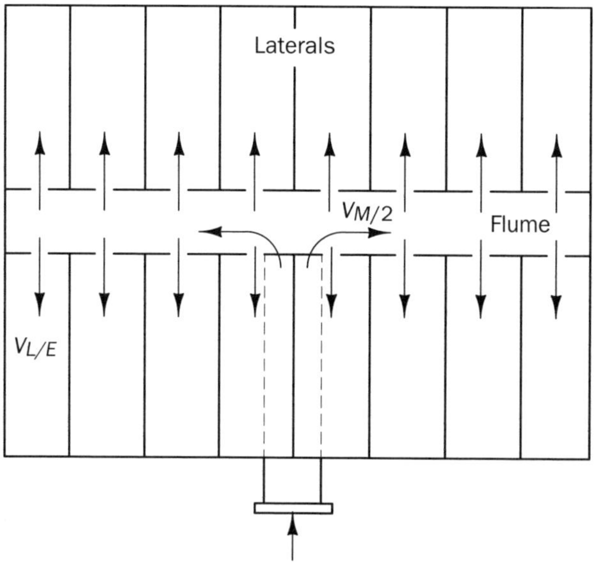

NOTE: The use of a center fed middle flume essentially reduces the lateral length in half, which requires a large flume to compensate for the reduction in velocity ratio.

FIGURE 3-14. Center-fed middle flume

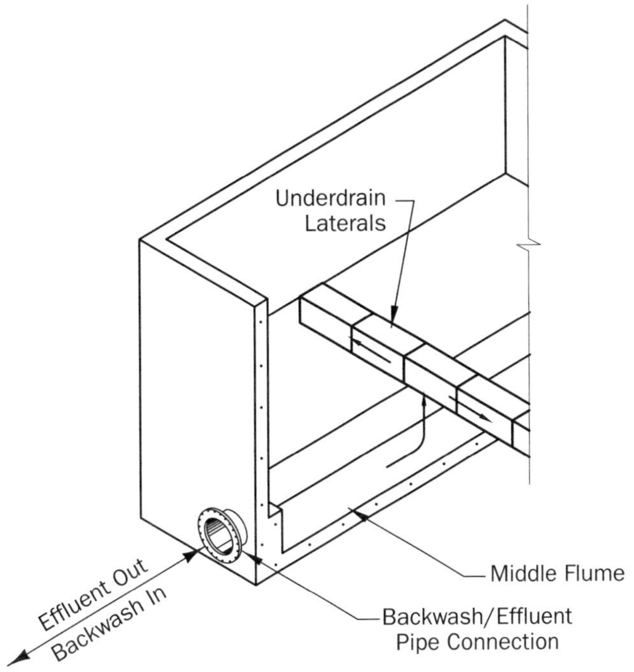

FIGURE 3-15. End-fed middle flume arrangement

the air and water to be fed in two directions, which reduces the velocities of both to manageable levels.

End-Fed Middle Flume

An end-fed middle flume (as shown in Figures 3-15 and 3-16) may have a low $V_{L/E}/V_M$ ratio, unless the flume size is increased. End-fed middle flumes or those with low $V_{L/E}/V_M$ ratios typically occur in plant rehabilitations and in long filters with low head loss underdrains (Figure 3-17). An end-fed front or side flume will also require a large flume.

End-Fed Front or Side Flume

End-fed front flumes are not common. End-fed side flumes are typically used on large filters, which are more nearly square than rectangular. Because of the size of the filter and the high backwash flow, the flume is generally tapered. The intent is to maintain a nearly constant cross velocity ratio ($V_{L/E}/V_M$) along the flume as the backwash flow is diverted to the various laterals. The space above the flume is typically used for influent and backwash waste connections.

The design of an end-fed flume has several other unique requirements, especially if air scour is used. Both concurrent and sequential air scour can be accomplished with a tapered end-fed flume. In both cases, the flume will have a backwash/effluent piping connection, an air piping connection, an air vent connection, and a penetration to

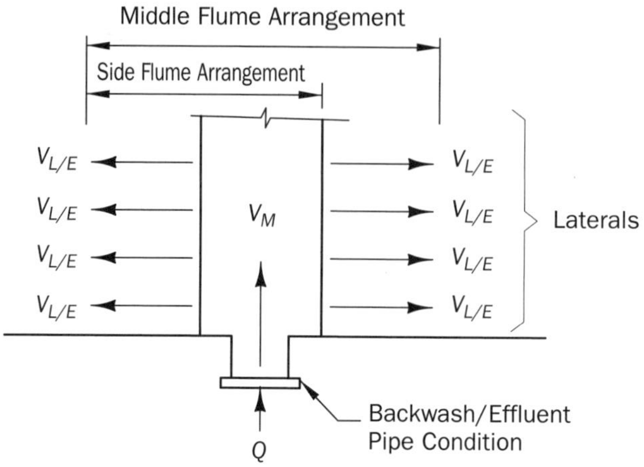

NOTE: In the case of an end-fed middle or side flume, the entire backwash flow is routed through the flume entrance, which results in a high initial velocity (V_M). For a middle flume, the flow through the laterals and V_1 are divided in half. The result is a much lower cross velocity in the front of the filter, which is less desireable. To counteract that effect, the flume must be larger.

Side Flume $\dfrac{V_{L/E}}{V_M \uparrow} = \downarrow$ (Decreases) – Requires a large flume to compensate

End-Fed Middle Flume $\dfrac{V_{L/E}/2 \downarrow}{V_M} = \downarrow$ (Decreases) – Requires a large flume to compensate

FIGURE 3-16. Middle or side flume cross velocity

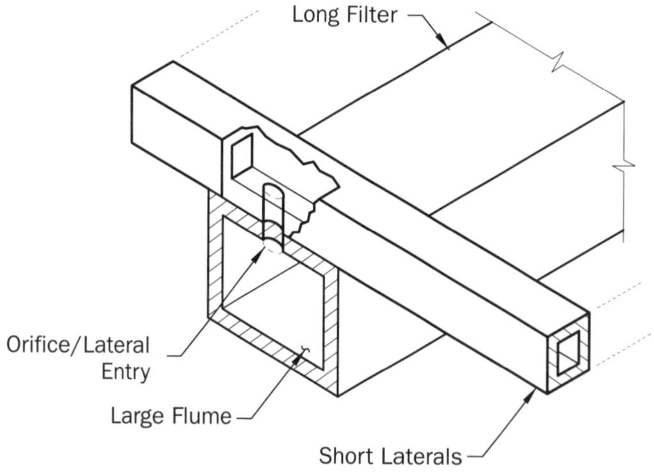

FIGURE 3-17. Flume for low head loss underdrain

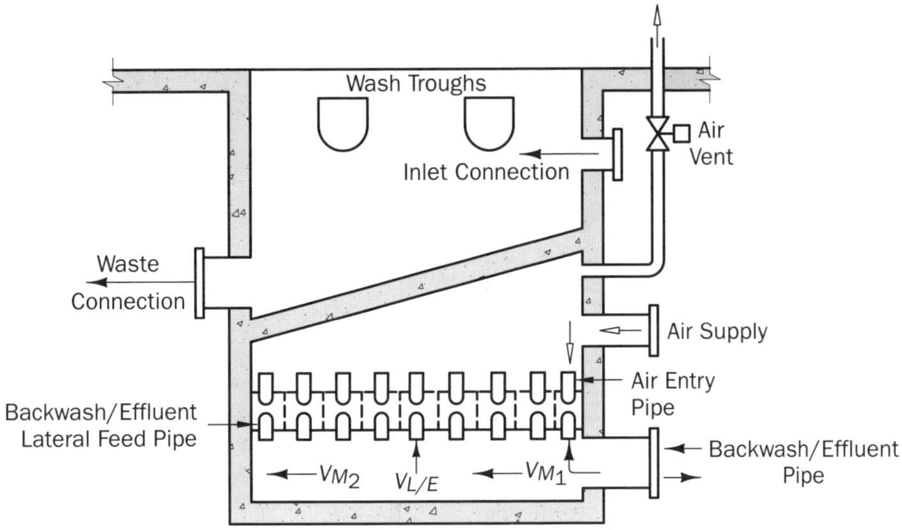

NOTE: One of the purposes of a tapered side flume is to maintain a nearly constant velocity (V_M) ratio along the length of the flume as backwash water is diverted to the various laterals. In that way, the cross velocity ratio ($V_{L/E}/V_M$) is also nearly constant.

$$V_{M2} \cong V_{M1}$$

FIGURE 3-18. Side flume

each lateral, as shown in Figure 3-18. The specific requirements of sequential and concurrent air scour are noted in the following section.

Sequential Air Scour

Many end-fed side flumes were originally designed and built for hydraulic-only backwash. The floor of the flumes in those filters was typically on the same level as the filter itself, as shown on the left side of Figure 3-19. The connection from the flume to the individual laterals was commonly a circular pipe or a rectangular cutout. The penetrations could not be as large as the full cross section of the underdrain for structural reasons.

Filters originally built in this manner for hydraulic-only backwash may be converted to sequential air scour by installing an air supply connection and an air vent connection. Concurrent wash requires a flume with a floor lower than the filter, as discussed in the next section.

The only limitation of using sequential wash is that the dual parallel lateral underdrain (if used) would have to be equipped with a porous plate media retention system in lieu of support gravel. Chapter 6 should be referred to for a more complete discussion of this issue.

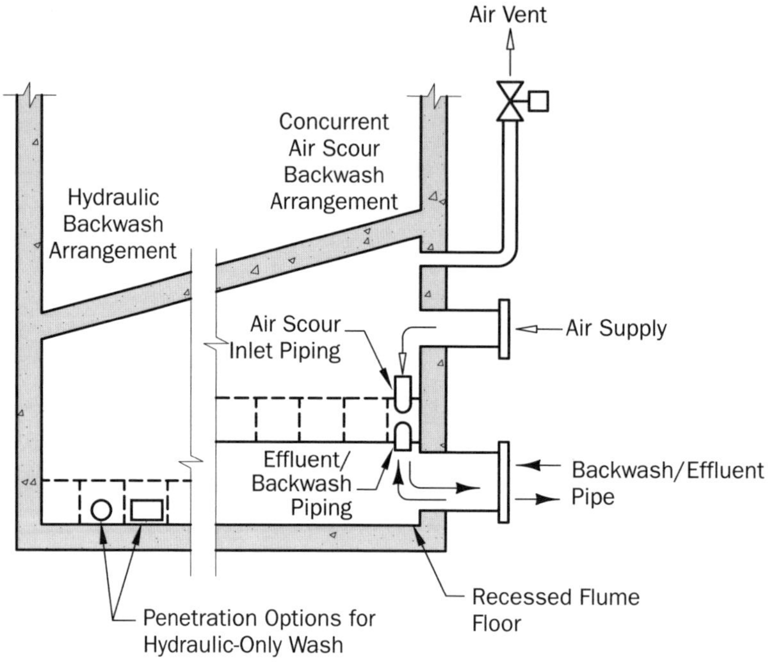

NOTE: The underdrain is typically at the same level as the flume floor for hydraulic-only applications.

FIGURE 3-19. Tapered flume alternatives

A sequential air scour cycle would be as follows:

1. Drain the filter down to a point below the lip of the wash trough. The level should be low enough to allow time for the air to be purged out of the system, before the water weirs over the lip of the wash trough, after the hydraulic backwash is begun.
2. Begin air scour for a set period of time.
3. End air scour and open the air vent for a set period of time.
4. Begin hydraulic backwash for a set period of time and close the air vent.
5. End backwash and return to service.

Concurrent Air Scour

An existing side flume could be rehabilitated for concurrent wash by installing an air header within the flume, as shown in Figure 3-20. The benefit would be an increased cleaning efficiency of the backwash cycle and a reduction in the amount of water used for that purpose.

Designing a tapered end-fed flume with concurrent air scour presents a number of interesting challenges.

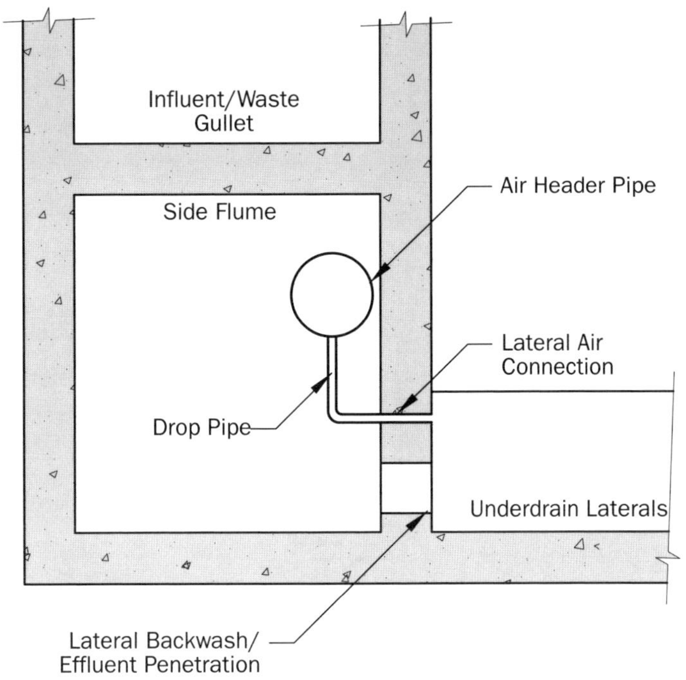

FIGURE 3-20. Side flume air header pipe alternative arrangement

During air scour, an air pocket is built up within the underdrain in the top half of the block. Chapter 6 has a more in-depth discussion of this issue. Because the complete cross section of the underdrain cannot normally be extended into the flume, separate penetrations must be constructed for the effluent/backwash water and for air, as shown in Figures 3-19 and 3-21. Because the effluent/backwash connection is smaller than the cross section of the underdrain, there will be a head loss associated with the water traveling from the flume into the underdrain. As a result, the air/water interface in the flume will be higher than in the underdrain. The air pipe must be extended above that point for proper operation.

A balance must be struck between the head loss required for the air to enter the underdrain and for the low-rate backwash water to enter the underdrain. The intent is to balance the pressure requirements so that the air/water interface is established at a reasonable location. In addition, the air entry pipe should extend above the air/water interface an adequate distance to allow for a higher low-rate backwash flow, if desired. When the concurrent step is completed, an air vent valve is opened to vent the remaining air from the flume in a similar manner to that of the sequential wash sequence.

It should be noted that the floor of the flume must be recessed below the filter floor to allow the backwash/effluent feed pipe to be turned down. The intent is to prevent a whirlpool from forming, which could cause unstable conditions in the flume. Another reason for the recessed area is to allow the backwash/effluent header to penetrate into the flume below the air/water interface during concurrent wash. If the header is too

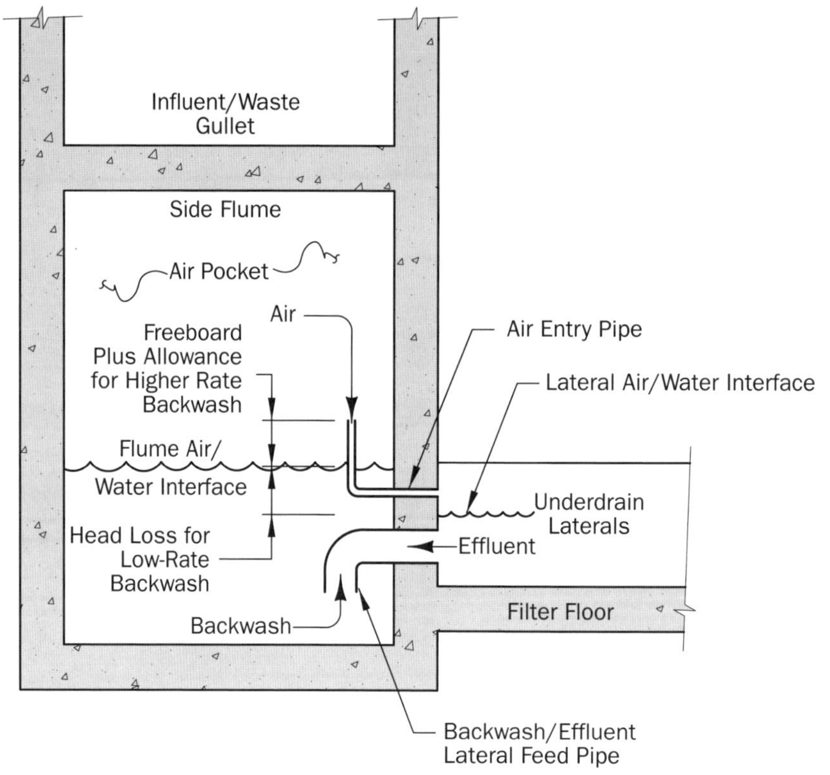

FIGURE 3-21. Side flume cross section

high, it will result in open channel flow in the pipe and could also cause unequal air and water distribution in the first few laterals.

Maldistribution Testing

It is difficult to determine maldistribution by measuring the actual water velocity over an underdrain in a flooded filter. As a result, common practice is to measure the underdrain pressure during a backwash instead, because pressure is related to flow.

To measure pressure, piezometer taps are first installed in the final or secondary compartments of the underdrain laterals. A hole is drilled and tapped at each location, and threaded 1/4-in. hose barb fittings are installed. Suggested locations for pressure taps are shown in Figures 3-22 and 3-23, depending on the flume arrangement. Pressure taps should be installed near the flume entry and at the far corners of the filter. Depending on the entry velocity, point *C* in the figures may have higher pressure than average because of the upward deflection of the incoming water.

Once the piezometer taps are installed, route 1/4-in. clear vinyl tubing from each sample point to a piezometer board containing a scale for each sample tube. Adjust the board and scales to the waterline during backwash in a location well away from a trough, as shown in Figure 3-24.

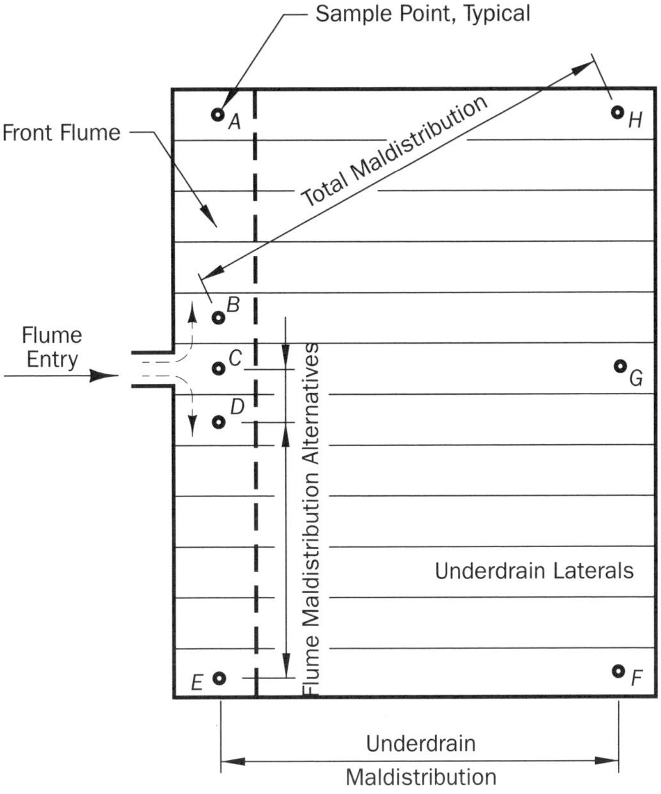

FIGURE 3-22. Front /middle flume pressure tap locations

Pressure readings representing the underdrain head loss are measured as the height of water in each tube above the waterline during a backwash. Maldistribution can be calculated as follows:

$$\text{Velocity Head } \frac{V^2}{2g} \text{ is proportional to pressure } (P)$$

$$V \text{ is proportional to } \sqrt{P}$$

$$\text{Maldistribution} = \frac{(P_{highest}{}^{1/2} - P_{lowest}{}^{1/2})}{P_{highest}{}^{1/2}} \quad \text{(Eq 3-9)}$$

$$\text{Maldistribution (\%)} = [1 - \sqrt{P_{highest}/P_{lowest}}] \times 100$$

Using the suggested locations for piezometer taps, the maldistribution for the flume and the underdrain laterals can be measured and calculated independently. Also, the total maldistribution can be measured between the points as shown on Figures 3-22 and 3-23. Total maldistribution is usually measured between a point close to the entry and the point farthest away.

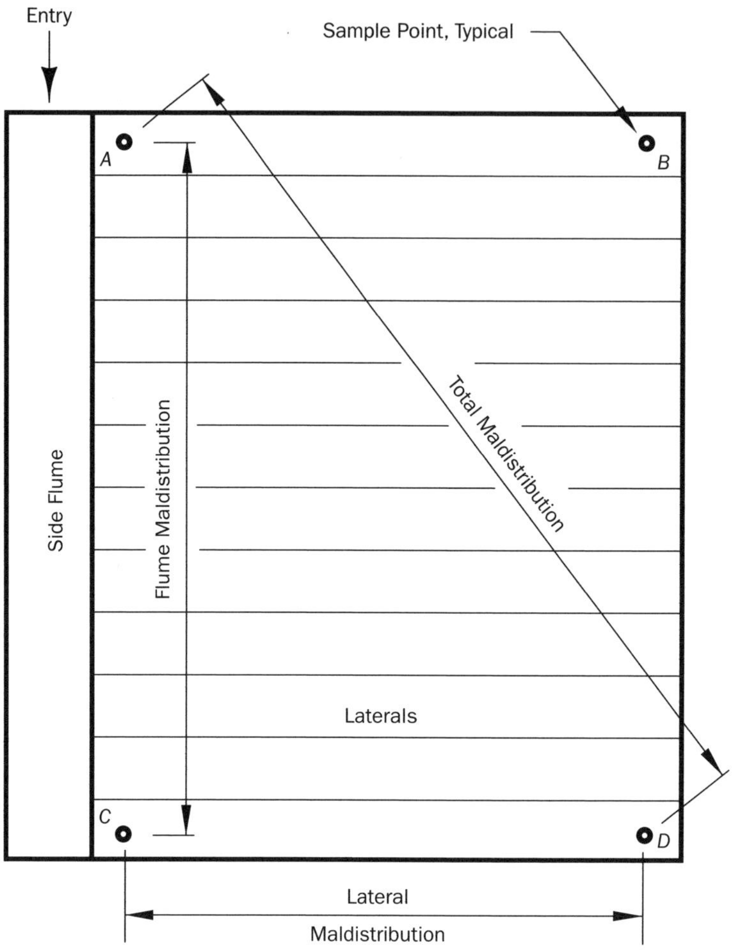

FIGURE 3-23. Middle or side flume pressure tap locations

For false deck–type underdrains, if measurements are to be taken, it is best for the piezometer taps to be cast or fabricated into the false deck during construction. It may be difficult to thread a piezometer tap into the cap of a nozzle, for example.

As previously mentioned, the goal for the filter installation is to have a total maldistribution of less than ±5 percent. For underdrains with relatively high maldistribution, large flumes will be required to reduce that component of the total maldistribution.

Summary

Assuming the same lateral geometry, the lateral velocity (V_L) will go down with shorter laterals. Short laterals rely primarily on the head loss of the final (or secondary) orifices when using dual parallel lateral underdrains. In that case, the flume must be sized to produce a much lower flume velocity (V_M) to maintain the desired cross velocity ratio and to ensure a low flume maldistribution component.

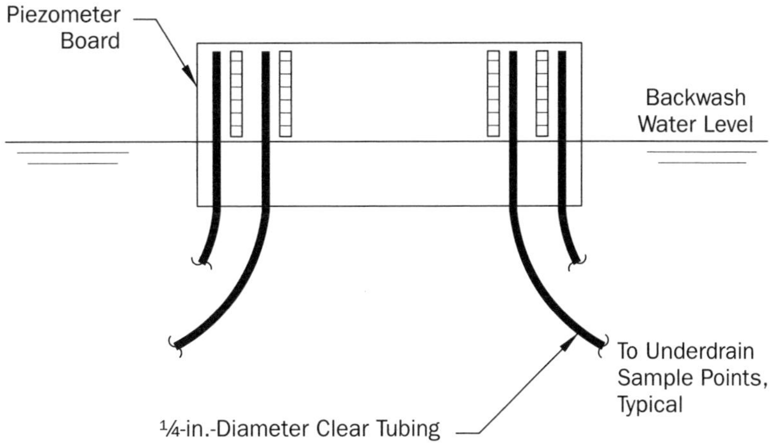

FIGURE 3-24. Maldistribution pressure measurements

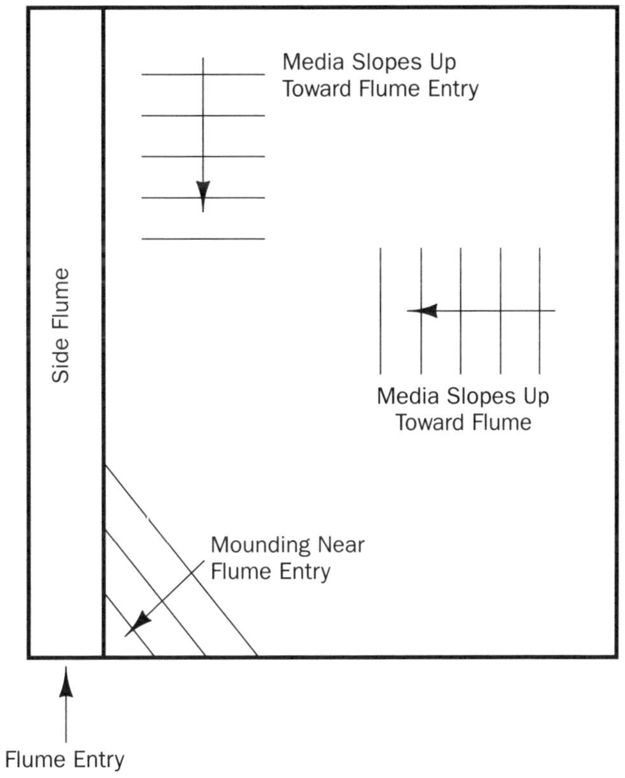

FIGURE 3-25. Potential flume problems

Problem	Probable Cause	Proposed Solution
Unequal backwash distribution to the laterals/nozzles	• High flume velocity (low cross velocity)	• Add orifice at entry to all laterals
First few laterals starved during backwash	• High flume entry velocity	• Add baffles at flume entry
Mounding near flume entry	• High flume entry velocity	• Add baffles at flume entry
Media surface sloped up toward flume	• High lateral entry velocity	• Enlarge entry orifice to each lateral (if possible)

NOTE: Once a filter is built, it may be very difficult to improve maldistribution caused by flume or lateral entry problems.

FIGURE 3-26. Filter hydraulics troubleshooting guide

The type of flume used will also have a major impact on the filter hydraulics. A front flume with a center feed results in the smallest overall flume cross section for small- to medium-sized filters. Middle flumes with a center feed are the best for long filters, while large side flumes are commonly used for large separate filters.

The recommended rules of thumb for flume or plenum design are as follows:

Backwash entry velocity $\leq$ 4 ft/sec

Equalized flume velocity $\leq$ 2 ft/sec

Cross velocity $(V_{L/E}/V_M) \geq 2.3$

Total underdrain head loss $\geq$ (5–6) × (flume entry velocity head)

If these conditions cannot be met, additional head loss may have to be taken with an orifice at the entry to each lateral. Figure 3-25 shows potential flume problems and Figure 3-26 shows a filter hydraulics troubleshooting guide.

CHAPTER 4

Filter Support Gravel

Introduction

Both gravity and pressure filters use support gravel. The governing specification for support gravel is ANSI/AWWA B100 *Standard for Filtering Material.* However, there are a number of direct media retention underdrains that do not use support gravel, as discussed in Chapter 5.

Graded gravel in layers is used to provide separation between the media and the underdrain. The bottom layer of the gravel is intended to protect the underdrain orifices from being plugged. The gravel bed is built in layers of smaller and smaller grain sizes to form a barrier layer, which is designed to support the media. According to the ASNI/AWWA B100 standard and the author's experience, the layers of gravel should be in 2:1 size gradations. For example, the layer below should be twice the size of the layer above. Gravel grain size is discussed in greater detail in other sections of this chapter. A typical single-taper gravel design for hydraulic backwash is shown in Figure 4-1.

The life of filter media generally depends on keeping the media clean and the support gravel intact. Cleaning the media and support gravel is dependent on having an efficient backwash system. During backwash, the media may be fluidized or agitated. However, it is critical that the support gravel not move or be agitated. Support gravel is laid very carefully in distinct layers of different sizes. If the layers begin to move, the support capability of the gravel will be greatly reduced or lost.

Disturbed single-taper gravel, as shown in Figure 4-2, is referred to as having become upset or blown. A hole has been blown through the layers, allowing them to become intermixed. When that happens, the media is no longer supported properly and can leak through the gravel and into the underdrain. There are a number of potential causes of upset or blown gravel, as discussed in other chapters.

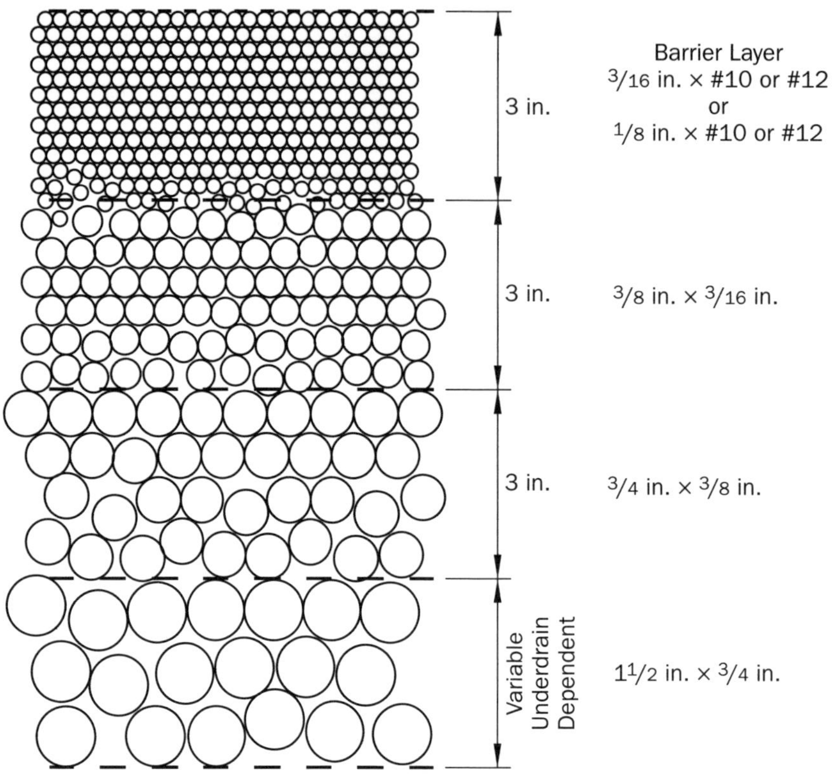

FIGURE 4-1. Typical single-taper support gravel design for hydraulic backwash

Once the support gravel becomes upset, the filter media will begin to deteriorate to the point where it must be replaced. The period of time involved in this deterioration process is highly variable. Physical symptoms that occur as a result of upset gravel include mudballing of the media and the formation of craters where media is being lost into the underdrain. When the gravel is upset, the filter effluent quality may deteriorate and filter runs may be shortened.

To ensure a long life for support gravel and the filter media, it is important to design the filter and auxiliary equipment properly and to understand the specifications used to obtain and install the gravel. Comments related to support gravel issues are given in the following section. Design considerations on other related equipment are covered in other chapters.

Physical Characteristics

Filter support gravel is a coarse aggregate composed primarily of silica. In addition, gravel must have adequate strength and durability to maintain its integrity over a long period of time. Another component is calcium, which contributes to acid solubility. Lower calcium content is recommended, as discussed in the next section. Some of the

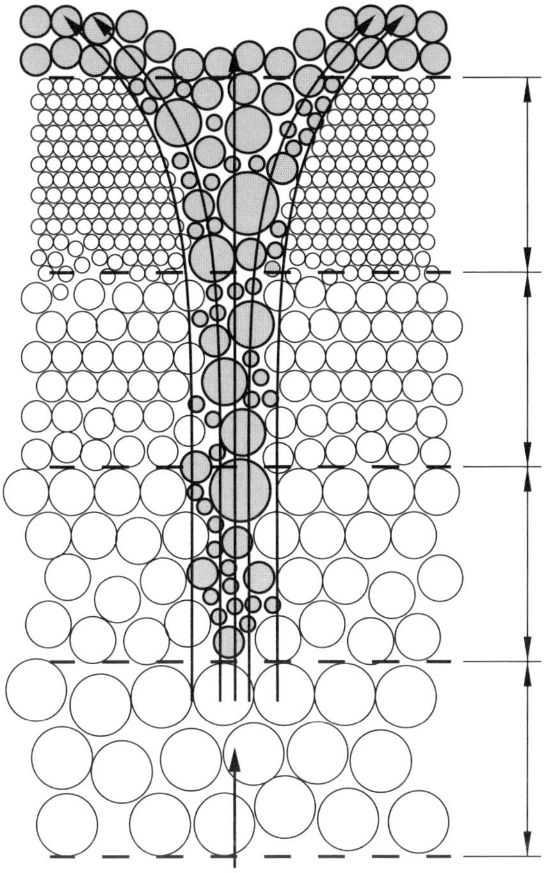

FIGURE 4-2. Upset single-taper support gravel

required physical characteristics for filter support gravel in the ANSI/AWWA B100 standard are as follows.

Acid Solubility

The ANSI/AWWA B100 standard limits acid solubility of gravel to less than 5 percent for No. 8 or smaller, less than 17 1/2 percent for 1 in. to No. 8, and less than 25 percent for 1 in. and larger gravels. Most silica gravel has a much lower acid solubility. It is important that acid solubility be low. Decomposition of the gravel caused by an acidic water condition would lead to a gradual movement within the various layers and a weakening of its support capability. If the water in the filter is expected to be acidic, a special effort will be necessary to ensure that low solubility gravel is obtained.

Fractured Faces

Crushed gravel tends to compact tightly and resist the flow of water. For filter gravel, it is important that the aggregate be rounded and smooth. Rounded gravel does not pack tightly and allows for the free passage of water. The ANSI/AWWA B100 standard sets a limit that less than 25 percent of the gravel should have more than one fractured face.

Specific Gravity

The minimum specific gravity in the ANSI/AWWA B100 standard for gravel is 2.5. Most gravel typically has a higher specific gravity of 2.6 or 2.65.

Bulk Density

A typical bulk density for filter gravel is approximately 100 lb/ft^3. With a specific gravity of 2.65, a solid cubic foot of gravel would weigh approximately 165 lb. Using a bulk density of 100 lb/ft^3, the void volume would be almost 40 percent. The void volume is an indication of its ability to pass water. A lower percentage would indicate more tightly packed gravel with less ability to pass water. Significantly higher numbers are probably not physically attainable.

High Density Gravel

As stated previously, the life of a filter bed is dependent on keeping the media clean and the support gravel stable. High density gravel is occasionally used as the barrier layer in order to provide additional stability to the gravel bed. High density gravel is usually composed of either garnet or ilmenite, with a minimum specific gravity of 3.8. Garnet gravel may not be available in sizes larger than the barrier layer (1.4–2.5 mm). Ilmenite is available in larger sizes but is rarely used for that purpose because of economic reasons and because much of it is manufactured from a crushing operation, which means it might pack too tightly.

The use of high density gravel does add stability to the gravel bed. However, in the author's opinion, the 2:1 sizing requirement between layers should also be maintained for a conservative design.

Design

The design of the support gravel and the number of layers used depends on the underdrain and the type of media.

Single-Taper Design

Filters using a hydraulic-only backwash typically use a single-taper gravel design with the coarsest layer on the bottom and the finest layer (barrier layer) on the top (Figure 4-1). The thickness of the bottom (coarsest) layer is dependent on the underdrain. For a discussion on the thickness of the bottom layer, refer to the section on Layer Thickness later in this chapter.

Backwash rates for single-taper gravel designs normally range from 10 to 25 gpm/ft^2, depending on the media types. At rates above 30 gpm/ft^2, the barrier gravel will normally begin to fluidize, which will cause disruption of the gravel layers. Some experiments have been made using ilmenite for all sizes of gravel used. Ilmenite gravel is very

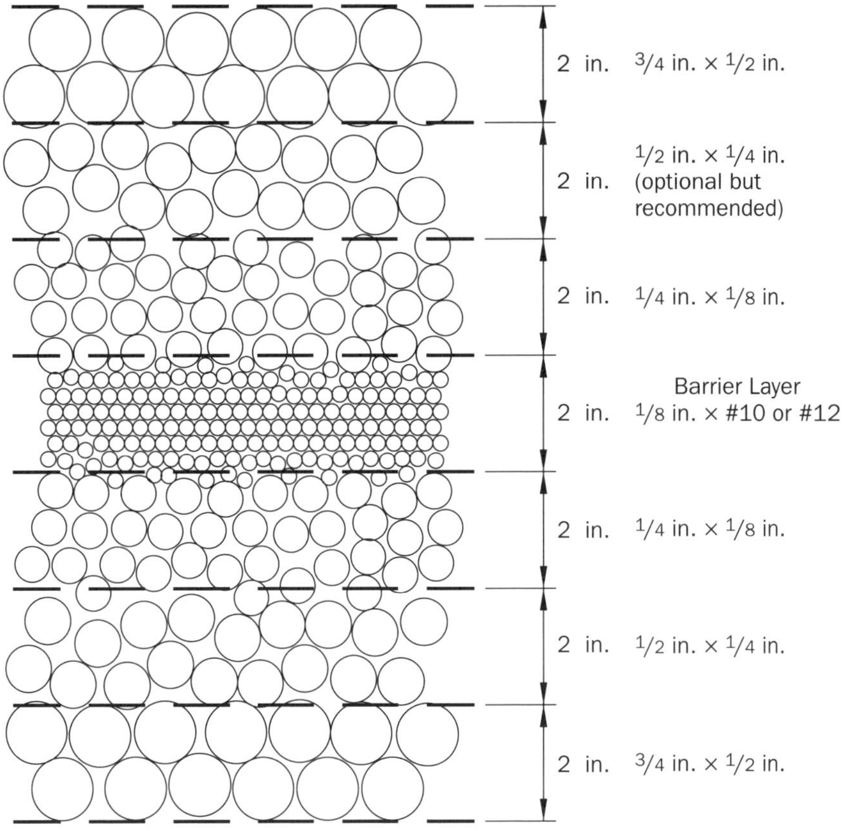

NOTE: 3-in. layers recommended

FIGURE 4-3. Typical double-taper support gravel design for air scour

stable because of its high density and will greatly extend the life of the filter. However, it is rarely used because of economic issues.

Double-Taper Design

Some air scour underdrain designs use a double-taper (reverse-graded or hourglass) configuration of support gravel, as shown in Figure 4-3. With a normal single-taper gravel design, the barrier layer is on top. Air coming up through the gravel from the underdrain will tend to lift the barrier gravel and push it to one side. Once the barrier gravel is disrupted, the remaining layers will become upset quickly.

In the double-taper design, the barrier gravel is usually the third or fourth layer down from the top. The intent is to sandwich the barrier layer between larger layers of gravel to keep it from moving. The larger gravel on top has large enough void spaces between the individual grains that the bubbles from normal air scour can easily pass through without disturbing them. If the upper layer remains intact, the rest of the layers normally will also.

To improve the stability of the standard hourglass design, garnet gravel has been successfully used as the barrier gravel.

Fluidizing Media

In filtration mode, normal dual media or multimedia will penetrate through the upper layers of double-taper gravel, down to the barrier layer. As much as 1 in. to 1 1/2 in. of media will penetrate into the gravel. During air scour and the various stages of backwash, the media will blow back out of the gravel. As with air scour, the head loss of blowing the media out is not sufficient to move or disturb the larger gravel on top. When the filter is returned to service, the media will again penetrate the gravel to its former position.

Nonfluidizing Media

Coarse media (substantially larger than 0.5 mm) may pack into a normal double-taper gravel design too tightly and require too high of a head loss to blow out during backwash. If that is the case, the upper layer of gravel could be lifted out of the way, resulting in disruption of the entire bed.

If a coarser media is to be used, testing may be required to determine the proper support gravel design. In some cases, the 1/8 in. × No. 10 gravel of the normal double-taper design has been eliminated, and the 1/4 in. × 1/8 in. gravel may become the barrier layer. The upper layers may also change.

It should be noted that many pressure filters use single-taper gravel and a hydraulic-only backwash. If air scour is used in a pressure filter, it may use an air grid above the gravel, in the media, instead of passing the air through the underdrain and support gravel.

Grain Size

The bottom layer of gravel is normally sized to be large enough to not obstruct orifice holes in the underdrain. Above the bottom layer, each succeeding layer is normally sized to be half the size of the layer below. The layers are sized to sit on the layer below, while supporting the layer above. The 2:1 size gradation from one layer to the next is designed to provide long-term stability. If the size difference becomes larger than 4:1, the aggregate in the upper layer can fall through the layer below, even when using garnet support gravel. However, the size difference between the barrier layer of gravel and the filter media may be as much as 6:1, 8:1, or even 10:1 for 0.5-mm filter sand or finer garnet sand. Wet sand "bridges" allow a much higher ratio than for the gravel. However, for coarser media, the support requirements may be closer to the 2:1 or 4:1 ratio.

Figure 4-4 illustrates poor gravel sizing between the barrier layer (garnet gravel) and the layer of silica gravel below.

Filter media must be installed wet on top of gravel in order for it to remain in place and be supported by the barrier gravel properly. In addition, once installed, the media should never be allowed to dry out, or it will begin to sift down through the support

FIGURE 4-4. Improper gravel gradations (garnet gravel on left, first layer of silica gravel on right)

gravel and foul the underdrain. For that reason, it is recommended that media installed in a filter not be allowed to stay dry for more than 2 days.

Each size of gravel is normally specified within a 2:1 size range such as $^1/_2$ in. × $^1/_4$ in. The reason for the range is because of physical limitations of the production process. The ANSI/AWWA B100 standard states that not more than 8 percent can be finer or coarser than the specified range. Quality control measures should be implemented at the job site to ensure that the gravel meets the specifications.

Quality Control

Specifications

Because of the manufacturing process used, a given size, such as $^1/_4$ in. × $^1/_2$ in., of support gravel is normally produced within a range.

Production

Once a gravel layer is sized, the normal tendency is to assume that the bulk of the material will be the average of the range. However, that is not always the case. It is not uncommon for the gravel to be primarily on one end of the range. For example, with $^1/_4$ in. × $^1/_8$ in. gravel, the average would be $^3/_{16}$ in., or 4.76 mm. If the gravel was more coarse, the aggregate might be closer to $^1/_4$ in., or 6.35 mm.

TABLE 4-1: Garnet gravel

Screen	Size
No. 8	2.05–2.45 mm
No. 8–12	1.40–1.55 mm
No. 12	1.40–1.70 mm

Having the gravel on the coarse side is not bad, as long as the other layers are also coarse. The silica gravel is normally provided by one supplier, and it is likely that all the layers will be either on the coarse or fine side. However, relative sizing of the layers should still be evaluated.

When garnet gravel is used, it must be compatible with the silica gravel. Garnet gravel is also produced in a range that is related to the screen size used to produce the material. The size range is much narrower than for silica gravel.

The size ranges from one manufacturer for garnet gravel are shown in Table 4-1.

Specifications for garnet gravel should be written for one of the sizes listed in Table 4-1. If the range is too large, such as 1.40–2.45 mm, the garnet gravel could easily be on the small end at 1.40 mm, with the silica gravel in the layer below on the coarse end at 6.35 mm. Both materials would be within their respective specification, but the two are incompatible.

The size ratio between the two gravels is calculated in the following equation:

$$\frac{6.35}{1.40} = 4.5:1 \quad \textbf{(Eq 4-1)}$$

The desired ratio is 2:1 to ensure stability between the two layers. With a ratio of greater than 4, the gravel design will be unstable, and the media bed will most likely not last as long as desired.

Summary

It is important that the procurement specifications for support gravel be written with the appropriate size ranges to ensure stability. In addition, submittals should be evaluated to ensure that the materials to be provided are all within the 2:1 size ratio. Another evaluation should be made of the materials delivered to the job site. The materials should visually appear to meet the desired 2:1 size ratio. If there is any question, a sieve analysis should be conducted on the materials at the job site before installation.

Media/Support Gravel Compatibility

In addition to the gravel layers being compatible with each other, the support design must be compatible with the media. For example, it is recommended that coarse monomedia not be placed over support gravel designed for fluidizing media. In

TABLE 4-2: Typical filter gravel gradations for fluidizing media in air scour applications

Layer No.	Gravel Size	Location
6	3/4 in. × 1/2 in.	Top Layer
5	1/4 in. × 1/8 in.	
4	1/8 in. × No. 10 (or 1/16 in.)	Barrier Layer*
3	1/4 in. × 1/8 in.	
2	1/2 in. × 1/4 in.	
1	3/4 in. × 1/2 in.	Bottom Layer

*2.0 to 2.5 mm garnet gravel optional

TABLE 4-3: Recommended gravel gradations for fluidizing media in air scour applications

Layer No.	Gravel Size	Location
7	3/4 in. × 1/2 in.	Top Layer
6	1/2 in. × 1/4 in.	
5	1/4 in. × 1/8 in.	
4	1/8 in. × No. 10 (or 1/16 in.)	Barrier Layer*
3	1/4 in. × 1/8 in.	
2	1/2 in. × 1/4 in.	
1	3/4 in. × 1/2 in.	Bottom Layer

*2.0 to 2.5 mm garnet gravel optional

addition, the barrier layer of gravel must be sized to properly support the media to be used.

Upset conditions will be created if either the gravel layers are incompatible with each other or if the barrier gravel is not specifically designed for the media to be used. If there is any question, pilot studies should be conducted for compatibility.

Layer Design

The gravel gradations listed in Table 4-2 represent typical design for air scour support gravel. (Refer also to Figure 4-3.) It should be noted that Layer No. 1 is repeated as Layer No. 6, and Layer No. 3 is repeated as Layer No. 5. Layer No. 2 is not normally repeated by most manufacturers. However, repeating Layer No. 2 would result in a more stable gravel design, and it is recommended, as shown in Table 4-3.

Layer Thickness

The thickness of individual layers of gravel depend on the media and the type of underdrain used. In addition, there are differences in many manufacturers' standard designs.

Minimum Layer Thickness

The foreword of the ANSI/AWWA B100 standard states that:

> "… the thickness of each layer of gravel should be at least three times the maximum particle size of gravel in the layer, but not less than 3 in. in any case."

Three times the maximum particle size for the gravel design shown in Table 4-3 would have been 3 in. × 3/4 in., or 2 1/4 in. According to the standard, a minimum layer of 3 in. would have been used regardless.

The ANSI/AWWA B100 standard allows a tolerance of ±8 percent of the gravel to be outside the specified size range. Particles larger and smaller than the specified sizes can therefore exist. Some handpicked pieces selected on one project were at least 1 1/2 in. long when the required size range was 3/4 in. × 1/2 in.

The upper two layers of the gravel design from Table 4-3 and Figure 4-3 are shown in Figure 4-5. On the left side, the upper layer of gravel is drawn as if all the grains of rock were 5/8-in. diameter, which would be the average size. In reality, the gravel is not spherical and has some aggregate, which is larger and smaller than the specified range, as mentioned previously. The gradation for the upper layer on the right side of Figure 4-5

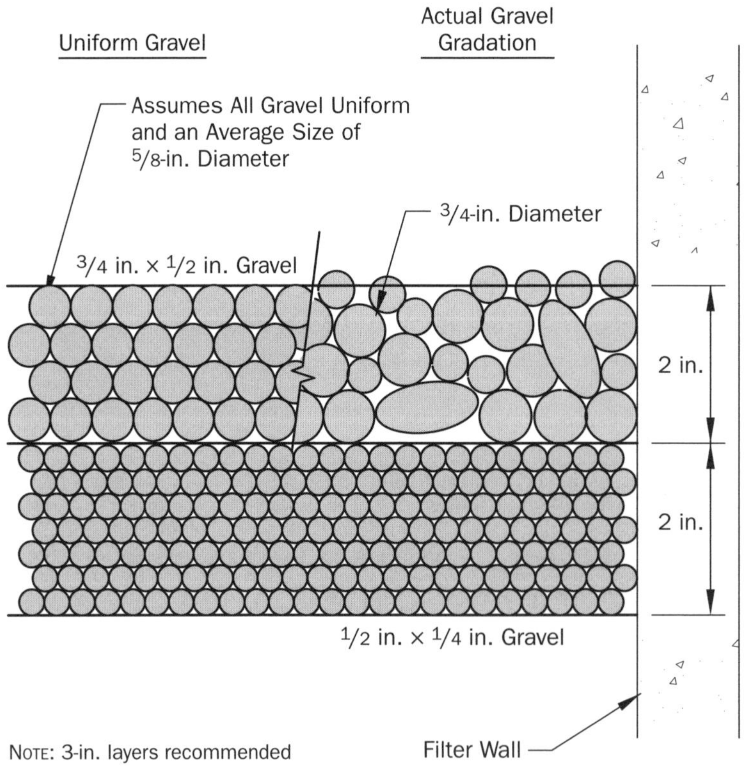

FIGURE 4-5. Filter gravel gradations

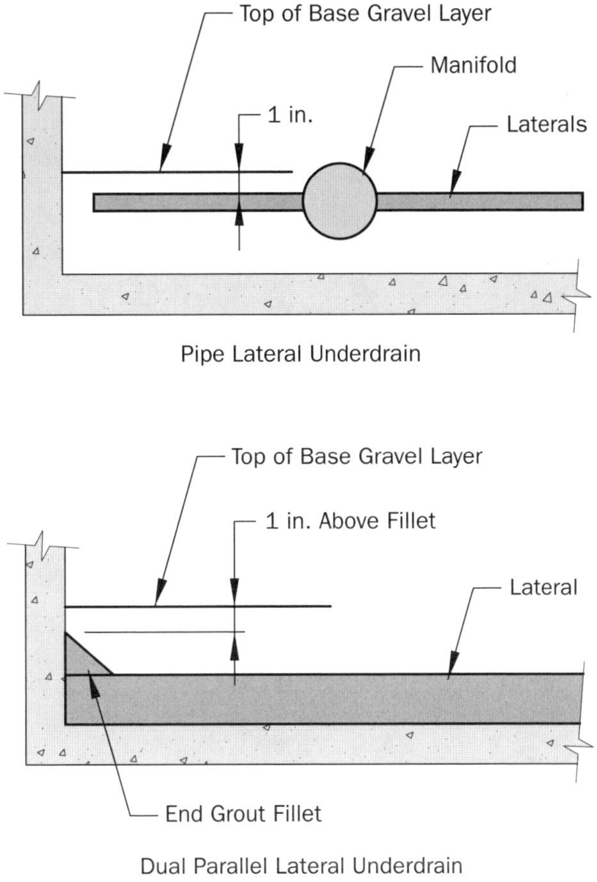

FIGURE 4-6. Base gravel layer depth

is more realistic, showing some smaller and some larger grains of rock. With a layer only 2 in. thick, the larger grains of rock will tend to weaken the layer. In places, a 2-in. layer of $^3/_4$ in. × $^1/_2$ in. gravel may be only two grains of rock deep. For this reason, it is recommended that the minimum depth for any layer be at least 3 in., as suggested by the ANSI/AWWA B100 standard.

Base Layer Thickness

The depth of the lower, or base, layer is dependent on the type of underdrain used. For example, the base layer is recommended to be at least 1 in. higher than the laterals in a pipe lateral underdrain or tepee/folded plate underdrain (Figure 4-6). For a dual parallel lateral underdrain, the recommended base layer depth should be at least 1 in. higher than the fillet at the end of the lateral. The reason is that the low head loss of this layer allows the backwash water to spread out and around the laterals or end fillets before the smaller layers are reached. The areas above these obstructions will receive a more normal backwash.

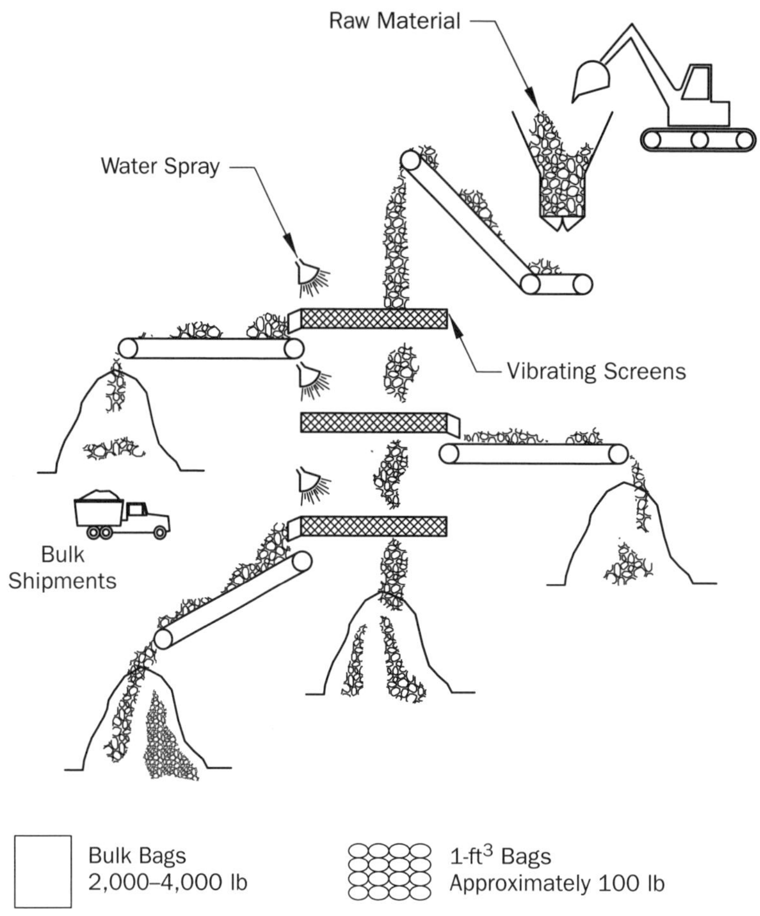

FIGURE 4-7. Gravel production

Methods of Production and Handling

Because of its weight and the expense of shipping, gravel is generally produced regionally for conventional filter plants. There is probably not enough business to support a large number of filter gravel suppliers for conventional plants. There are many producers of gravel; however, it is important to note that only a few of them are set up to produce material in the sizes required for filters.

Production

Gravel must be carefully screened to produce the necessary sizes required for use in filters. Graded gravel is typically produced by passing the raw materials through a series of vibrating screens, as shown in Figure 4-7. A water spray is also used to help move the gravel through the screens and to wash off some of the dirt and other contaminants.

Conveyors take the materials from the various screens for individual handling and packaging.

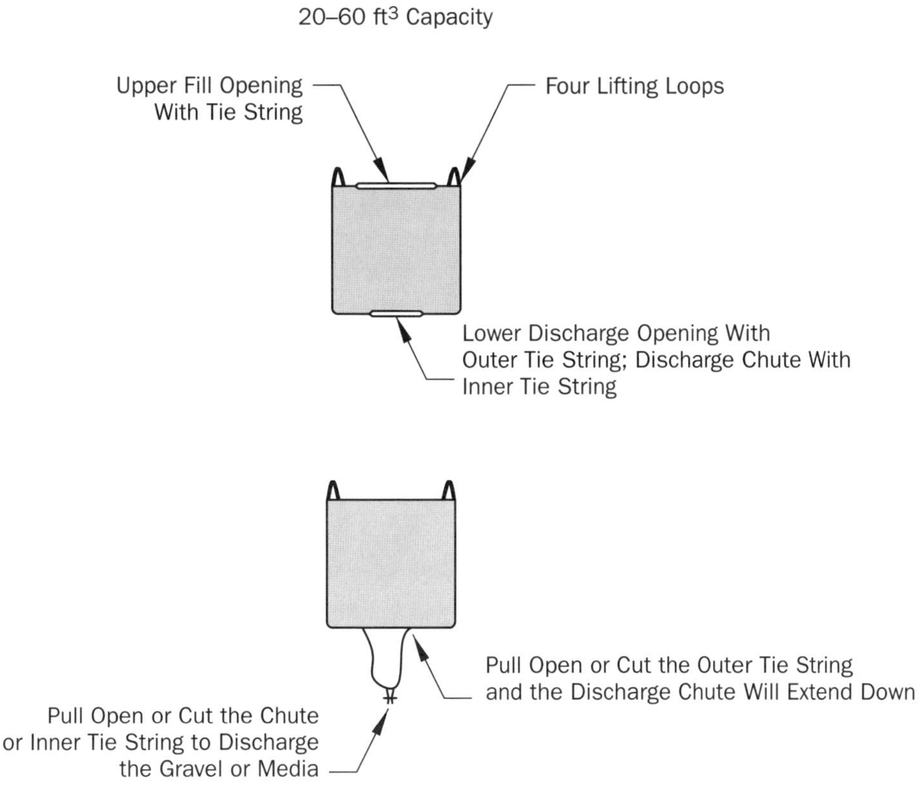

FIGURE 4-8. Bulk bags

Bagged Shipments

Gravel manufacturers who produce materials for conventional filter plants generally package it in either bulk bags (2,000–4,000 lb) or 1-ft^3 bags (100 lb). Bulk bags (superbags) are usually made of woven polypropylene and can be 3 ft to 4 ft square and 2 ft to 6 ft tall. One-cubic-foot bags for gravel are also usually made of woven polypropylene, while those for filter media may be made of paper or thin plastic sheeting. The smaller bags are normally used for pressure filters and smaller gravity filters, while the bulk bags are used for larger gravity filters that are accessible by a crane.

Both the bulk bags and the 1-ft^3 bags are normally made from a woven polypropylene material. The 1-ft^3 bags may either be sewn or closed with a wire tie.

The bulk bags have four lifting loops: an upper *fill opening* with tie string, a lower *discharge opening* with an outer tie string, and a *discharge chute* with an inner tie string (Figure 4-8).

It should be noted that the four lifting loops of a bulk bag should never be slung from a single lifting hook. A specialized lifting bar designed for the weight to be handled is recommended, per Figures 4-9 and 4-10.

FIGURE 4-9. Installing superbag of gravel

Bulk Shipments

Support gravel for large slow sand filters is commonly produced in bulk. The gravel manufacturer screens each size material into separate large piles for shipment by truck to the job site (Figure 4-7).

The piles of screened gravel are sometimes very large. On one project, the pile produced by the manufacturer for oversize gravel contained 3,500 tons. However, the project required only approximately 40 tons, or four truckloads of that size material. When the trucks were loaded, the gravel was taken from the outside of the pile, which is normal. Because of the nature of the materials, when gravel is produced into a large pile, a natural classification occurs where the outer edge is coarser than the interior of the pile. Therefore, when only a relatively small quantity of material is needed out of a large pile, the material obtained will most likely be coarser than the specified size

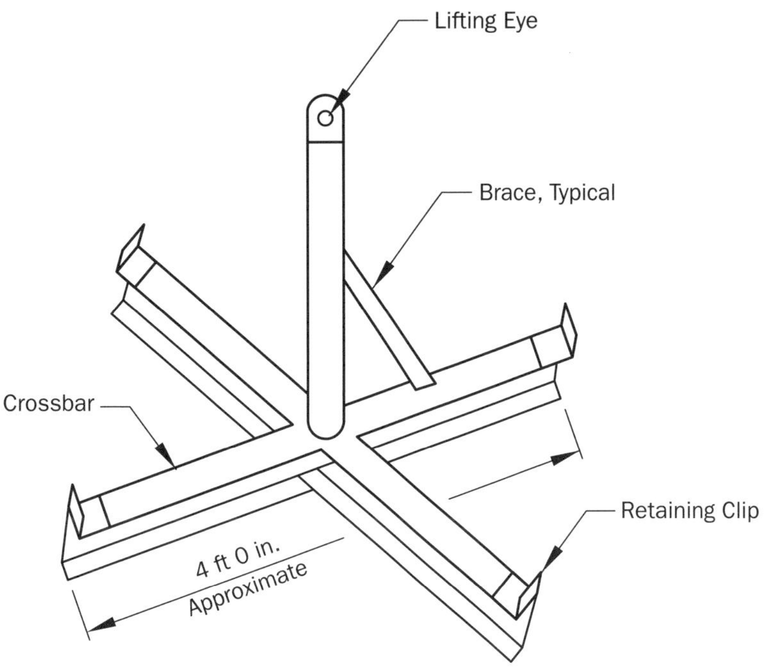

NOTE: Design to handle weight of bulk bags to be used (2,000–4,000 lb)

FIGURE 4-10. Bulk bag lifting bar

range even though the composite sample of the pile meets the specification. Quality control and methods of handling are, therefore, major issues in bulk shipments.

Bulk shipments also present a unique problem in that the gravel may classify in the truck because of vibration and bumping during transit. With adequate vibration, the finest gravel will settle to the bottom, with the coarsest on top. For this reason, all the gravel in a given truck should go in the same filter bay. It is very difficult to split a truckload between two filters and keep the proper gradation.

Classification during shipment of material in bags is not normally a problem, because material from one bag tends to mix with others during installation. The volume in one bag is usually too small to create much of a problem.

On-Site Quality Control

Gravel supplied by manufacturers who specialize in filter materials is seldom a problem. However, gravel supplied in bulk shipments from manufacturers unfamiliar with filters could potentially vary from the specifications. As a result, it is helpful to have a series of sieves on hand for testing purposes at the job site. A manual testing procedure is shown in Figure 4-11, with typical results in Table 4-4.

The material sampled in Table 4-4 contains 12.5 percent of its total weight, which is smaller than the specified range. As a result, this sample would fail. No more

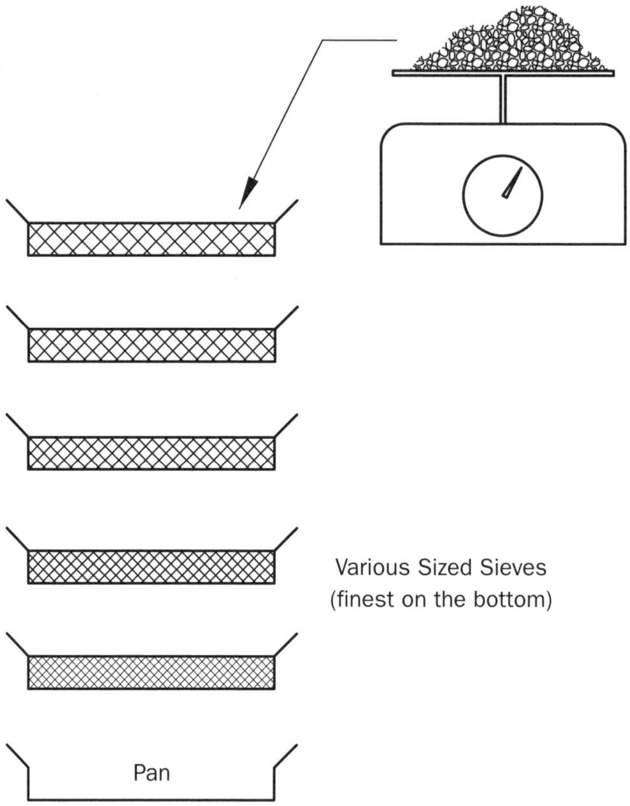

1. Weigh the gravel sample to be tested.
2. Select a series of sieves for the range to be tested.
3. Pass the sample of gravel through the various sieves. Shake the stack of sieves manually or with a machine.
4. Develop a table of the percentage of material passing each sieve.
5. Compare the results with ANSI/AWWA B100 and the project specifications.

FIGURE 4-11. Sieve testing gravel or media

TABLE 4-4: Typical gravel sieve results*

Screen Size	Weight Retained	% Retained	% Passing
1 in.	—	0	100
3/4 in.	8 oz	5	95
1/2 in.	120 oz	75	20
3/8 in.	12 oz	7.5	12.5
1/4 in.	20 oz	12.5	—
Pan	—	—	—

*Specification: 3/4 in. × 3/8 in.; sample weight: 10 lb (160 oz)

than 8 percent above or below the specified range is allowed by the ANSI/AWWA B100 specifications.

Test results obtained from manual on-site testing cannot be considered as laboratory-grade results. However, on-site testing can determine if there is a potential for a problem and whether or not materials should be rejected or further laboratory tests conducted. For laboratory tests, the American Society for Testing Materials standard specifications used are:

- ASTM E11, Standard Specification for Wire Cloth and Sieves for Testing Purposes
- ASTM C136, Standard Test Method for Sieve Analysis of Fine and Coarse Aggregates
- ASTM C702, Standard Practice for Reducing Samples of Aggregate to Testing Size
- ASTM D75, Standard Practice for Sampling Aggregates

Volume Calculations

When specifying the quantity of gravel required for a filter, the basin dimensions and layer depths are normally given along with each size of gravel used. However, the actual volume required is not normally specified. The reason is that gravel layers overlap, as shown in Figure 4-12. In other words, the gravel in Layer B occupies some of the voids between the aggregate in the top of Layer A, creating an overlap. As a result, more gravel is needed in Layer B than would ordinarily be calculated.

Another reason for additional gravel is that both the filter bottom slab and underdrains are typically laid slightly out of level, because of errors normally associated with the construction process. Therefore, to maintain the minimum required depth at the

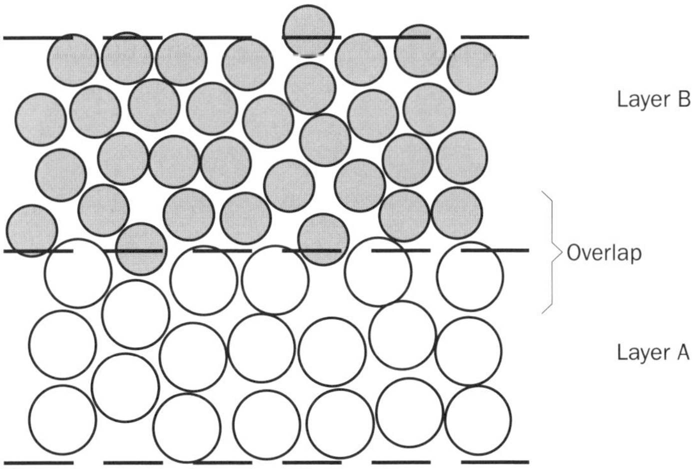

FIGURE 4-12. Typical overlap of gravel layers

shallow point in the filter and to ensure a level base layer surface, an excess volume of gravel is normally required.

To compensate for the overlap problem in the upper layers, it is recommended that at least 8 percent extra gravel be provided of each size. Depending on the distance that the floor and underdrains are out of level, excess volumes ranging from 15 to 20 percent may also be required for the base layer. The intent is to maintain the specified layer thickness at the highest point of the floor or underdrain. The extra material will also allow for a small amount of wastage at the job site. It is recommended that additional material be specified as noted previously, above the calculated amount, to ensure that the proper depths are obtained in the filter.

Installation

The following guidelines are recommended for the installation of support gravel.

Inventory

The inventory list should be checked to determine if there is enough gravel on-site to do the job. The contractor or manufacturer should be notified of any shortages and the required shipping schedule.

Shipment

The gravel should not be received at the job site more than 30 days in advance of installation. If the gravel is on-site for more than 30 days, the bags must be covered with plastic for UV protection. Normal bulk bags and 1-ft^3 bags will decompose with prolonged exposure to sunlight.

Handling Superbags

The manufacturer's instructions should be referred to regarding handling and use of superbags.

NOTE: Superbags require a four-point pickup when lifted by a crane. A single point pickup can break a superbag. Figure 4-10 illustrates the recommended lifting bar.

Height Calculations

White, yellow, or orange chalk (or an environmentally acceptable spray paint) and a straight edge should be used to mark the wall in one location, indicating the height of the gravel for each lift (or layer). When leveling or checking the gravel height, water should be run to the mark for each lift. The finished height should be as specified or as discussed previously.

Media Load Chart Calculations

The amount of gravel required for each lift from the area of the filter should be calculated, and the required depths using the overage percentages discussed previously should also be calculated. A gravel load chart should be prepared for this purpose (Figure 4-13).

Material Type	Total Inventory Quantity	Depth	Amount Needed per Filter
Gravel			
$1^1/_2$ in. × $^3/_4$ in.	________ lb or ft^3	________ in.	________ lb or ft^3
$^3/_4$ in. × $^3/_8$ in.	________ lb or ft^3	________ in.	________ lb or ft^3
$^3/_8$ in. × $^3/_{16}$ in.	________ lb or ft^3	________ in.	________ lb or ft^3
$^3/_{16}$ in. × #10	________ lb or ft^3	________ in.	________ lb or ft^3
Garnet Gravel	________ lb or ft^3	________ in.	________ lb or ft^3
Other	________ lb or ft^3	________ in.	________ lb or ft^3
Media			
Greensand	________ lb or ft^3	________ in.	________ lb or ft^3
Carbon	________ lb or ft^3	________ in.	________ lb or ft^3
Garnet Sand	________ lb or ft^3	________ in.	________ lb or ft^3
Silica Sand	________ lb or ft^3	________ in.	________ lb or ft^3
Anthracite	________ lb or ft^3	________ in.	________ lb or ft^3

FIGURE 4-13. Typical media load chart

Support Gravel Installation

Gravel Bag Weight

The quantity of gravel in each bulk bag (if used) should be checked, as it varies widely. An approximation of 100 lb/ft^3 can be used for this purpose. Each gravel bag should be marked with its weight. If a partial bulk bag is required to complete a layer, the bag weight should be divided by its height. The resultant weight per inch can be used to calculate the amount from that bag to be used.

Dividing a bulk bag can be difficult. Gravel can be removed through the top opening manually using buckets, or the discharge chute can be retied.

Retying the discharge chute is possible, but it is not easy. The first half of the bag should be emptied slowly per the "Bag Unloading Procedure" outlined in a following section. When approximately half of the bag has been unloaded, the crane should be stopped and a strong cord should be wrapped around the unloading chute. The cord should be tightened to the extent possible, and the bag raised gradually with the crane while tightening the cord further. When the unloading chute is completely closed off, the cord should be secured and the bag removed from the filter.

Dividing a bulk bag takes practice. Therefore, it is recommended to experiment with the first bag of that lift. It may take several attempts to divide a bulk bag successfully.

Leveling

Once the gravel for a given layer is in the filter, the water level in the filter should be adjusted to the mark on the sidewall established previously, and the gravel raked level using the water surface as a guide. Long boards can be used to "rough screed" the gravel level. However, steel rakes are usually required for final leveling to the water mark. It should be noted that it is very difficult, if not impossible, to level gravel accurately without using water.

The theoretical standard for leveling gravel is for the finished surface of a layer to be level within ±1/2 in. of the nominal grain size of that layer. For example, 3/4 in. × 1/2 in. gravel should be leveled within ±3/8 in.

For larger sizes of gravel, the theoretical standard for leveling can be closely approximated. However, for 1/2 in. and smaller size gravel, the theoretical standard is impossible to achieve. The smaller sizes of gravel tend to be somewhat "soupy," and they cannot be leveled to any high degree of accuracy. Even walking boards will leave an unacceptable impression. The field inspector, working with the contractor's personnel, should get the layer leveled to the extent possible and proceed to the next layer. However, it is important that there be no large depressions or mounds. The goal should be for the smaller gravel to be leveled within ±1/2 in. as a minimum.

Walking Boards

After the first layer of gravel has been leveled, any personnel working on the filter must walk on boards to prevent making depressions in the media. It is recommended that the boards be a minimum of 3/4-in. plywood, 1 ft wide × 2 ft long.

Verification of Quantities

The calculated amount of gravel for a given layer should be installed and raked flat. If the layer is short, the calculations and bag weights should be checked. The manufacturer should be notified of any anticipated changes to the quantity of material required.

Final Gravel Layers

The remaining gravel should be installed to the specified marks on the wall in a similar manner using water as a leveling guide.

NOTE: After the first layer, *do not throw the gravel with a shovel.* It can gouge a hole in the layer below. The gravel should be pushed with a rake or "screeded" with a board.

Bag Unloading Procedure

Bulk Bags/Dry Method

The superbag should be lowered to within 15–18 in. of the bottom. When the first (outer) drawstring is pulled, the onloading chute will come down. It should not be more than 6 in. from the bottom (Figure 4-14). Next, the second drawstring should be pulled. A small amount of gravel will come out of the bulk bag until it forms a pile sufficient to choke off the flow. The bag should be raised very slowly to allow it to empty gradually. In no case should the gravel be allowed to free-fall onto the bare underdrain or any previously installed layer of gravel. If that occurs, all previously installed material will have to be removed and new material installed. Whenever emptying bags, each bag should be throttled to prevent it from discharging at full speed.

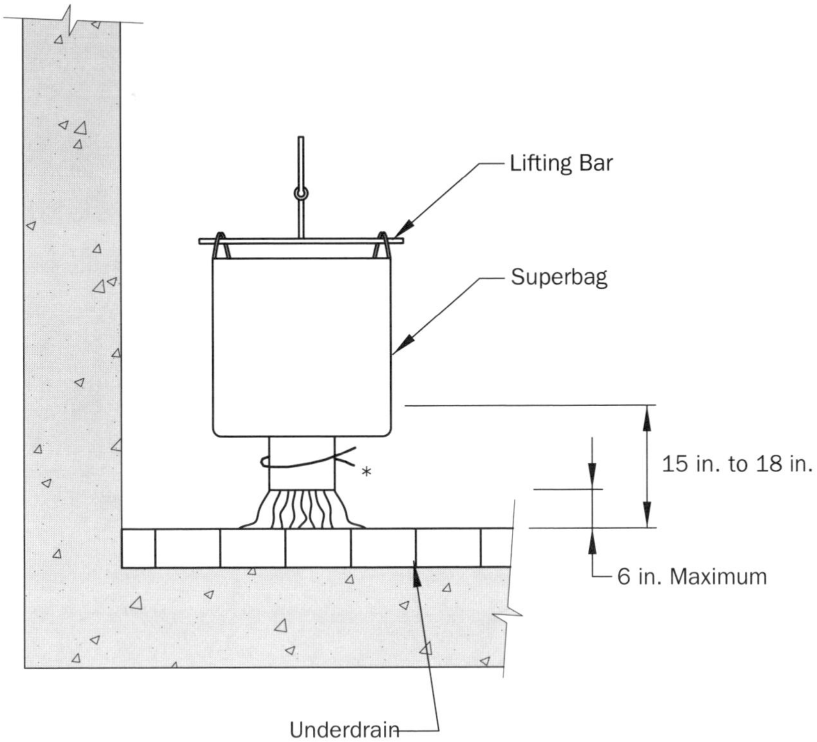

*Use rope to choke off discharge chute while raising slowly to split gag (if required).

FIGURE 4-14. Emptying a bulk bag

If the bag cannot be lowered close to the bottom, it must be emptied slowly onto a splashboard (Figure 4-15). The media will hit the splashboard and spread without disturbing the layer below. The splashboard should be a minimum 4 ft × 8 ft sheet of plywood, $^5/_8$ in. or $^3/_4$ in. thick. It should be raised on one end while the other end sits on a baseboard to prevent it from digging into and disrupting the layer below. It should make approximately a 30° angle to the horizontal. Care should be taken to count all boards and implements to ensure that they are removed when the work is complete. The use of a splashboard is not recommended unless the bulk bag cannot be lowered to the bottom of the filter.

After a pile is established by emptying one bag, others can be emptied on top of it. If a splashboard is being used, as shown in Figure 4-15, several bags may have to be emptied to make an adequate size pile. After that, the splashboard may be removed and other bags emptied on top of the pile. Care should be taken when removing the splashboard and digging up the baseboard that the layer below is not disturbed. The gravel can then be spread out from that point. Several piles may be established to reduce the labor and time required for raking.

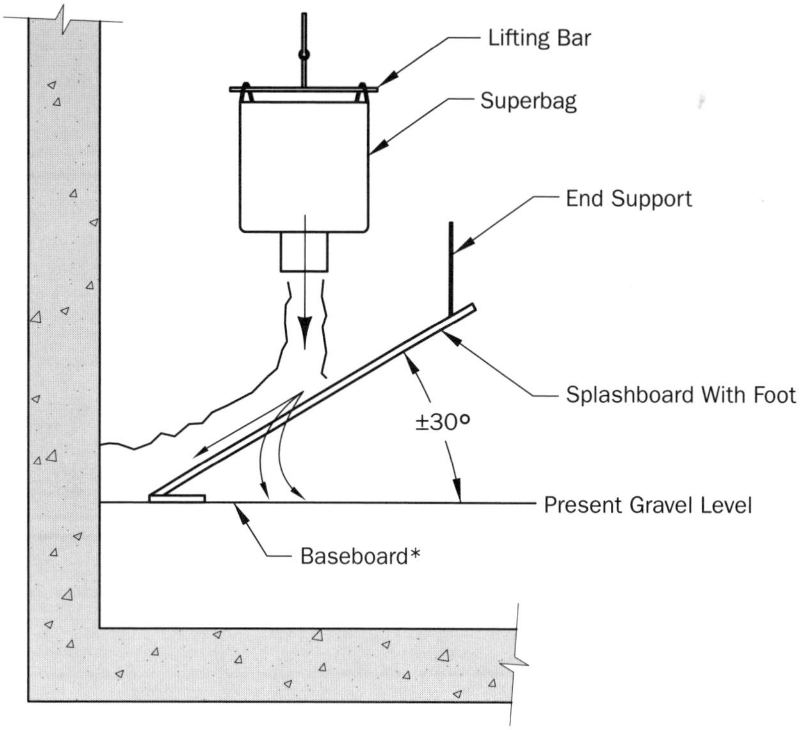

*Uncover and remove baseboard carefully so as not to disturb the layer below.

FIGURE 4-15. Laying gravel or media with a splashboard

NOTE: When installing gravel or media, all workers in the filter should wear dust masks.

Splitting Bulk Bags

When installing media with 1-ft^3 bags, the exact number of bags required should always be installed. However, when using bulk bags, a given layer of support gravel or media may only require a partial bag to provide the desired thickness.

To split a bag, it should be lowered close to the bottom and the discharge chute opened as shown in Figures 4-8 and 4-14. A rope should be wrapped tightly around the discharge chute as shown in Figure 4-14. The bag should be raised gradually and the rope pulled tightly until the flow of gravel or media is stopped.

Once the discharge chute is secure, the bag can be lifted out of the filter and stored until needed. Because the discharge chute is extended, great care must be taken when setting the bag down. Any sideways movement while the chute is in contact with the ground could tear out the bottom of the bag.

Splitting a bag is not easy. Several attempts may be necessary to do it properly. Each size of gravel or type of media will act differently. As a result, it is recommended

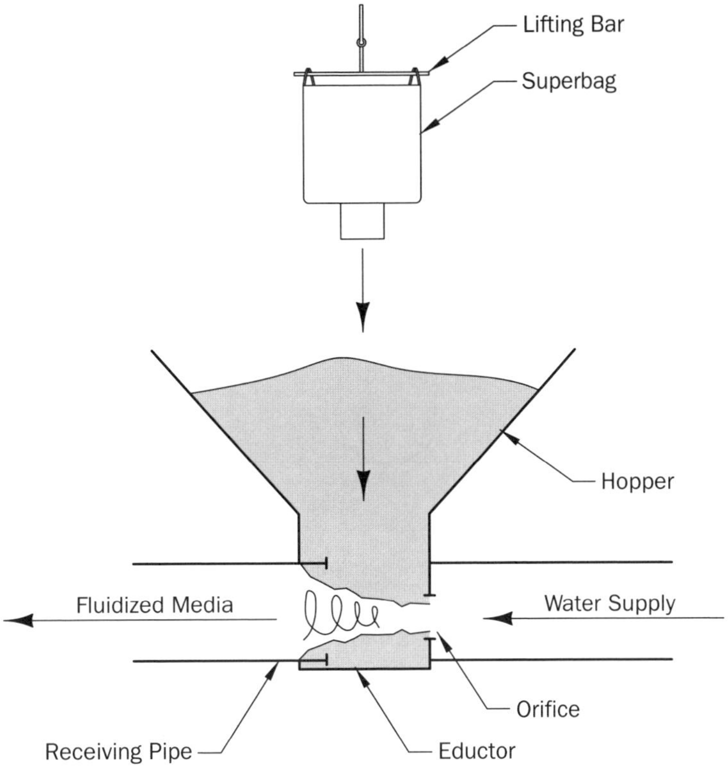

FIGURE 4-16. Eductor loading system

that the installers practice on the first bag for the layer in question in order to have the best chance of splitting a bag properly.

Installation (Eductor Method)

An eductor may be used to "sluice" the gravel (if required) in with water (Figure 4-16). For this method, a minimum of 4 ft to 5 ft of water is also required in the filter. In addition, it is necessary to keep the discharge hose close to the surface of the water or slightly submerged. The hose discharge must also be horizontal or parallel to the water surface to prevent the new material from disturbing the layer below. Using this method is not recommended except as a last resort.

One-Cubic-Foot Bags

The installation procedure for gravel in 1-ft^3 bags is similar to that for bulk bags, except that it is more labor intensive. The bags are initially raised to the top of the filter individually by a conveyor, or on a pallet with a forklift. They are then lowered into the filter one at a time by hand. The bags should be set on a plywood board to prevent damage or fouling of the underdrain or the layer below.

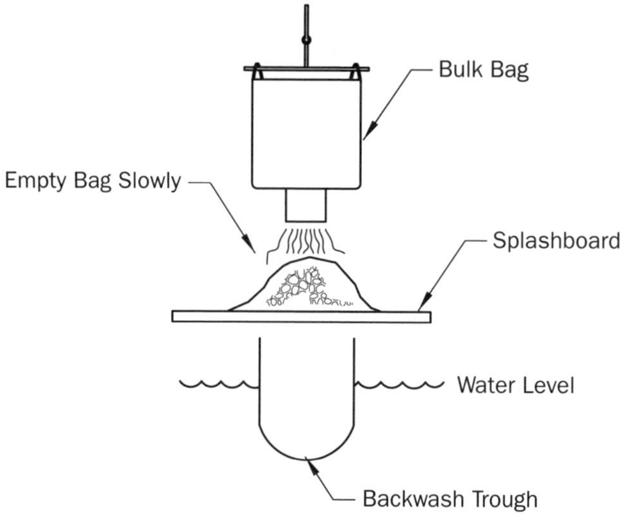

NOTES:
1. Verify load capacity of troughs and weir boards (may only work for small quantities).
2. May have to span two troughs or use the sidewall of the filter instead.

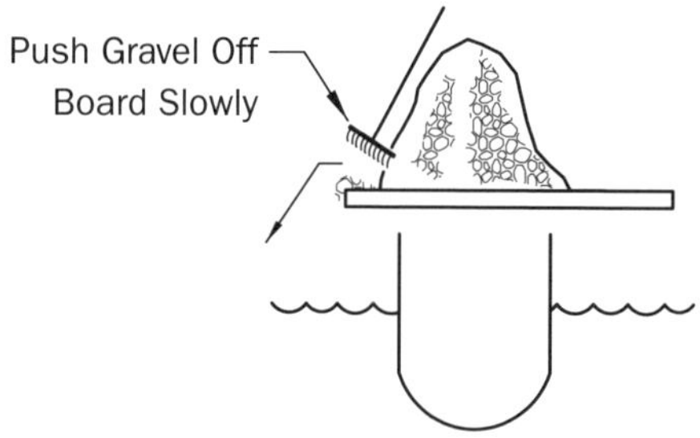

FIGURE 4-17. Wet method of installing gravel

After untying or cutting open a bag, the open end should be lowered close to the underdrain or layer below, and then emptied gently. As with any method of installing gravel, the intent is to protect the underdrain and not damage the layer below (if any).

Raking and leveling each layer will be similar to the procedure previously outlined.

Wet Method

There are occasions when the wet method of installing gravel may be used, such as in pressure filters or gravity filters where a bulk bag will not fit between the troughs.

The wet method involves filling the filter with water almost to trough level. Construct a splashboard over one trough or between troughs if structurally feasible (Figure 4-17). A bulk bag should be emptied onto the board slowly and the gravel pushed off the board gradually. If 1-ft^3 bags are used, the gravel should be trickled slowly into the water.

It is very important that large amounts of gravel *not* be dumped quickly into the water. Large amounts of gravel dumped suddenly tend to hit the bottom with a piston-like effect, disrupting the layer below.

Once the calculated amount of gravel is layered, the water level must be lowered to the mark previously established for that layer. The gravel is then raked level, as previously discussed. After that, the filter is refilled for the next layer.

The wet method is not commonly used unless necessary. It requires the use of a considerable amount of water and usually takes longer than the dry methods.

Dry Method

Support gravel may be installed using either the wet or dry method. The wet method will be similar to that previously outlined. However, the dry method is slightly different.

Using the dry method, one worker lowers a 1-ft^3 bag of gravel down into the pressure filter, where another worker receives it. The bag should be set down on a board or pile of gravel being installed for the current layer. The bag is then untied or cut and emptied gently. All strings or wire ties should be collected and removed from the filter. Leveling can then proceed as outlined previously.

It should be noted that the shape of a pressure filter can result in a significant unswept area in the bottom of the filter, below the underdrain. Over a period of time, solids can collect in this area and potentially cause process problems. It is recommended that this unswept or "dead" area be filled with concrete, unless it is not compatible with the process (Figure 4-18). Chapter 10 offers a more in-depth discussion of pressure filters.

Cleaning

During installation of the gravel, any foreign objects, such as wood debris or strings, etc., should be removed manually. After it is installed, the gravel can be backwashed for cleaning. However, it is usually simpler to perform that task at the same time as backwashing and cleaning the media.

On completion of installation, care should be taken to remove all tools, walking boards, tie strings, etc., before the workers leave the filter.

Field Testing

Sieve analysis for the actual material produced will be shipped with the material. Sieve tests may be conducted at the job site to check for general conformity to the specifications. If the material is questionable, installation should be stopped until independent lab tests can verify that the specifications have been met.

Installation Tolerance

The normal tolerance used for installing support gravel is $\pm^1/_2$ in. of the nominal size. It is not practical to achieve this tolerance with smaller sizes. However, it is possible to come reasonably close to it with the coarser layers. The author normally uses

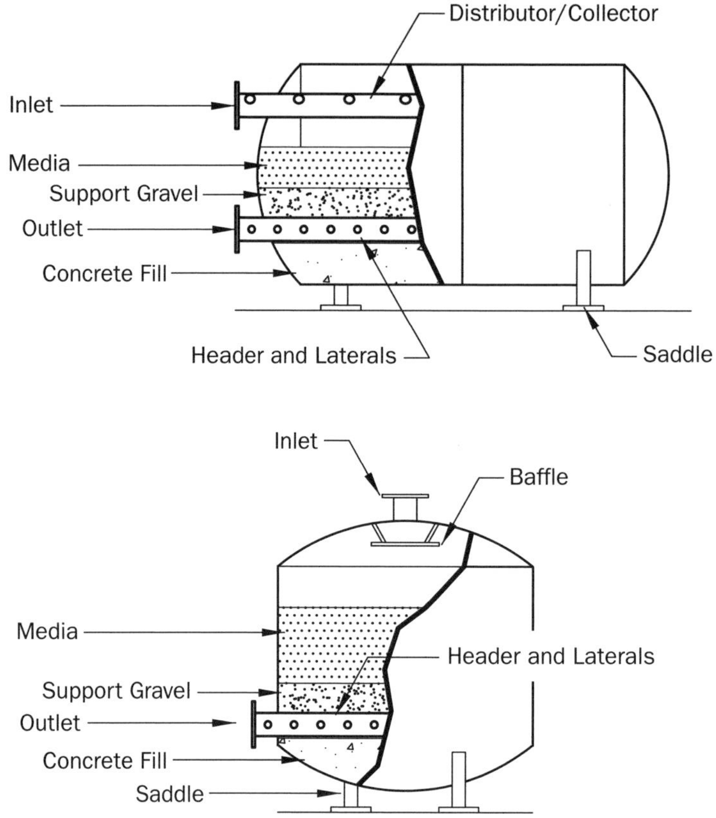

FIGURE 4-18. Typical horizontal and vertical pressure filters

an installation tolerance of ±$^1/_2$ in. of each layer separately and the final depth of gravel overall as a conservative estimate if the installation is done properly.

Operational Comments

Sidewall Effects

The sidewall of a filter is the weakest point. Within the bed, the gravel grains tend to interlock with each other to some extent. However, at the sidewall there are periodic void spaces not found inside the filter, as shown in Figure 4-19. For the same reason, full-scale filters will usually provide better service than pilot columns with a smooth plastic sidewall.

Media Loss

The life of a filter is normally dependent on keeping the media clean and keeping the gravel intact. If the gravel becomes upset, sand and high density garnet sand will be lost through the underdrain into the finished water. Leakage downward can be analyzed by sample ports installed on the bottom of the effluent manifold or by inspecting the clearwell. If abnormal media loss is occurring, it may mean that the gravel is disrupted.

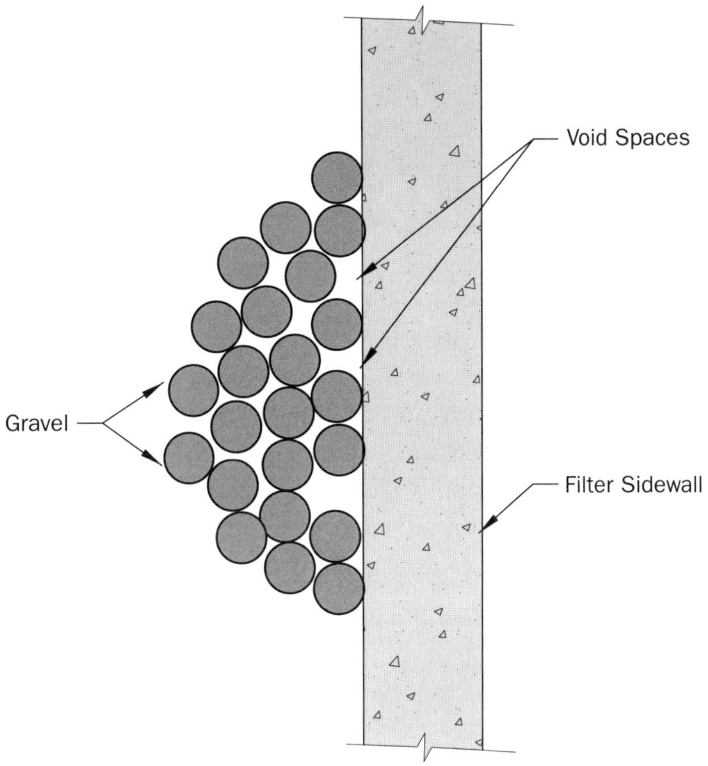

NOTE: The sidewall is the weakest area of a filter caused by the void spaces on the sidewall.

FIGURE 4-19. Sidewall effects

Anthracite breaks down with time and is usually lost over the top of the trough weirs in the backwash wastewater. Anthracite is usually not lost down through the bed, unless the underdrain is ruptured or there is a major failure of the gravel layers. High backwash rates, excessive surface wash, or air scour still occurring when the backwash water is weiring over the trough also contribute to the loss of anthracite.

Design Issues

Repeated backwashes over a 15- to 20-year period will disrupt gravel very slowly. However, uncontrolled air and hydraulic shock can upset a gravel bed in one backwash cycle. Once a gravel bed becomes upset, the only recourse is to clean out the entire filter and fill with new material.

During air scour (if used) and water backwash, the media expands and rolls and is agitated. It is important that the gravel does not move at any time. If it does move, it loses its ability to support the media. Therefore, it is necessary to have a very stable gravel design to ensure long life.

Because of the expense involved, it is highly desirable to design the backwash system to eliminate the potential for uncontrolled air or hydraulic shock. These issues are discussed separately in Chapter 7.

Direct Retention Underdrains

Most underdrain types do not have 100 percent coverage of the filter surface. Many underdrains leave spaces on the sides and ends, while others have grout fillets to cover the ends of their laterals. When support gravel is used, backwash water can travel laterally through the gravel, to some extent, and up the walls of the filter (Figure 4-20). Even though the underdrain does not cover the complete area of the filter, the media can still be cleaned. However, with direct retention underdrain types, the media is installed directly on top of the laterals, with no support gravel. Therefore, the backwash water tends to go straight up, without diffusing to the side (Figure 4-21 and 4-22). The resulting unswept area may not get properly cleaned and could result in the accumulation of mud over those areas. Air scour, if used, tends to mitigate this problem.

If a direct media retention underdrain is used, the unswept areas should be minimized, and air scour is strongly recommended.

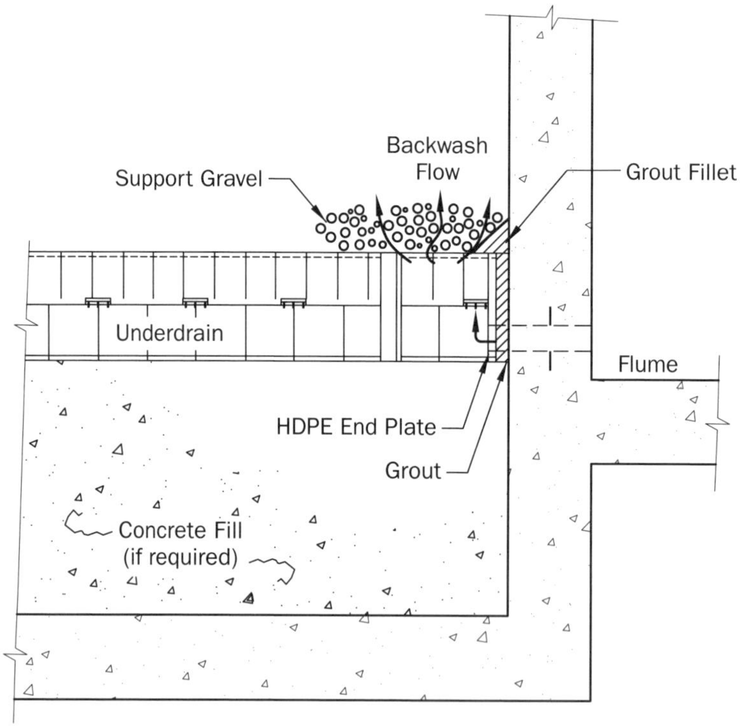

NOTE: Backwash water rising out of the underdrain can move laterally through the gravel because of low head loss.

FIGURE 4-20. Water movement in gravel

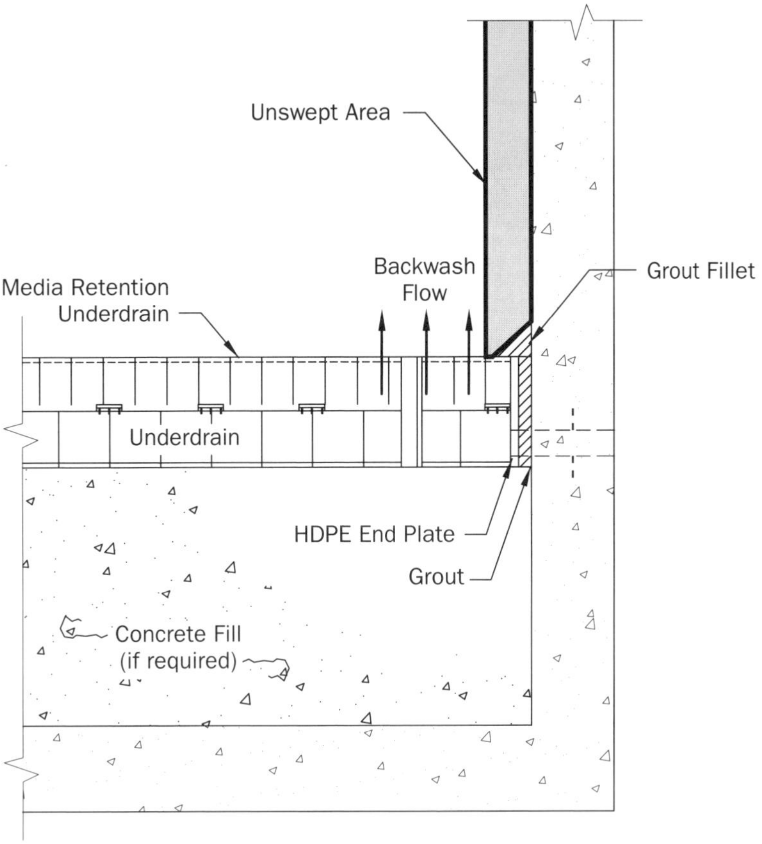

FIGURE 4-21. Edge condition

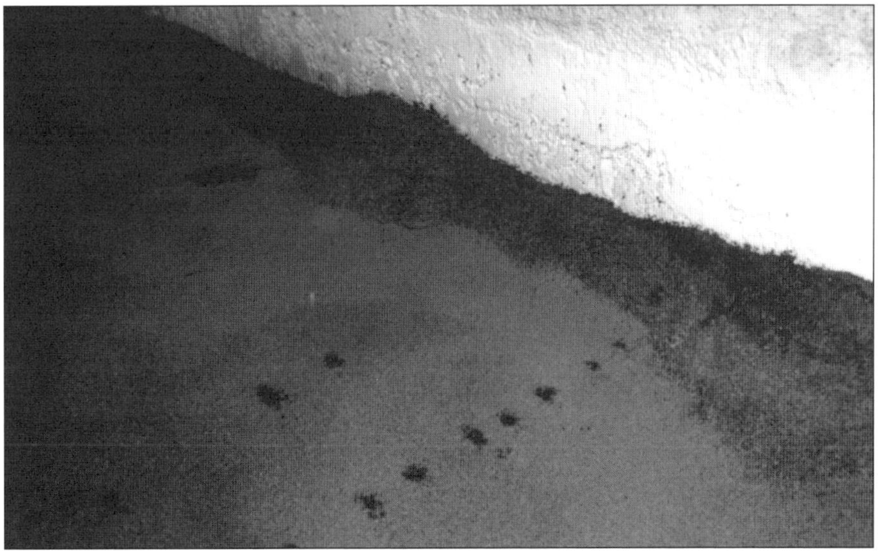

FIGURE 4-22. Photograph showing edge condition in filter

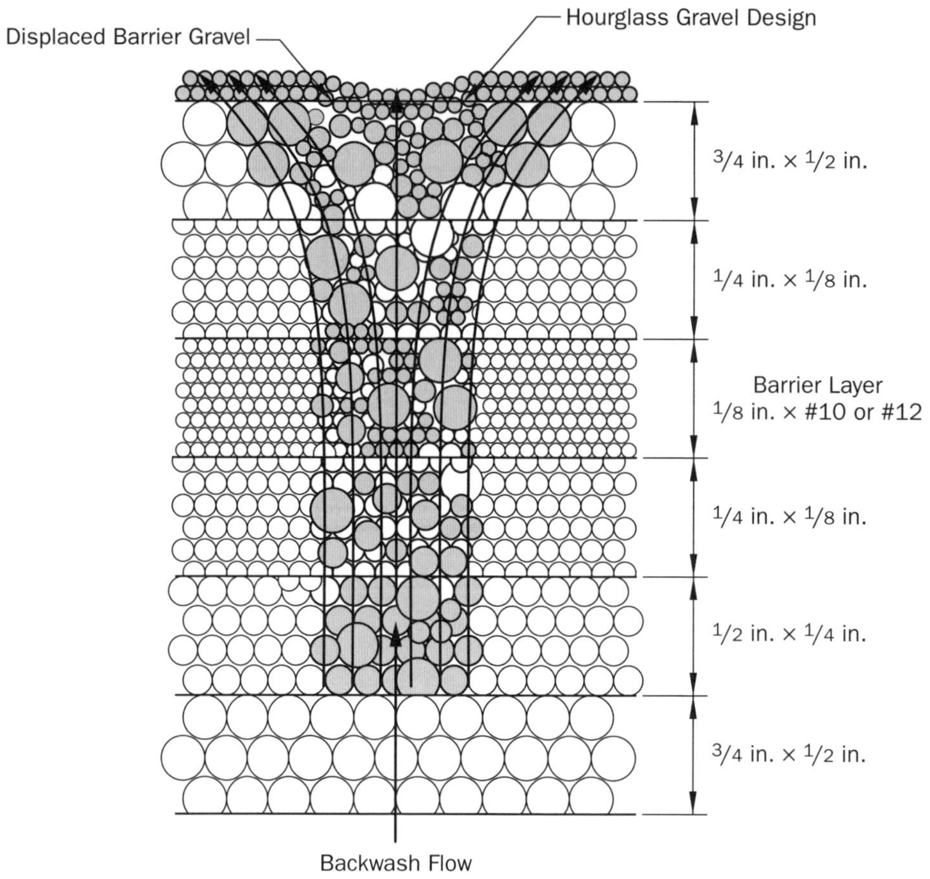

FIGURE 4-23. Disruption of hourglass support gravel

Gravel Stability

The life of the filter media is dependent on the life of the gravel. The stability of the top layer is critical to gravel life. If the gravel begins to move, the filter will be in a failure mode. Once the support gravel becomes upset or intermixed, the performance of the filter is compromised, and it will deteriorate over time. If the media and support gravel are not replaced, it could eventually result in deterioration of effluent quality, short filter runs, and potential clogging and catastrophic failure of the underdrain. The time involved is variable and depends on the status of other plant equipment and processes.

Upset Gravel

Upset or blown gravel is characterized by an intermixing of the layers, as shown in Figure 4-23. In that figure, a typical hourglass support gravel design is illustrated. With this gravel design, an upset or blown condition is usually characterized by a depression and/or the presence of barrier gravel on the gravel surface. Single-taper gravel that has been blown is shown in Figure 4-2. An upset condition with single-taper gravel is typically characterized by having a depression and/or gravel from a lower level on the surface.

A filter can be blown with one occurrence of either hydraulic shock or uncontrolled air. Once this occurs, the condition is irreparable and will continue to degrade. From an operational point of view, blown filters typically have shorter and shorter filter runs, which are terminated by turbidity breakthrough as opposed to terminal head loss.

A blown filter is usually described as having a hole blown through the gravel layers. In reality, there is not always an actual hole, unless there has been a catastrophic failure. Usually there is an area in which the gravel layers have become intermixed. Such an area is a *weak spot* that will appear soft when probed. Because of the weakness, the support function has been compromised. Higher than normal backwash flow will go through this area. In addition, there will also be higher than normal flow in filter mode, which is usually accompanied by a loss of media.

Media that has penetrated the support gravel and entered a pipe manifold underdrain is shown in Figure 4-24.

As a result of the higher than normal backwash flow through a blown area, the backwash flow in adjacent areas is lower, resulting in inadequate cleaning and significant mudballing.

Blown areas often have a "rat hole" through the media or a depression on the media surface, with mudballing around the sides.

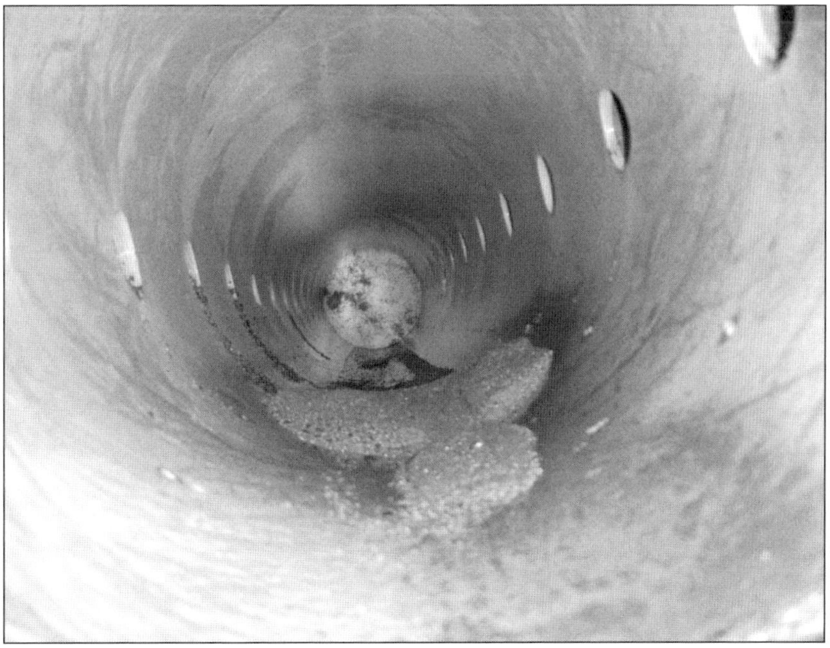

FIGURE 4-24. Media in pipe manifold

Definition of Upset Gravel

Gravel layers that have become intermixed will have begun to lose their support capability. To determine the condition of the gravel, a hole can be dug down through the media. The exposed surface of the gravel can be examined for intermixing.

Manual excavation is time consuming and only exposes a small area. Probing the gravel will allow the surface condition of the gravel to be evaluated. An intact gravel bed will have a firm and relatively level surface, whereas upset gravel will be uneven.

There is no exact definition in the available literature the author is aware of regarding what constitutes upset gravel in regard to surface level. However, any measurable difference to the recommended installation standard of ±1/2 in. represents a movement of gravel and is a problem. Measurements of ±5/8 in. could arguably be a result of differences in measuring techniques or installation tolerances. However, by the time the gravel becomes ±3/4 in. or ±1 in., it is definitely a serious problem. By that point, one-third to one-half of the upper layer has moved, which will leave little or no support for the layer above (Figure 4-25).

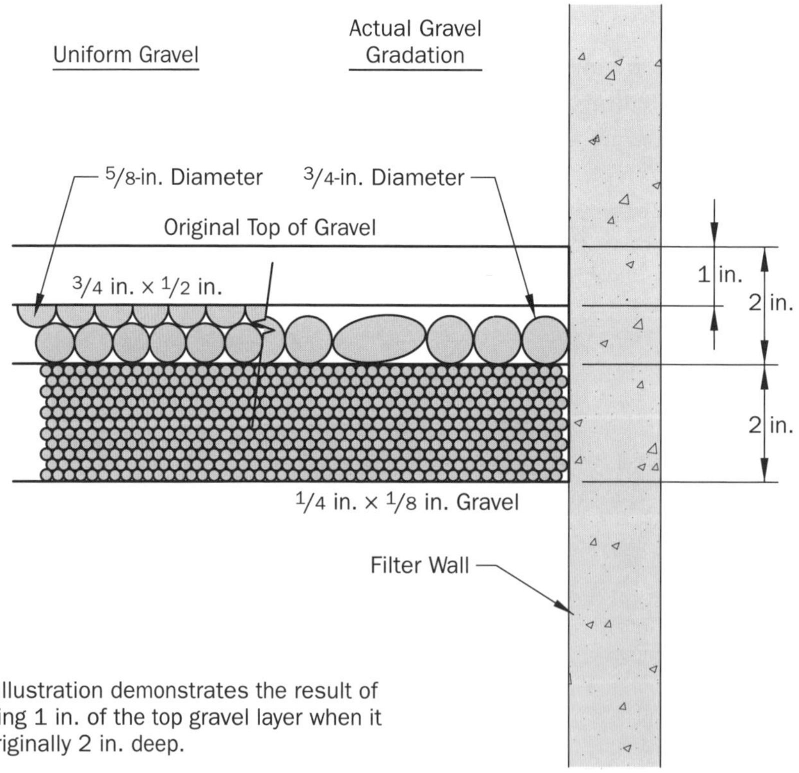

FIGURE 4-25. Upset filter gravel

One of the effects of using narrow layers of gravel, 2 in. thick for example, is that there is no room for error. Therefore, using layers with a minimum thickness of 3 in. is again recommended because it gives a greater safety factor.

Gravel stability measurements are made by probing, as discussed in the following section.

Probing the Gravel

It is recommended that a filter using fluidizing media be probed periodically to determine the condition of the gravel surface. The information obtained can be used as an early warning of potential problems.

To probe the filter, a length of 1½-in. or 2-in. PVC pipe is recommended, with a coupling glued on one end and a scale marked on the side, starting at zero on the bottom of the coupling. The probe should be lowered gently into the media during a high-rate backwash until the gravel can be felt (Figure 4-26). Care should be taken that a hole is not punched into the gravel with the probe. The surface of the gravel

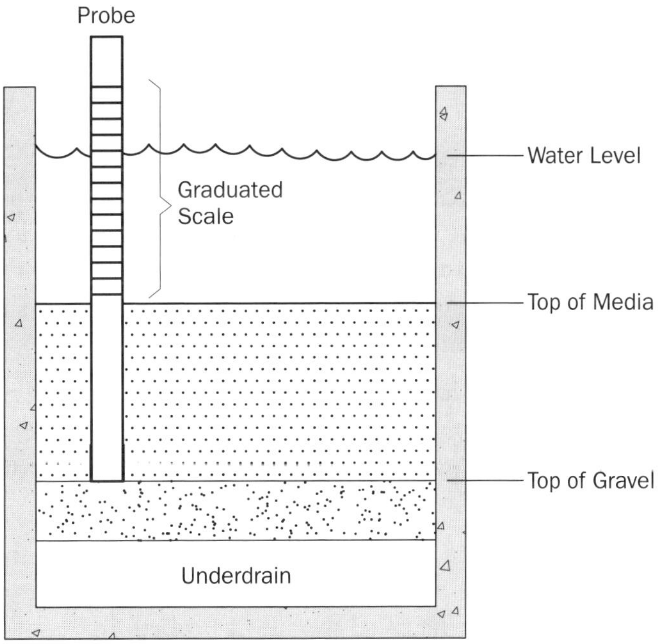

NOTES:
1. A typical probe would consist of 1½-in. or 2-in. PVC pipe with a coupling on one end and a scale marked on the side.
2. Probe the filter during a high-rate backwash, approximately 16–20 gpm/ft^2.
3. Probe very gently with a slight tapping motion.
4. The gravel surface should be reasonably flat and firm, not mushy. Be careful not to probe or tap too hard or holes can be punched in the gravel.

FIGURE 4-26. Probing gravel

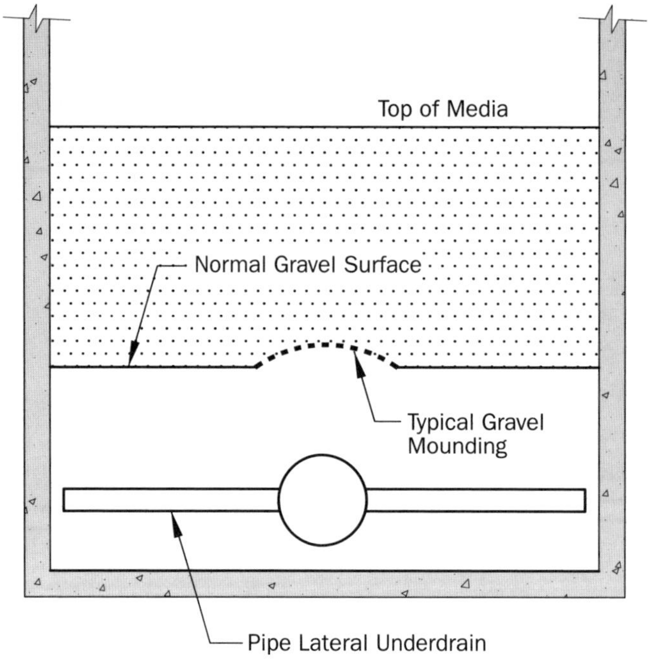

FIGURE 4-27. Gravel mounding with pipe lateral underdrains

should be gently touched and the probe rotated slightly to ensure that the surface of the gravel has been reached. The scale on the probe can be used to measure the water depth to the top of the gravel. Record the depth at that location and compare the results with depths measured in other places in the filter. If the results vary more than $\pm^1/_2$ in., a problem may be developing and an investigation should be made into the cause so that the problem can be resolved.

It should be noted that a slight mound may develop over the top of an embedded manifold in a pipe lateral underdrain or any unswept areas in other underdrain types (Figure 4-27). Support gravel and media will tend to shift from areas of high energy to low energy as a result of repeated backwash cycles.

Before beginning the probing, a backwash of the filter should be observed. Weak spots should be located, such as places where there are bursts of air or heavy plumes of solids. The weak spots identified should be probed. Intact or mounded gravel will feel solid, while upset gravel areas will feel soft and mushy, and it may be difficult to find a solid base.

Typical results from probing three different filters are shown in Figures 4-28 to 4-31. These filters had dual-media and hourglass support gravel, with an air scour backwash system.

The information in Figure 4-28 is a result of observations of the top of the media, probing the gravel, and excavation of an area (upper left) that appeared to be upset in Filter A. As a result of these investigations, this filter was shown to be definitely upset.

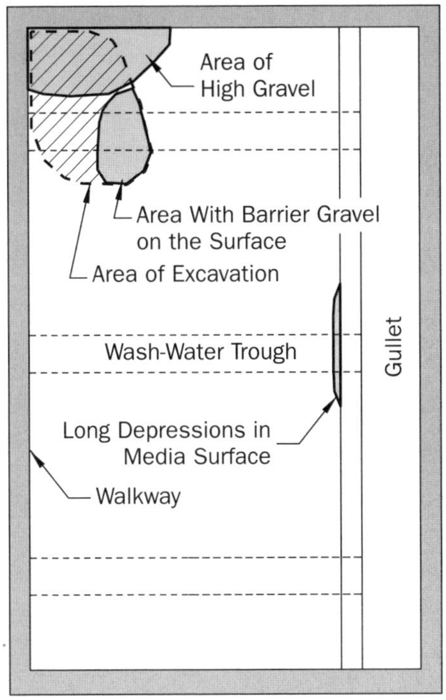

FIGURE 4-28. Filter A gravel condition

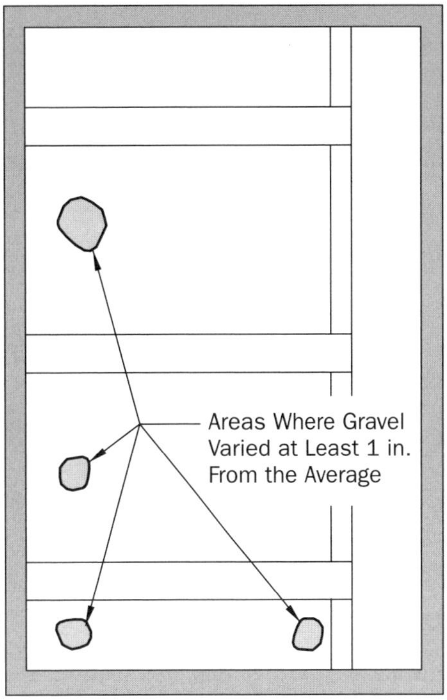

FIGURE 4-29. Filter B gravel condition

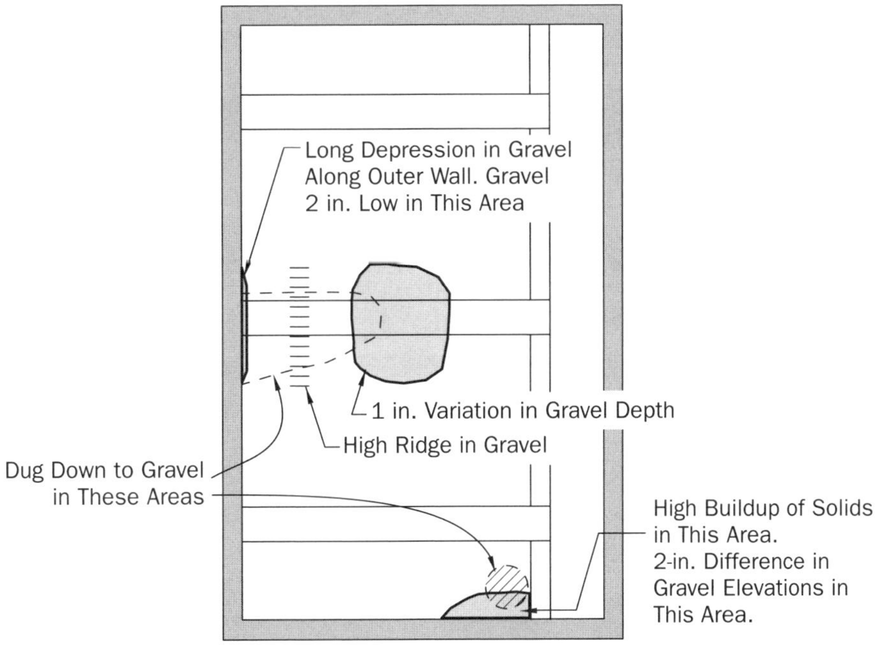

FIGURE 4-30. Filter C gravel condition

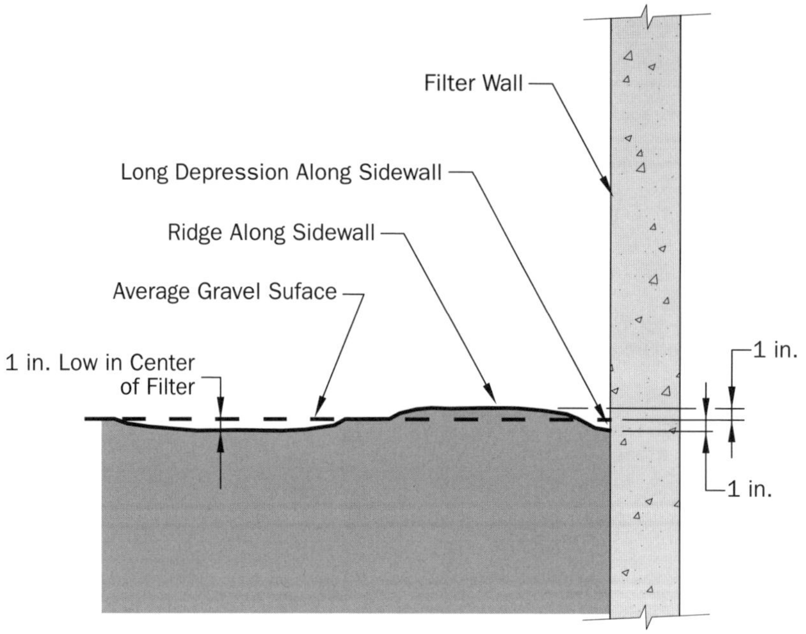

FIGURE 4-31. Filter C gravel section

Media was being lost into the clearwell down through a long depression along the sidewall on the right-hand side. In addition, the barrier gravel had blown through the surface in the upper left of the filter, creating a mounded area in the corner, which was not being cleaned effectively.

The results of probing Filter B are shown in Figure 4-29. Four areas were located where the surface of the support gravel varied at least 1 in. from the average. These areas represent potential locations of upset gravel, and this filter is probably in an early stage of gravel failure.

Filter C (Figure 4-30) was shown to be in a similar condition to Filter A. Media was lost down through a depression along the sidewall (Figure 4-31). There was also a large area of mounded gravel adjacent to the depression. All these factors indicate that the filter gravel is in an advanced stage of failure.

It is recommended that filters be probed at least once per year. The data should be retained for comparison with future probings. If any variations occur regularly, the plant should be analyzed to determine if improvements or modifications are warranted.

Summary

- Use 3-in. layers of support gravel as a minimum.
- The use of support gravel requires a deeper filter than a direct media retention underdrain. However, gravel is preferred by the author in some cases, with the possible exception of rehabilitating shallow filters.

Problem	Probable Cause	Proposed Solutions
Upper gravel (blown or intermixed)	• Hydraulic shock • Uncontrolled air • Air shutdown problem (refer to Chapter 6) • Contamination (leaves, pine needles, plastic, etc.)	• Resolve problems and completely rebed the filter
	• Installation problems	• Completely rebed the filter
	• Poor gravel design	• Completely rebed the filter with a revised gravel design

NOTE: The life of the filter depends on the support gravel remaining intact. Once the gravel becomes upset, the only solution is to completely rebed the filter. Because that is a relatively expensive and time-consuming task, great care should be taken to ensure that the plant is consistently operating properly and that potential problems are quickly eliminated.

Cribbing a small localized area and repairing the gravel has been practiced occasionally; however, it is not recommended. Cribbing (if done) should only be considered as a very short-term fix until the filter can be rebedded.

FIGURE 4-32. Support gravel troubleshooting guide

- Quality control during manufacturing and installation of support gravel is critical.
- The support gravel will be in failure mode if its surface is measured to be ±$^3/_4$ in. to 1 in. out of level.
- Once filter support gravel is determined to be in failure mode, steps should be taken to identify and resolve the problems that are causing those conditions. The media and support gravel should then be replaced as soon as possible before the conditions become worse. If the gravel and media are not replaced, it could eventually result in deterioration of effluent quality, short filter runs, and potential catastrophic failure of the underdrain.
- A support gravel troubleshooting guide is included in Figure 4-32.

CHAPTER 5

Filter Media

Introduction

The selection of the size and type of filter media used on a given project is usually determined by process considerations. Pilot studies may be used to select the media size or, in the case of a plant expansion, the same materials may be used as in the original plant. This discussion begins after the media is selected and continues to the specification and procurement stages of the various materials. Ultimately, the effect of the physical parameters of filter media on troubleshooting the plant operation will be demonstrated.

A brief discussion of granular media will help in better understanding how the specifications are written and what they mean to the operation of the plant. A general overview of the physical characteristics of filter media is also provided, so that the basic concepts can be understood and used in the troubleshooting of filters. A definition of terms used in this chapter is given in Figure 5-1.

Physical Characteristics

Theoretically, perfect filter media would have high solids storage, low head loss, consistently high removal efficiency, and very high operating rates. However, the granular medias currently in use are limited by physical constraints and the manufacturing processes used to make them. Granular media, such as anthracite, sand, and garnet, are produced to meet specifications, such as effective size and uniformity coefficient.

Because it is not possible to produce granular media to exact requirements, a range is given. For example, a typical specification for anthracite might be as follows:

Uniformity Coefficient (UC)	Less than 1.3
Effective Size (ES)	0.95 mm – 1.05 mm
Specific Gravity (SG)	1.60 ± 0.05

Effective Size	That point where 10% of the grains are finer than the rest (D_{10})
Fluidization	The state where the upward drag of the backwash water is balanced by the weight of the media, causing the grains to be suspended in the flow. Fluidization is characterized by a rolling and churning of the media.
Incipient Fluidization	That point where the increasing backwash flow causes fluidization to begin in the upper levels of a media layer
Complete Fluidization	That point where the D_{90} grains (90% finer) are fluidiized
Coarse Fraction	The D_{90}–D_{100}-size media grains
Uniformity Coefficient	A nondimensional factor used to depict the range of particle size in the portion of the bed which does the bulk of filtration $UC = \frac{D_{60}\text{ (60\% finer)}}{D_{10}\text{ (10\% finer)}}$

FIGURE 5-1. Filter media definitions

Any material within these parameters and the general requirements of the ANSI/AWWA B100 standard (latest edition) would then be acceptable. These and other physical characteristics are discussed in the following sections.

Effective Size

To calculate the effective size, a sieve analysis is conducted on the media using a series of pans with calibrated screens in the bottom, per ASTM C136. It is the same procedure used in Figure 4-11 of the previous chapter on support gravel. The pans are stacked on top of each other, with the screens decreasing in size from the top down. A measured amount of media is added to the top pan, and the stack is shaken for a specific period of time. The amount of media retained on each screen is weighed, and the results are plotted on a log probability or semi-log paper, as shown in Figure 5-2. The effective size is defined as the screen size that passes 10 percent of the material (10 percent finer or D_{10}).

The effective size selected will impact the support gravel design, or the design of any direct media retention system used. In the case of a dual- or multimedia design, the various medias need to be compatible. Specifically, intermixing between the sand and the anthracite should not be more than about 4 in. to 6 in. Intermixing will be discussed in greater depth later in this chapter.

Formulas have been developed to match the sizes of one media to another. However, because of the manufacturing processes involved, it is very difficult to accurately predict the exact effective size of any given media. Therefore, exactly matching two medias is extremely difficult. From a practical point of view, it is assumed that the various medias used in dual media or multimedia are half the size of the layer above. For example, 1-mm anthracite would be on top of 0.5-mm sand. The anthracite might be specified as 0.95 mm to 1.05 mm in effective size, while the sand could be specified as a 0.45-mm to 0.55-mm effective size. The two medias will vary in size slightly because

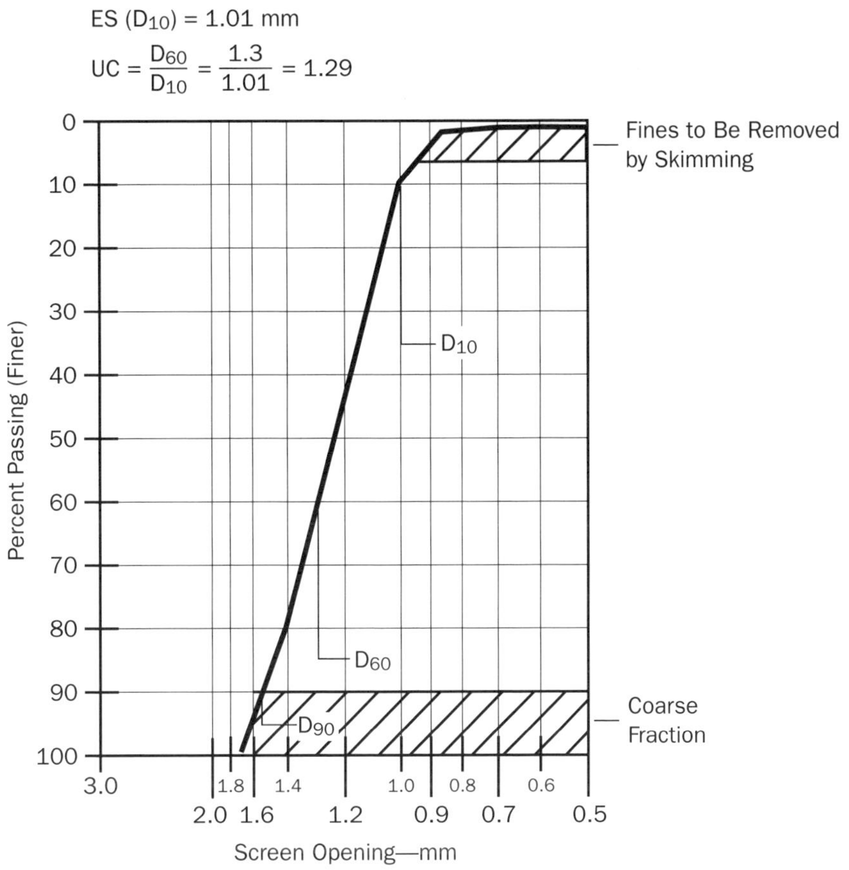

FIGURE 5-2. Anthracite sieve analysis

of production methods, and there will most likely not be exactly a 2:1 difference in size ratio between them.

In the case of multimedia, garnet sand would be in the range of 0.25 mm to 0.3 mm in effective size and would be below the silica sand. The exact size would depend on the material currently produced at the mill. There are several suppliers of filter anthracite and many different suppliers of filter sand. However, there are only a very few suppliers of garnet sand, and the users are dependent on using the material currently produced by the manufacturer.

The effective size of the media affects the unit filter rate, the effluent quality, and the backwash rate necessary to clean the media. Finer media will have a higher effluent quality but shorter filter runs. On the other hand, finer media will have lower backwash requirements. If the backwash rates are not properly matched to the effective size of the media, either loss of media from high backwash rates or improperly cleaned media from a too low backwash rate can result. The exception is when nonfluidizing, coarse monomedia is used, which has very low backwash rates.

Uniformity Coefficient

The uniformity coefficient is the screen size that passes 60 percent of the material (D_{60}) divided by the effective size (D_{10}), as shown in Figure 5-2. As shown in Figures 5-2 and 5-4, there is a wide range of particle sizes in a given material. The fines can be removed by skimming, but nothing can be done about the coarse fraction except to improve the screening processes used in producing the material and lower the UC. An example of the D_{90} particle size from high UC anthracite (UC 1.7) is shown in Figure 5-3. Figure 5-4 illustrates the difference in the amount of coarse particles in anthracite with different UCs. The definition of complete fluidization used in this handbook is the point where the D_{90} grains (Figures 5-2 and 5-4) are agitated. As Figure 5-4 illustrates, the D_{90} size varies dramatically with UC. Therefore, the backwash requirements also vary.

Many plant operators try to achieve an established percent expansion of the media during backwash. To accurately determine the proper expansion to be used, a pilot study is recommended. By using a clear pilot column, the backwash rate can be increased, while observing the depth of fluidization within the anthracite. In a dual or multimedia filter, the anthracite will be the controlling media for determining the proper backwash rates. As the backwash rate is gradually increased, the fluidization begins at the surface, with the finer grains, and moves deeper in the bed. Once 90 percent of the anthracite is fluidized, the backwash flow rate and water temperature can be recorded and the amount of expansion calculated. The backwash rate and the amount of expansion obtained can be used to verify that the proper backwash rate and

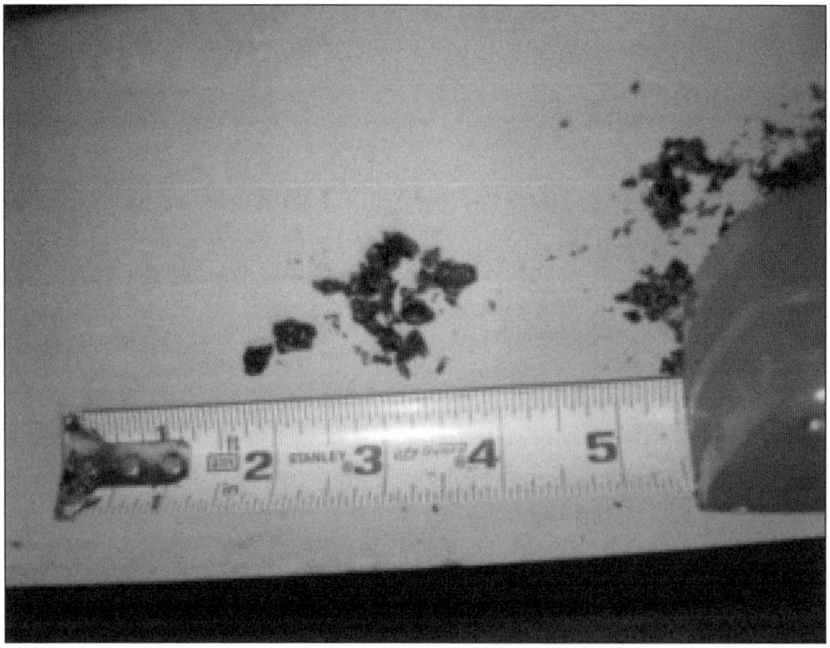

FIGURE 5-3. High uniformity coefficient anthracite

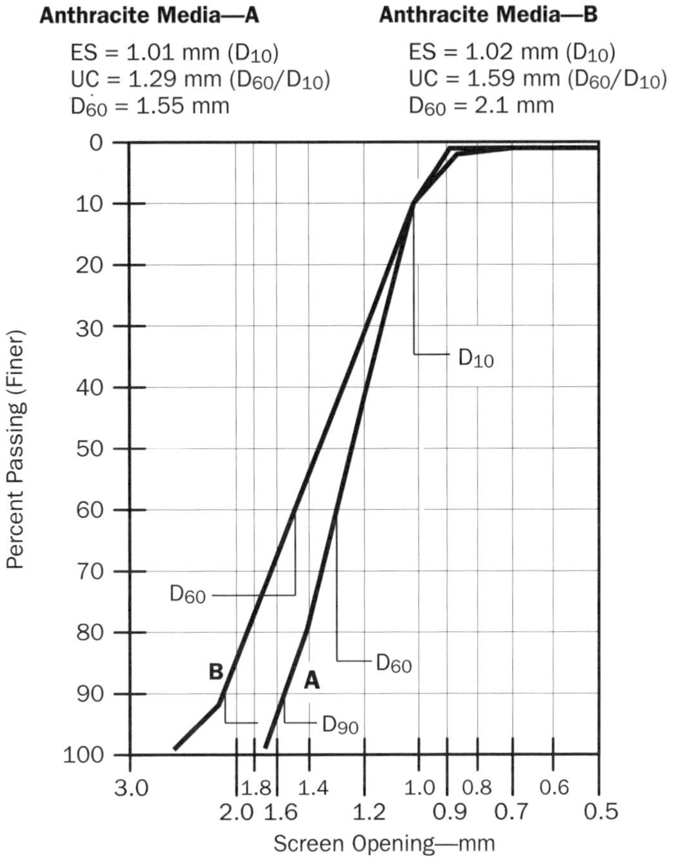

FIGURE 5-4. Uniformity coefficient comparison

expansion in the plant are achieved. It should be noted that a large pilot column should be used to reduce or eliminate any plug effect in the media that may occur. A column diameter of at least 200 times the effective size of the media is recommended. For 1-mm anthracite, the pilot column diameter is calculated as follows:

$$1 \text{ mm} \times 200 = 200 \text{ mm}$$

$$\frac{200 \text{ mm}}{25.4 \text{ mm/in.}} = 7.9 \text{ in.} \rightarrow \text{Use 8-in.-diameter pilot column}$$

Once the D_{90} grain size has been fluidized, there is no additional increase in agitation or solids removal with an increase in the backwash rate. The D_{90} fluidization rate is as high as the backwash flow needs to be. Higher rates simply separate the grains further apart, without any corresponding increase in shear or agitation.

Specific Gravity

The specific gravity is normally 1.65 for anthracite or 2.65 for sand. Slight variances have little impact on the performance of coarse monomedia. However, in a dual- or multimedia filter design, the specific gravity of each of the materials is important, not only in promoting intermixing at the interface between the two materials, but also in limiting the total amount of intermixing that occurs. Intermixing will be discussed in the section, Fluidizing Media, in this chapter.

Specific gravity is also of special concern when anthracite is used in dual- or multimedia filter designs. If more than one form of raw feedstock is used in the production of filter anthracite, the average specific gravity could be as specified, even though a wide range of different specific gravity materials are present. If that is the case, the heavier, higher specific gravity material would sink further into the silica sand, resulting in a much higher amount of intermixing than desired. Therefore, it is recommended that specifications for procuring filter anthracite require that only one source of material be used in the production of the material.

Acid Solubility

The acid solubility test is a method for determining the potential loss of material from contact with chlorine, coagulants, and naturally occurring chemicals. It is of most concern when the raw water or treatment processes result in the filtration of low-pH water. Tertiary filters in wastewater plants are a special concern because of the potential for aggressive water quality.

If the water in a filter is aggressive, it could lead to the degradation of the support gravel and also the silica sand, either of which could in turn result in creating an upset filter condition. Therefore, if the process is expected to have aggressive water, particular attention should be paid to obtaining low acid solubility silica sand and support gravel. Acid solubility is less likely to be a problem with filter anthracite.

Friability

The term *friable* relates to the ease in which a material crumbles. The more friable a material is, the easier it crumbles. Less friable material is usually tougher and more resistant to crumbling. Very hard material may also be very brittle. Therefore, toughness is desired more than hardness.

The author is not aware of any direct relation between friability and any other normal physical characteristic of anthracite. Friability may be more affected by differences in porosity and grain structure than hardness or percent carbon.

Friability is an issue primarily concerning anthracite. If the anthracite has a high friability, it will break down more quickly during normal operation in a filter, resulting in a more significant amount of fines, with a higher loss of media than would be expected because of normal attrition. It is recommended that a friability of less than 4 percent be specified when using the paint can shaker test.

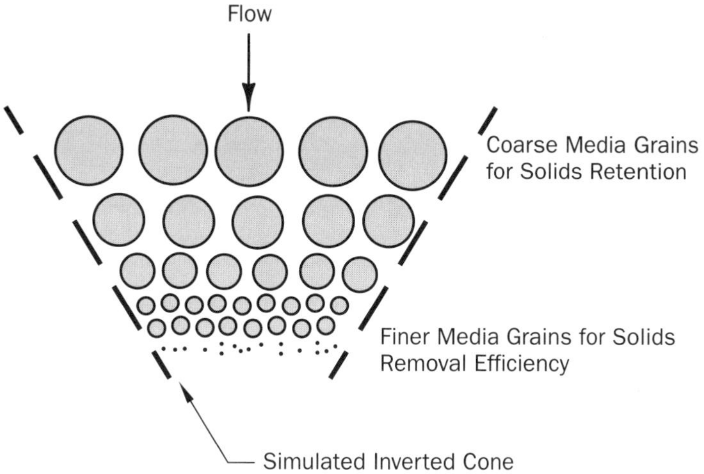

NOTE: The ideal filter media would have a smooth transition of grain size with the coarsest on top and the finest on bottom, simulating an inverted cone.

FIGURE 5-5. Ideal grain distribution for filter media

Other Characteristics

There are several other physical characteristics of filter media that can be evaluated, including hardness and percent carbon for anthracite and alkaline solubility. However, these issues are not often of great concern when troubleshooting filters.

Media Types

The ideal filter media would have coarser media grains on the surface for solids retention and increasingly finer grains deeper in the bed for solids removal efficiency, as shown in Figure 5-5. The ideal media could then be characterized as filtering through a cone or funnel, with solids removal efficiency becoming greater toward the bottom.

There are two basic types of filter media that predominate in the industry: fluidizing and nonfluidizing designs. The approach of each type and their relation to the ideal is discussed in the following sections.

Fluidizing Media

When an individual layer of media is completely fluidized down to the D_{90} level, the media grains classify in a series of layers, with the finest on top and the coarsest on the bottom. The result can be compared to an inverted funnel, with the finest material on top and the coarsest on the bottom, as shown in Figure 5-6. The resulting grain distribution is exactly opposite from that in the ideal media described previously.

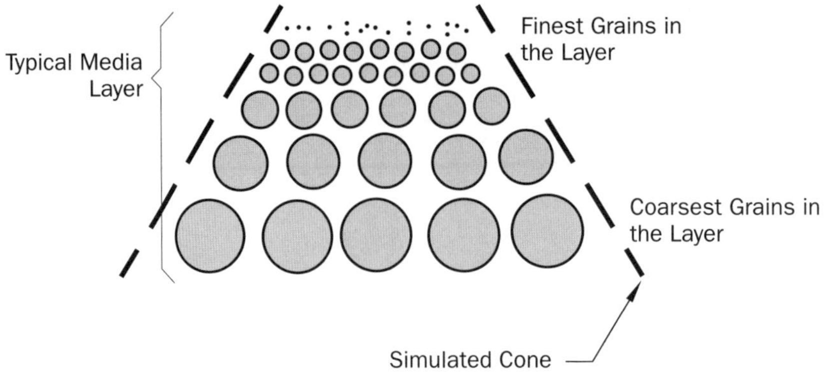

NOTE: Whenever a fluidizing backwash is used, the grains in each layer classify with the finest on top and the coarsest on bottom, simulating an inverted cone.

FIGURE 5-6. Actual fluidizing media grain distribution

Multiple Media Layers

To more closely approximate the ideal grain distribution, two or more different media are installed in layers in the same filter, as shown in Figure 5-7. Dual media uses anthracite over silica sand, while multimedia adds a third layer of garnet sand underneath. The various medias are in decreasing effective size and increasing specific gravity so they stay in their relative location during backwashing. The effective size and specific gravity of the various materials are adjusted so that there is a small amount of intermixing between types at the interface. Otherwise, each layer is separate and distinct. The intent is to provide a relatively smooth transition from one media type to the smaller one below, while minimizing any sudden increase in head loss that might occur from an abrupt change to a smaller media. It is desirable to promote the penetration of solids so that the larger media on top collects the heavier solids, while the finer solids penetrate to the next media type below. In that way, the entire depth of the media can be used for the accumulation of solids, while the finer material on the bottom produces the final effluent quality.

Intermixing

The intermixing zones between media types are typically in the range of 4 in. to 6 in. deep. A narrower intermixing zone could result in a rapid increase of head loss at that point. On the other hand, a wider intermixing zone could result in decreased solid storage of the upper layer and decreased solids removal efficiency of the lower layer. The exception is garnet sand, in which as much as 50 percent or more of the material may mix with the silica sand above.

Pilot studies can be conducted to determine the amount of intermixing required if representative samples of the media for a given project can be obtained in advance. An alternative is to pilot test a sample of the materials delivered to the site. In addition, acceptance of new media could be conditional on achieving the proper amount of

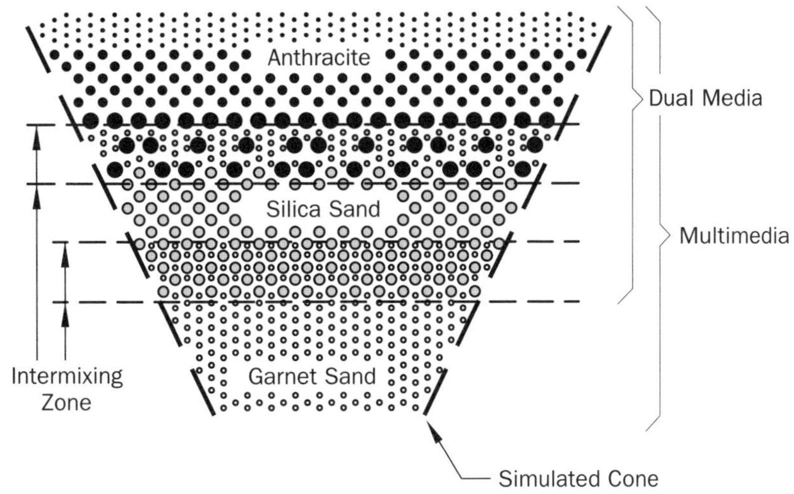

NOTE: The combination of anthracite, silica sand, and garnet sand simulates the ideal grain distribution of a cone.

FIGURE 5-7. Dual/multimedia grain distribution

intermixing. Other recommendations for procurement specifications are included in the section, Physical Characteristics.

Nonfluidizing Monomedia

Nonfluidizing monomedia is typically composed of anthracite or coarse sand (fine gravel) with an effective size in the range of 1.3 mm to 3 mm.

The theoretical distribution of nonfluidizing monomedia is that all the grains are very nearly equal throughout the depth of the bed, as shown in Figure 5-8. Therefore, the incoming raw water receives the same removal efficiency at any depth. In addition, because of the coarse nature of the monomedia, the bed will typically hold a high volume of solids.

Coarse monomedia is nonfluidizing because the backwash rates to achieve fluidization are normally too high to be practical. As a result, coarse nonfluidizing monomedia is backwashed with a combination of a low-rate water wash (typically 6–12 gpm/ft^2) combined with air scour. The air scour is essential in keeping coarse monomedia clean. The section on Fluidizing Media discusses backwash for these types of media.

When using a coarse sand or fine gravel, the actual grain distribution of monomedia is close to the theoretical distribution, shown in Figure 5-8, because the media does not fluidize and therefore it does not classify. The manufacturing process for coarse sand or fine gravel is such that there are typically few fines or oversized grains. The size of the media grains is usually restricted to a fairly narrow band.

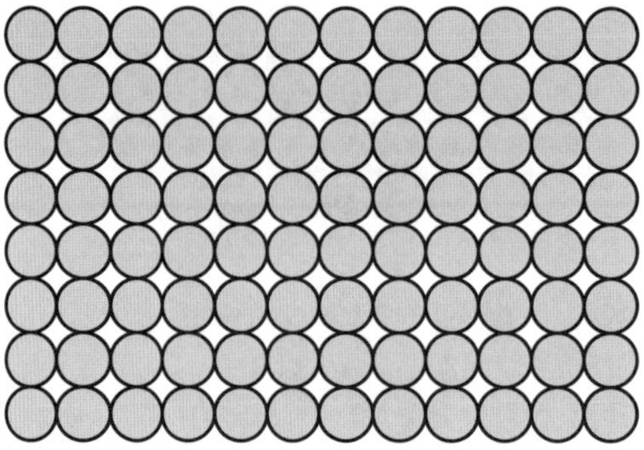

Coarse Monomedia

NOTE: The actual grain distribution of coarse monomedia is close to the theoretical distribution because the media does not fluidize or classify.

FIGURE 5-8. Nonfluidizing monomedia theoretical distribution

On the other hand, coarse anthracite will have fines. However, because the backwash used is nonfluidizing, few of the fines will come to the surface at any one time, so that the overall grain distribution will remain relatively uniform.

There are advantages and disadvantages to using either anthracite or a coarse sand or fine gravel. Anthracite is usually a better filter material because of the irregular surface but will tend to break down with time and require periodic skimming and replacement. Coarse sand or fine gravel is usually more rounded and, therefore, will have a lower filtration efficiency than anthracite, but the grains are more durable and will last longer.

Typical Media Designs

Typical media designs are shown in Figure 5-9 along with the advantages of each. It should be noted that for those filter plants using dual media, the effective size of the materials used is on the smaller end of the range, whereas the larger end of the effective size range is typically found in wastewater applications. Anthracite used for nonfluidizing monomedia is often found in potable water treatment plants, while coarse sand or fine gravel is common to tertiary filters in wastewater treatment plants.

Grain Surface Area

The concept of dual or multimedia filter designs relies on a high surface area for filtration efficiency. In addition, the irregular shape of the anthracite grains also contributes to filter efficiency.

Coarse monomedia will have a much lower grain surface area per cubic foot of material, as shown in Figure 5-10. The difference in filtration efficiency is made up in part by having a bed depth of 4 ft to 6 ft, as opposed to the $2^1/_2$ ft to 3 ft common with dual or multimedia.

	Fluidizing: Dual Media	Fluidizing: Multimedia	Nonfluidizing Monomedia
Anthracite	18 in. to 24 in. ES 0.9 mm to 1.5 mm UC 1.3 to 1.4 SG ±1.65	16 in. to 18 in. ES ±10 mm UC 1.3 to 1.6 SG ±1.65	48 in. to 72 in. ES 1.3 mm to 1.5 mm UC 1.3 to 1.4 SG ±1.65
Silica Sand	10 in. to 12 in. ES 0.5 mm to 0.8 mm UC 1.3 to 1.5 SG ±2.65	8 in. to 10 in. ES ±0.5 mm UC 1.3 to 1.5 SG ±2.65	↑ or ↓
Coarse Sand (fine gravel)			48 in. to 72 in. ES 1.5 mm to 3.0 mm UC 1.3 to 1.5 SG ±2.65
Garnet Sand		3 in. to 4½ in. ES 0.25 mm UC ±1.6 SG 4.0	
Advantages	Lower Cost Good Turbidity Removal	High Surface Area Better Removal of Small Particles	High Solids Storage High Flow Rate (8 to 12 gpm/ft^2)

FIGURE 5-9. Typical media designs

Activated Carbon

Activated carbon is a high porosity material typically made from bituminous coal or coconut shells. Other materials are also used, but these are the most common. The high porosity is achieved by thermal activation at high temperatures in the absence of air. The high porosity of the grains of activated carbon makes it very efficient in adsorbing certain types of molecules that are held in place by weak physical forces. Activated carbon is typically used in a number of applications, such as taste-and-odor control and organics removal for potable water treatment.

In keeping with the rest of this handbook, it is not the author's intent to discuss process issues in detail, rather, to discuss the physical properties of media types and their relation to troubleshooting filters.

Adsorptive Capacity

The adsorptive capacity of carbon is measured by tests using carbon tetrachloride or iodine. For example, a typical iodine number on activated carbon would be between 800 and 1,000.

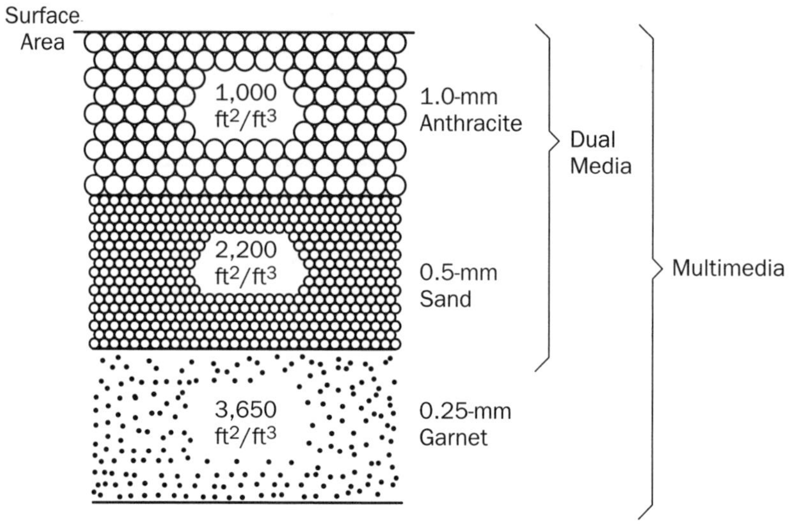

NOTES:

1. Typical mixed media grain surface area = 4,400 ft^2/ft^3 (based on the use of 16½ in. of anthracite, 9 in. of sand, and 4½ in. of garnet sand).
2. Dual- and multimedia designs use a layered approach to filtration where coarse solids are removed in the upper layer and the finer solids are removed at the bottom.
3. Multimedia uses a small particle size and a high surface area to improve the capture of small particles.

FIGURE 5-10. Grain surface area—dual/multimedia

Porosity

The porosity of activated carbon is illustrated by the difference in bulk density between the base material and the final activated carbon product. For example, the bulk density of crushed bituminous coal is in the range of 43 to 50 lb/ft^3, whereas the bulk density for activated carbon is 26 to 28 lb/ft^3.

Uniformity Coefficient

Because of the methods of manufacture, the UC of activated carbon is typically much higher than is used for filter materials, such as anthracite. A UC in the range of 1.9 to 2.1 is common.

One of the impacts of the higher UC is that there is a high percentage of fines in the material, which go to the surface after a backwash. When activated carbon is used for adsorption purposes, the fines are a benefit because of the very high surface area and adsorptive capability of the smaller grains. When used for a dual role of filtration and adsorption, the water quality coming into the filter must be very high, otherwise the filter runs will be short because of surface filtration.

Effective Size

Activated carbon is available in a number of size ranges, including 0.8 mm to 1 mm, which is typical for filter anthracite. However, as noted previously, the amount of fines is significantly higher than for other media types.

Backwash Rates

As with any other media type, the backwash rate is dependent on the bulk density, effective size, the UC, the water temperature, and the desired expansion of the media. As a result of its lighter weight, backwash rates are typically in the 10- to 14-gpm/ft^2 range. When adjusting the backwash rate, care should be taken that the rates used are not so high as to wash the media out into the backwash waste.

Application

Adsorption Only. Some plants have separate carbon adsorption filters after conventional treatment and filtration. In that case, the carbon adsorbers do not function as true filters. However, they are built like filters and the troubleshooting issues are similar.

Adsorption and Filtration. Other plants use activated carbon in a dual role of adsorption and filtration. In that case, it may be desirable to skim or blow off some of the fines to prevent surface filtration, and, as previously mentioned, the water entering the filter should be of a very high quality, which requires highly efficient clarification.

Some activated carbon filters operating in a dual role have bed depths similar to standard dual media, while others have depths up to 5 ft or 6 ft. In either case, the activated carbon typically has an effective size similar to filter anthracite (approximately 1 mm). In the dual role, there is also a layer of filter sand underneath. The sand may be 10 in. to 12 in. deep, similar to dual media, with an effective size of 0.45 mm to 0.55 mm. Even though activated carbon uses a relatively low backwash rate, it is sufficiently high to fluidize the sand.

Troubleshooting Issues

Cleaning. Filters functioning as adsorbers only should not accumulate significant fines. However, filters using activated carbon in a dual role will collect solids and may be hard to clean as a result of the low backwash rates. Because of this, air scour is recommended to assist in cleaning the bed. However, even with air scour, the operators may have to drain the filters periodically and hose down the surface to break up large accumulations of solids.

Polymers. Excess polymers or other coagulants entering the filter can plug the pores and interfere with the activated carbon's ability for adsorption. When using activated carbon in a dual role, the chemicals used must be very carefully fine-tuned.

Remaining Useful Life. There are a number of ways to determine the remaining useful life of activated carbon. Some plants use pilot columns, so that samples at different media depths can be withdrawn to determine the level of adsorption. The remaining adsorptive capacity of the activated carbon can then be determined. Another method is to take a sample of the activated carbon and send it to a lab for analysis. However, it will have to be a core sample of the complete depth of the bed in order to be accurate, as the activated carbon becomes spent from the top down.

From a filtration point of view, the activated carbon will be effective as long as it maintains its physical characteristics. If the carbon gets too old and has possibly exceeded its adsorptive capability, it may get soft and mushy. In that case, the activated carbon should be changed out as soon as possible.

Biological Fouling. Activated carbon will adsorb chlorine in addition to other compounds. As a result, carbon media is prone to biological fouling. In some cases, biological growth may be part of the process design. Backwashing with chlorinated water may be necessary to keep biological fouling under control, even though it affects the carbon's adsorptive capability.

ANSI/AWWA Standard B604

For additional information on carbon as filter media, refer to ANSI/AWWA Standard B604, Granular Activated Carbon. It should be noted that the author was a member of the original committee that generated that standard.

Support Requirements

Once the media design is selected for a given project, the supporting materials can be determined. The support requirements generally fall into two categories: those using a direct media retention underdrain, and those using support gravel.

Direct Media Retention Underdrain

Whether using a porous plate, slotted screen, perforated plate, or any other type of direct media retention underdrain, the device must be tailored to fit the media load. For example, there is a significant difference in supporting coarse monomedia, as opposed to multimedia with garnet sand. Some standard designs of direct media retention underdrains may support coarse monomedia or 0.5-mm silica sand but may not support 0.25-mm garnet sand.

Support Gravel

When support gravel is used, it must be tailored to the filter media design and the underdrain. There are two basic types of support gravel: a single-taper design, as shown in Figure 5-11, is commonly used for hydraulic-only backwash; while an hourglass-type design is commonly used for air scour applications, as shown in Figure 5-12.

In gravel designs of either type, the lower layer is sized to protect the underdrain. A recommendation for the size of gravel used for the lower layer can be obtained from the manufacturer of the underdrain that is used. Once the lower layer of gravel has been sized, the next layer above should be half the size of the lower layer. Each succeeding gravel layer above the underdrain should be half the size of the layer below. Chapter 4 has an additional discussion of filter support gravel.

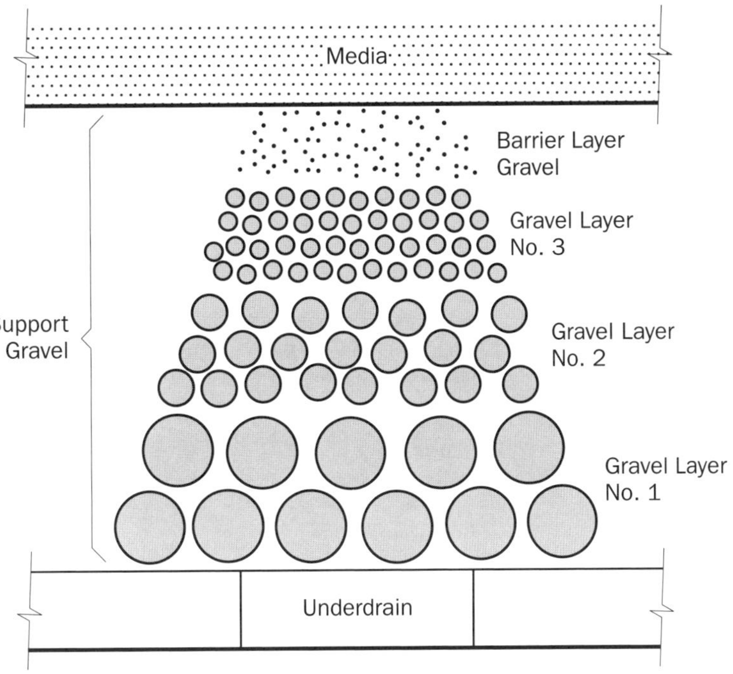

NOTES:
1. The layers should have not more than a 2:1 size difference. For example, the layer above should be half the size of the layer below.
2. Fluidizing media can be as much as 10 times smaller than the barrier layer because of the bridging effect of fine sand. Larger, nonfluidizing monomedia should be close to the 2:1 size difference because the bridging effect is less.

FIGURE 5-11. Single-taper gravel

The barrier layer is the one that is intended to support or retain the filter media. When using fluidizing media, the media grains in contact with the barrier layer can be as small as one-tenth the size of the barrier layer because of the bridging effect of fine sand. Larger, nonfluidizing monomedia should be closer to the 2:1 size ratio of support gravel, because the bridging effect is less.

Air Scour Applications

A typical hourglass gravel design commonly used for air scour applications with fluidizing media is shown in Figure 5-12. In this figure, the bottom layer of $^3/_4$ in. × $^1/_2$ in. gravel should adequately protect a dual parallel lateral underdrain.

In fluidizing applications with an air scour backwash and typical dual- or multimedia designs, the finer media grains penetrate completely through the upper layer of $^3/_4$ in. × $^1/_2$ in. gravel. In addition, the finer media grains also penetrate through the next gravel layer to some extent, while the barrier gravel allows no penetration of fine media, as shown in Figure 5-13. During air scour and backwash, much of the fine media blows back out of the gravel, as shown in Figure 5-14. Once the backwash cycle is completed

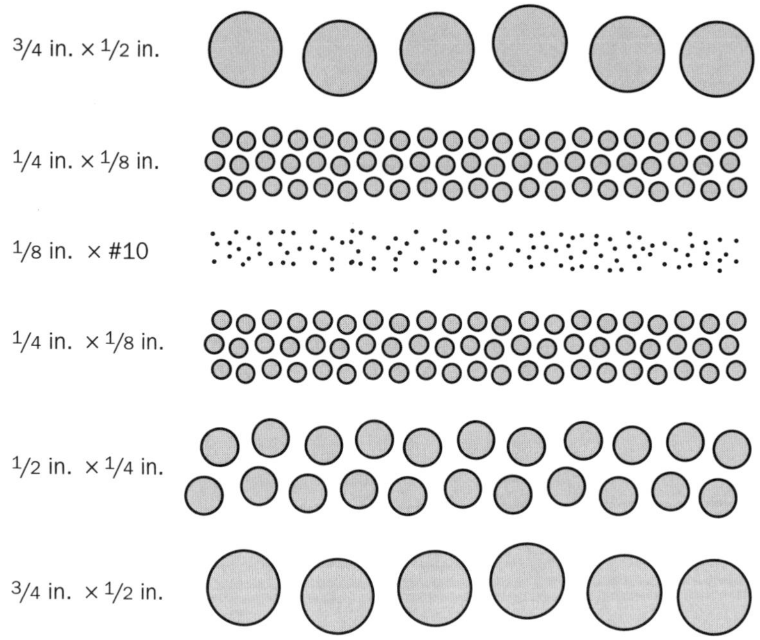

FIGURE 5-12. Hourglass support gravel

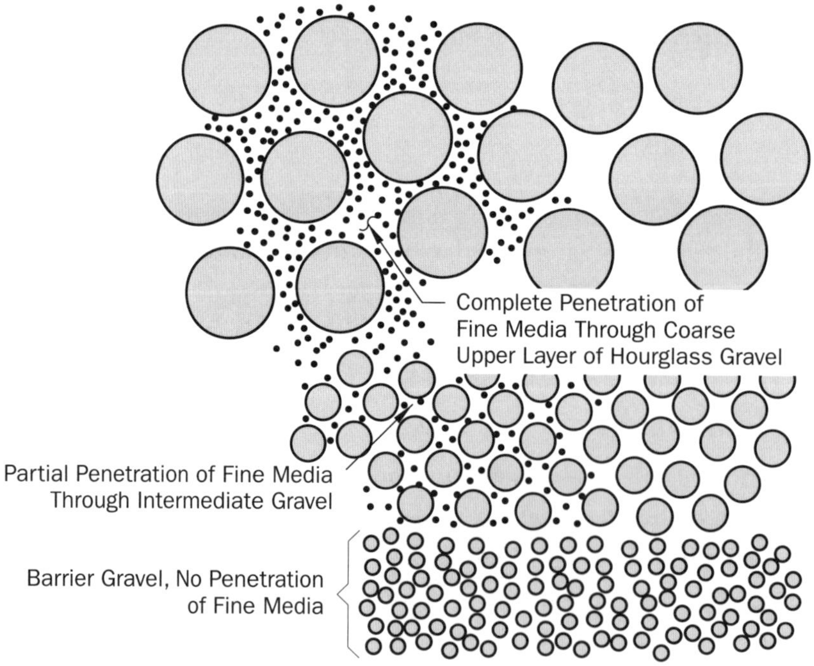

FIGURE 5-13. Penetration of fluidizing media in hourglass support gravel

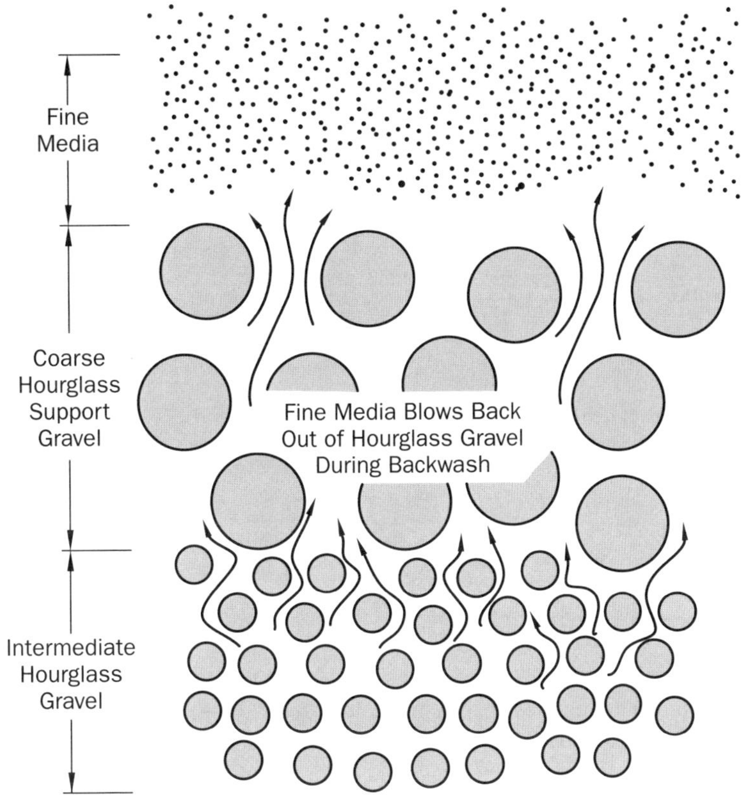

FIGURE 5-14. Fluidizing media during backwash

and the filter is returned to service, the media will again penetrate back down through the gravel. The cycle of movement of media through the gravel is made possible by the size difference between fine media and the $^{3}/_{4}$ in. × $^{1}/_{2}$ in. gravel. The short time it takes for the fine media to resume its position is another reason a filter-to-waste cycle is required in potable water treatment plants to reestablish filtration efficiency.

Nonfluidizing Monomedia

The use of coarse nonfluidizing monomedia presents unique support requirements when using support gravel. Hourglass gravel is sometimes used, but modifications may be required to the standard designs normally used.

In a similar manner to fluidizing media, coarse nonfluidizing monomedia will also penetrate through the top layer of $^{3}/_{4}$ in. × $^{1}/_{2}$ in. gravel in the normal hourglass configuration. Even though the barrier layer is the third layer down, little if any of the coarse monomedia will penetrate through the second layer of $^{1}/_{4}$ in. × $^{1}/_{8}$ in. gravel.

Coarse monomedia can vary in size from 1.3 mm to 3 mm, which is several times the size of the 0.5-mm sand or 0.25-mm garnet sand commonly used in fluidizing media. As a result of its larger size, the grains of coarse monomedia tend to interlock

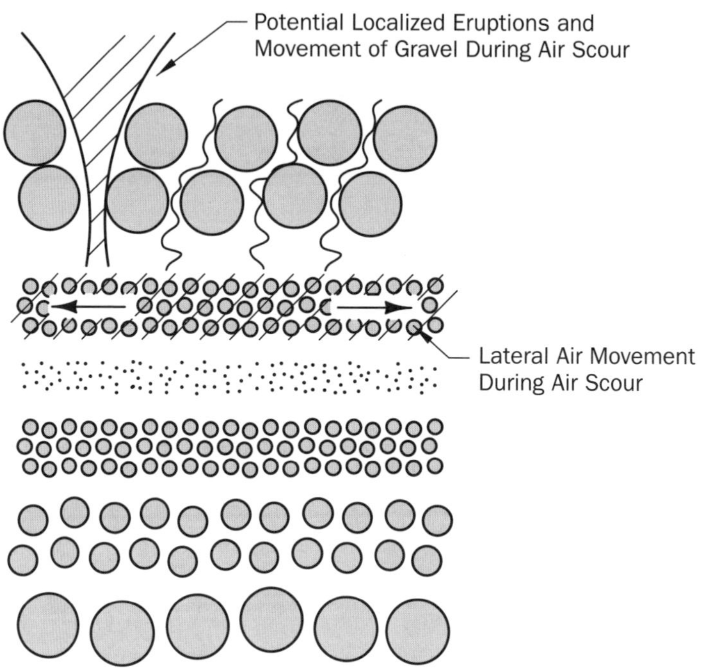

NOTE: Coarse nonfluidizing media (1.5 mm and larger) has little, if any, penetration through or into the 1/4 in. × 1/8 in. layer.

FIGURE 5-15. Standard hourglass support gravel with coarse nonfluidizing media

within and underneath the top layer of 3/4 in. × 1/2 in. gravel. (Figure 5-15.) Then, during air scour backwash, the monomedia tends to stay in place and not blow back out of the gravel. The result is to create a considerable resistance to the upward movement of air. The air tends to move laterally until it finds a weaker location, increasing the potential for localized eruptions and movement of the upper layer of gravel during air scour. Any movement of gravel could potentially result in loss of the entire support gravel bed.

When using coarse nonfluidizing monomedia, some manufacturers eliminate the top layer of 3/4 in. × 1/2 in. gravel in the normal hourglass design. The resulting modified hourglass design would then have only one layer of gravel on top of the barrier gravel. In some cases, the barrier gravel itself, as well as the other layers, may be modified for coarse monomedia bed designs. For specific recommendations on the support gravel design used, contact the underdrain and/or the media supplier.

Fluff

The manufacturing process used to produce filter media results in a wide range of particle sizes. The size variations common for anthracite are depicted in Figures 5-2 and 5-4.

When the material is shipped, whether in 1-ft^3 bags or bulk bags, the various particle sizes are mixed uniformly throughout the container. Different conditions occur once the material is installed in the filter, depending on whether or not fluidizing or nonfluidizing filter media is used.

When a given media layer is installed and given a fluidizing backwash, the different particles classify according to grain size. The smallest grains end up on the surface, with the largest on the bottom and a relatively uniform gradation in between. If the media is originally installed to a depth of 12 in., as shown in Figure 5-16, the gradation and the resulting height after backwash are as illustrated in Figure 5-17. As a result of the classification of particle sizes, the bed depth after backwash will be greater than the original 12 in. For anthracite, the gain in height, or *fluff*, can be as much as 7 to 10 percent of the original height.

It is typical to skim at least ½ in. for the removal of fines. For normal dual media or multimedia, the height of anthracite, for example, would normally be adequate to generate enough extra height caused by fluff for skimming purposes. However, because of potential losses of media at the job site, and for potential multiple skimmings, it is recommended that at least 5 percent extra of each filter material be provided. The depth of silica sand and garnet is not normally sufficient to obtain an adequate gain in height for skimming. An extended discussion of skimming is contained later in this chapter.

Fluidizing Media

Fluidization is the suspension and agitation of the media that is created by the upward flow of backwash water. Fluidization occurs when the upward flow of water creates sufficient drag on the media grains to overcome their weight. Fluidization of any given filter media is dependent on the effective size, the specific gravity, the UC, water temperature, and grain shape. For accurate information on fluidization, pilot studies are recommended. (The following information was developed by the author using a 12-in.-square pilot column.)

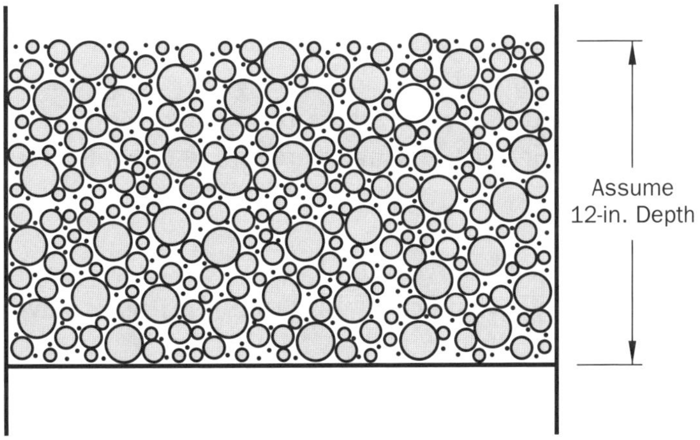

FIGURE 5-16. Media as initially installed

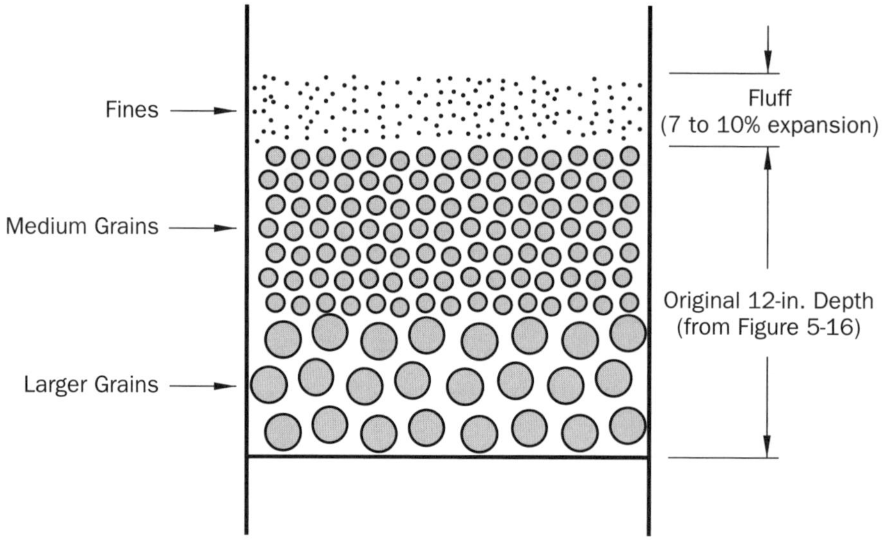

NOTE: The media grains, within a given layer, will classify by particle size during backwash with the finest on top and the largest on the bottom.

FIGURE 5-17. Media after backwashing

Silica Sand Media

A head loss versus backwash flow rate curve is shown for silica sand media in Figure 5-18. The curve represents a 12-in. bed of silica sand with an effective size of 0.5 mm and a UC of 1.34. It should be noted that the head loss rises steeply at flow rates up to approximately 7 gpm/ft^2. During this stage, the sand media is expanding but is not yet agitated. When the backwash flow rate reaches approximately 7 gpm/ft^2, incipient fluidization begins wherein the top surface of the sand begins to be agitated. As the flow rate is increased, fluidization begins to occur deeper in the bed. By approximately 8.0 to 8.5 gpm/ft^2, complete fluidization of the D_{90} level has been achieved.

At flow rates above the incipient fluidization rate, the curve flattens out, with very little increase in head loss for increases in the backwash flow rate. For sand-only media of this type, there would be no need for backwashing at a rate higher than 8.0 to 8.5 gpm/ft^2. Beyond that point, there is no corresponding increase in shear between grains to scrub solids off. The grains are simply expanded farther apart.

Anthracite

A graph of anthracite head loss versus backwash flow rate is shown in Figure 5-19. The graph represents a 12-in.-deep bed of anthracite, with an effective size of 1.0 and a UC of 1.22.

The resulting head loss versus flow curve is similar to that of sand, except that the head loss is lower. Incipient fluidization begins at approximately 8 gpm/ft^2, with complete fluidization (D_{90}) at approximately 16.5 gpm/ft^2.

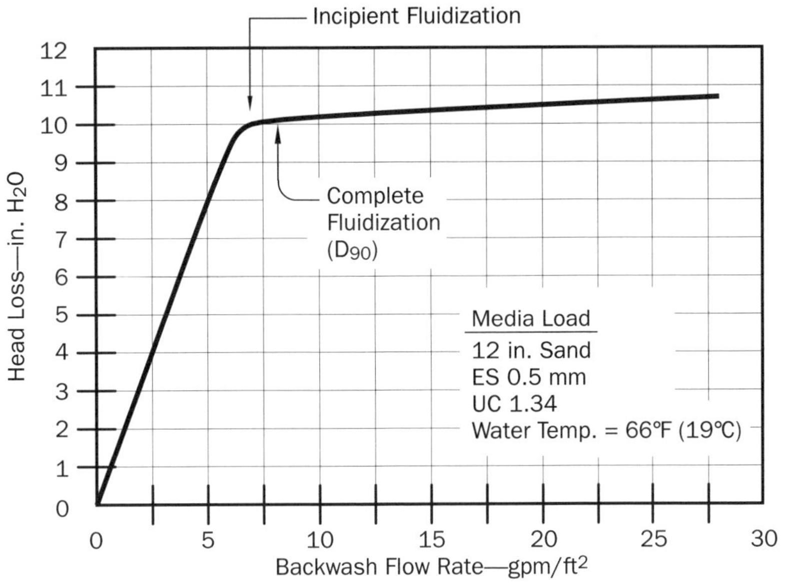

FIGURE 5-18. Sand media head loss versus backwash flow rate

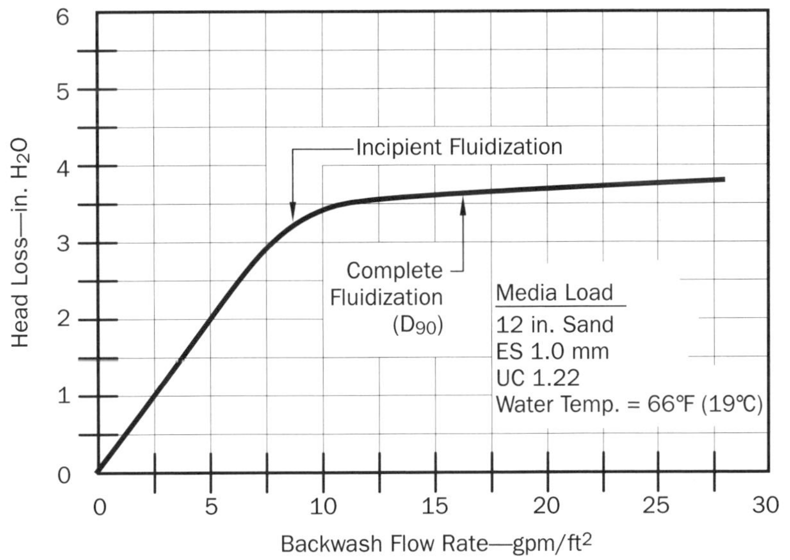

FIGURE 5-19. Anthracite head loss versus backwash flow rate

Anthracite achieves incipient fluidization at approximately the point where sand is completely fluidized. As with the sand, there is no need to fluidize anthracite above the D_{90} point. It only separates the grains farther apart without any increase in shear or head loss.

Dual Media

A head loss versus backwash flow rate curve for dual media is shown in Figure 5-20. The dual media curve uses 18 in. of anthracite instead of the 12 in. used in Figure 5-19, plus 12 in. of sand similar to Figure 5-18. The D_{90} fluidization is approximately the same point as for the anthracite-only curve. However, incipient fluidization begins slightly earlier.

The curves shown are for clean media only. When the media is dirty, there is an additional head loss spike that occurs between incipient and complete fluidization, which is caused by the head loss required to break up the encrustation of solids on the media surface. The amount of head loss spike that actually occurs is highly variable and is a function of chemical coagulation, solids loading, run time, and head loss. The actual spike can be several times higher than depicted in Figure 5-20.

During filtration, the flow distribution through the filter is dependent on the condition of the filter media. However, during backwash it is desirable that the underdrain control the distribution of the backwash water. A high efficiency of distribution of backwash water is desirable to ensure that all parts of the filter are backwashed uniformly with the same flow rate of water. Therefore, it is highly desirable that the head loss of

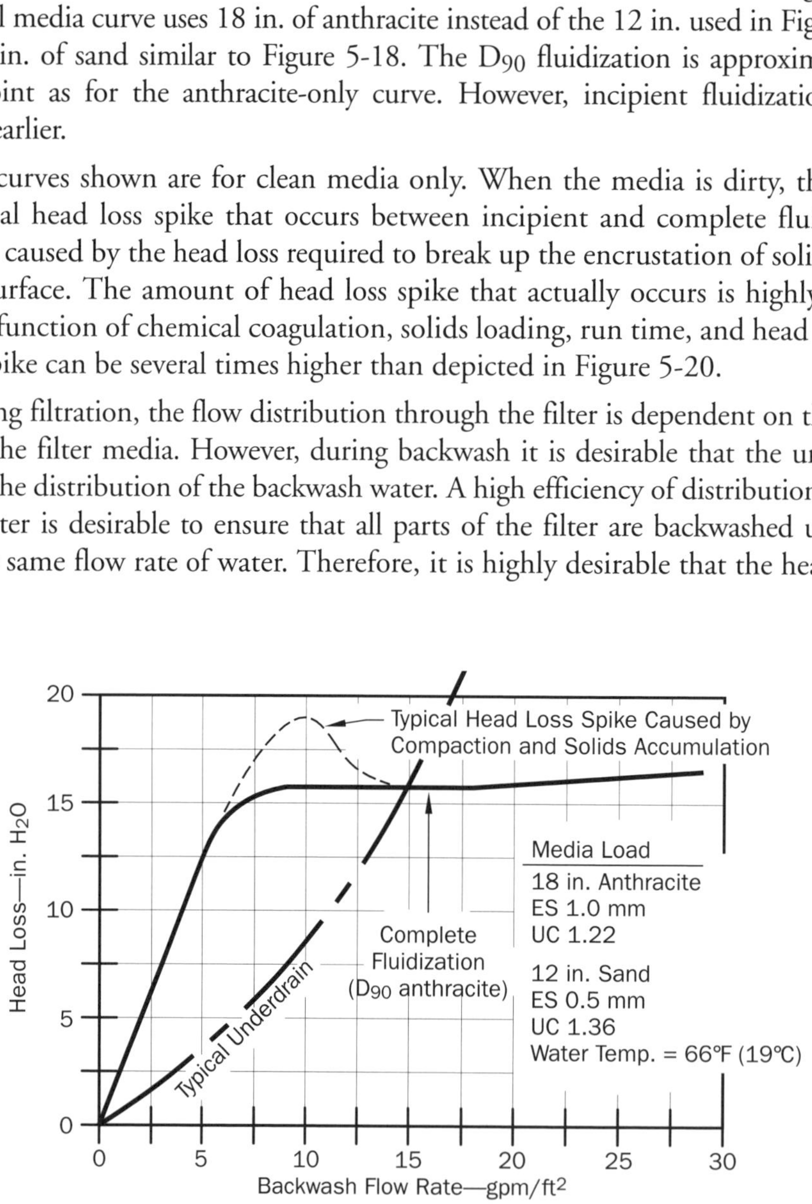

FIGURE 5-20. Dual media head loss versus backwash flow rate

the underdrain be greater than that of the filter media at the backwash rate used. In Figure 5-20, a typical underdrain head loss curve is overlaid on the graph to show the desired relationship between the underdrain and the filter media head loss.

Head loss typically controls flow distribution. If the underdrain does not have higher head loss than the filter media, it could be difficult to clean the media uniformly across the filter. It would then be easier for fouled media to redirect the flow of backwash water in the underdrain.

Backwash Sequence

Hydraulic-only backwash applications typically have only one high-rate backwash step. A low-rate water wash is sometimes used in air scour applications, which is followed by a high-rate wash to flush solids out of the filter and reclassify the media. During high-rate wash, the media becomes reclassified as part of the fluidization process.

As the backwash rate is ramped down, the media begins to rapidly settle into place. By the time the flow is reduced to the incipient fluidization rate (7 to 8 gpm/ft^2; Figure 5-20), the media has ceased to be agitated and has come to rest in the relative position established by fluidization. As the rate is reduced further, the media begins to recompact, but the individual grains stay in the same position relative to each other.

In some plants, the low-rate/high-rate sequence is followed by another low-rate wash. If the intent is to promote better classification, another low-rate wash step is unnecessary. Tests by the author have shown that the best reclassification occurs with a rapid shutdown of the backwash flow. However, that is not practical and causes other problems.

After a backwash, typically there are solids remaining in the filter that have been loosened but not flushed out. There may be some benefit in using a second low-rate wash step to flush out those solids. Chapter 7, Optimizing Backwash, has an additional discussion of this process.

Expansion and Compaction

After fluidized filter media is operational, it undergoes a continuous series of expansions and contractions. There is one set of conditions that occurs during normal operation, and another that occurs during backwash and air scour (if used). The media is slightly "fluffed up" after backwash and at the beginning of a filter run.

Air Scour

After being operational for a period of time, the filter media will begin to develop operational head loss from accumulated solids. The operational head loss will cause the media to compact. The amount of compaction varies widely with the type and depth of the media and could amount to several inches or more, as shown in Figure 5-21.

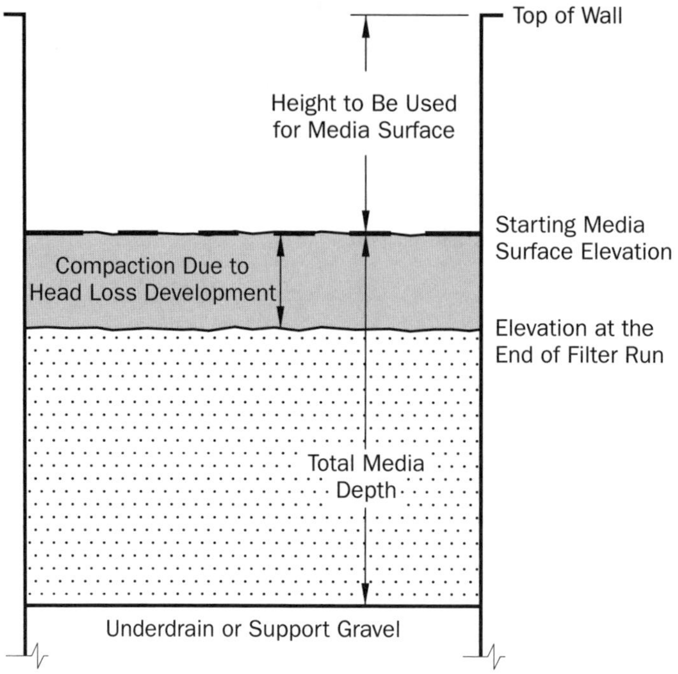

NOTES:

1. Backwashing and cleaning of media "fluffs up" anthracite and restores it to the starting elevation.
2. Height measurements should always be taken at the same point in the filter cycle, preferably after backwash.

FIGURE 5-21. Media expansion and contraction

Operational Expansion

Backwash and cleaning of filter media "fluffs up" the media and should restore it to the starting elevation. The expansion and contraction described previously will occur during each filter cycle.

Height Measurements

It is desirable to measure the height of the bed periodically to determine if media is lost. Media can be lost through a variety of reasons, as is described in following sections in this chapter. Media bed height can also increase with accumulation of calcium carbonate or manganese, etc.

As a result of the expansion and contraction that occurs in the filter bed, it is desirable to measure the height of the media at the same point in the operational cycle every time. For reference purposes, it is recommended that the height of the media be measured after backwash and before the filter returns to service, otherwise the elevation of the media is variable depending on the length of run and the amount of head loss development that has occurred. The height of the media should be measured to a

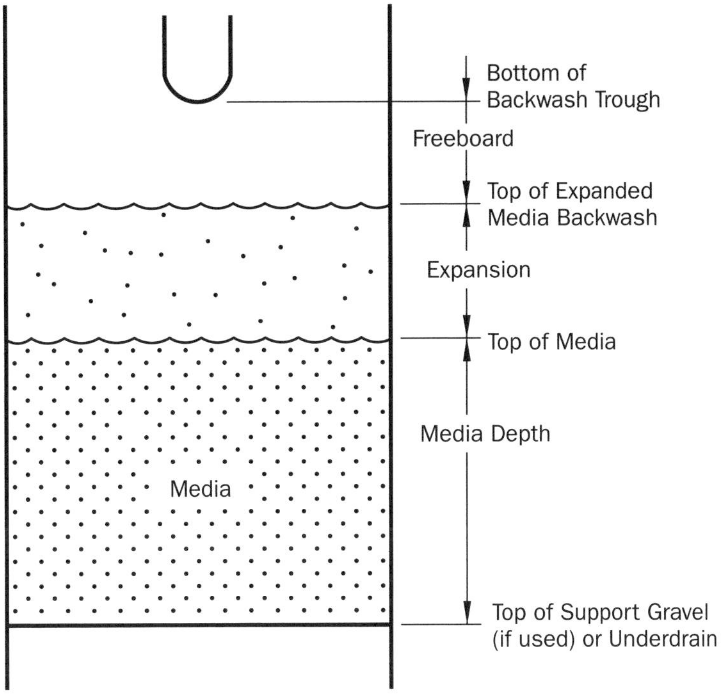

$$\text{Expansion (\%)} = \frac{\text{Expansion}}{\text{Media Depth}} \times 100$$

= 20–30% depending on the UC

Freeboard (Above Expanded Media) = 9 in.–12 in. min. recommended

FIGURE 5-22. Expansion and freeboard

fixed reference point, such as the top of the filter wall. The troubleshooting section in this chapter describes height measurement.

Backwash Expansion

The amount of expansion desired during backwash is the expanded height of the media during backwash (top of exposed media) less the height after backwash (top of media at rest) divided by the total media depth, as shown in the calculation in Figure 5-22. The actual amount of expansion desired is dependent on the effective size, the UC, specific gravity, water temperature and grain shape, and should be determined by pilot testing. An expansion of 22 to 30 percent is typical for most fluidizing medias.

Media Freeboard

Once the media load and the desired amount of expansion are determined, the minimum trough height can also be determined. *Media freeboard* is defined as the distance from the top of the expanded media to the bottom of the wash trough, as shown in Figure 5-22. The intent of media freeboard is to prevent the expanded media from being carried out of the filter with the backwash water. The minimum freeboard is recommended to be at least 9 in. to 12 in., and more is better.

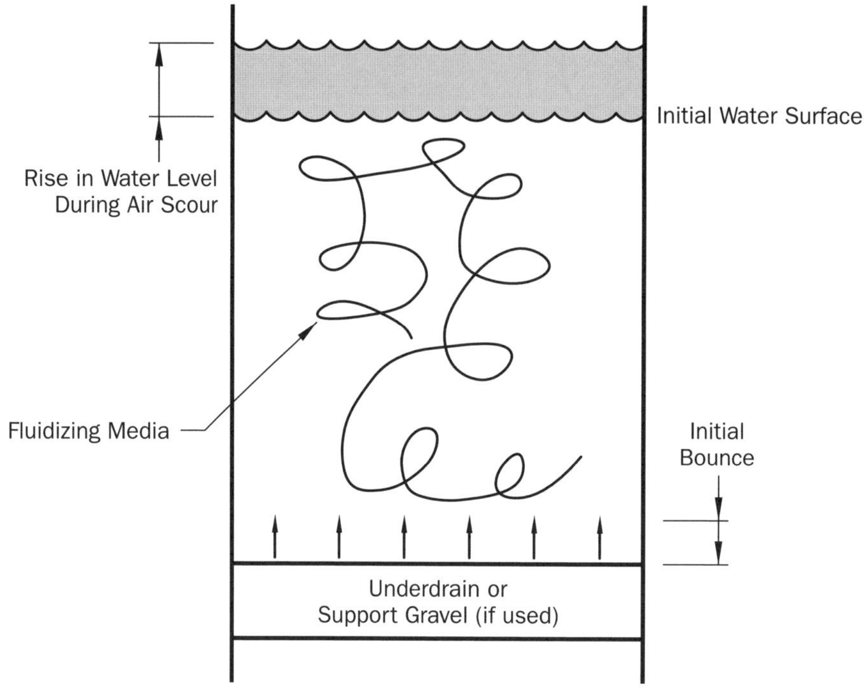

NOTE: Fluidizing media expands to the water surface and completely rolls and churns when air scour begins.

FIGURE 5-23. Initial expansion caused by air scour

It may be difficult to obtain the desired freeboard in older, shallower filters. In those cases, an investigation should be made to determine whether or not the troughs can be raised. If not, a *winged trough* design provided by some manufacturers may be used. The winged troughs are designed to shed expanded media and prevent its loss. For an in-depth discussion of media freeboard, the Media Loss section in this chapter should be referred to. Another option is to use a direct media retention underdrain to eliminate the height normally used for gravel.

Air Scour

Several things occur at the beginning of an air scour backwash. The first thing that occurs is the media bounces slightly, even if it is nonfluidizing coarse monomedia (Figure 5-23). The initial in-rush of air moving upward raises the media slightly until a steady-state condition occurs. This initial "bounce" may be as much as $^1/_4$ in. to $^1/_2$ in. or more. The "bounce" is another reason that the support gravel design (if used) must be very stable.

If a fluidizing media is used, the second thing that occurs is the media begins to roll and churn immediately on the introduction of air into the filter. In doing so, the media will expand to the water surface. The water surface rises perhaps 10 percent because of

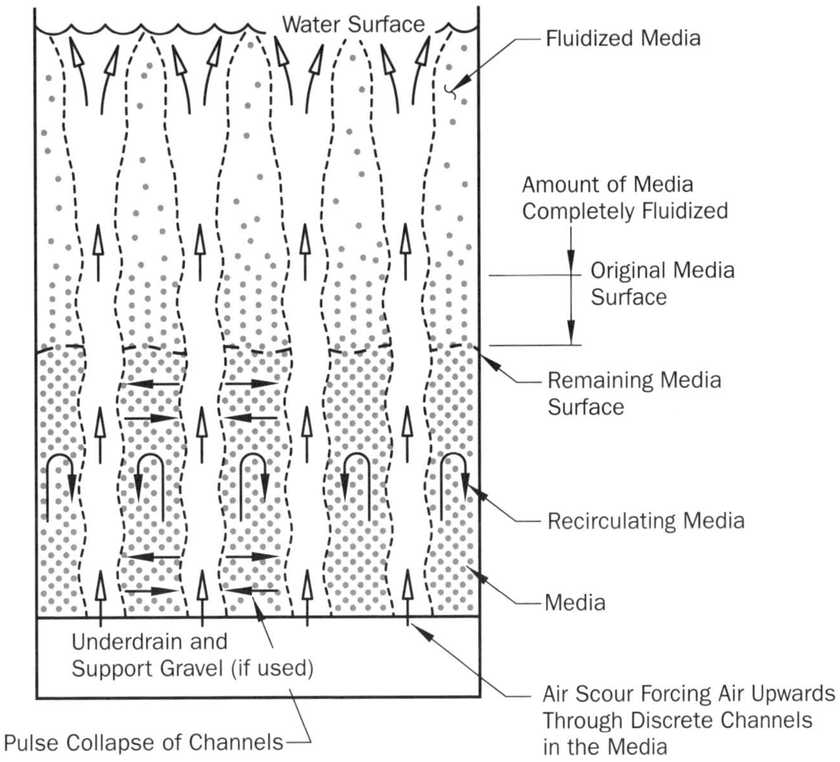

NOTE: This figure represents the condition of fluidizing media after the first minute of air scour.

FIGURE 5-24. Pulse collapse mode during air scour

the space occupied in the filter by the air at any one moment. During this step, dual or mixed media will become thoroughly intermixed. A final step of water-only fluidization is required to reclassify the various media types. After 30 to 45 sec, the rolling and churning ceases, and the media begins to compact, as shown in Figure 5-24. The upper few inches of the media will continue to be fluidized to the water surface, but the bulk of the media compacts with the air passing upward through discrete channels in the media. The media then begins what has been called a *pulse-collapse mode* of scouring whereby the air alternately pulses upward, then the media collapses somewhat until the next bubble pulses through. At the same time, the media begins a slow recirculation whereby the media slowly turns over.

With fluidizing media, it is the author's opinion that approximately 1 min is adequate for air-only scour. After that, a low rate of water wash should be added to the air to create a concurrent wash. During the concurrent wash cycle, the media again rolls and churns as it did in the beginning of the air-only cycle.

The agitation during the concurrent wash is the highest level of agitation that occurs during an air scour backwash. It is the author's experience that when 1 bed volume of concurrent water has passed upward through the filter, the media is clean at

Anthracite

Bulk Density (varies slightly) = 52 lb/ft^3 (dry)

Specific Gravity (varies slightly) = 1.65

Weight of a Solid Block = 1.65 × 62.4 lb/ft^3 (water)

= 103 lb/ft^3

$$\text{Void Volume} = \left(\frac{103\ lb/ft^3 - 52\ lb/ft^3}{103\ lb/ft^3}\right) \times 100$$

= 49.5% ≈ 50%

Sand

Bulk Density (varies slightly) = 100 lb/ft^3

Specific Gravity (varies slightly) = 2.65

Weight of a Solid Block = 2.65 × 62.4 lb/ft^3 (water)

= 165 lb/ft^3

$$\text{Void Volume} = \left(\frac{165\ lb/ft^3 - 100\ lb/ft^3}{165\ lb/ft^3}\right) \times 100$$

= 39%

FIGURE 5-25. Void volume

that point. The air can then be turned off and the water ramped up to a high-rate backwash to flush the loosened solids from the filter and reclassify the media.

Void Volume and Acceleration

In any bed design, the media obviously takes up some of the volume in the filter. The amount of space taken up by the media and the void volume remaining for the passage of water vary with the type and size of the media. Sample calculations of void volumes for anthracite and sand are shown in Figure 5-25. Typical values are used in these calculations for bulk density and specific gravity. For more accurate information, refer to the equipment suppliers for physical data on the material to be provided on the specific project.

For anthracite, the void volume is calculated to be approximately 50 percent, whereas for sand the void volume is only 39 percent. If a filtration rate of 5 gpm/ft^2 is used, as shown in Figure 5-26, the resulting downward velocity will be 0.67 ft/min above the media. Once the incoming water reaches the surface of the filter media, the water experiences an immediate acceleration. Because of the 50 percent void volume, the water will accelerate to twice the incoming velocity or 1.35 ft/min. However, because of the tortuous zigzag path the water must take around the grains, the actual velocity will be higher. As a result of the rapid acceleration, it is extremely important that the residual floc and incoming solids be tough enough to keep from being broken

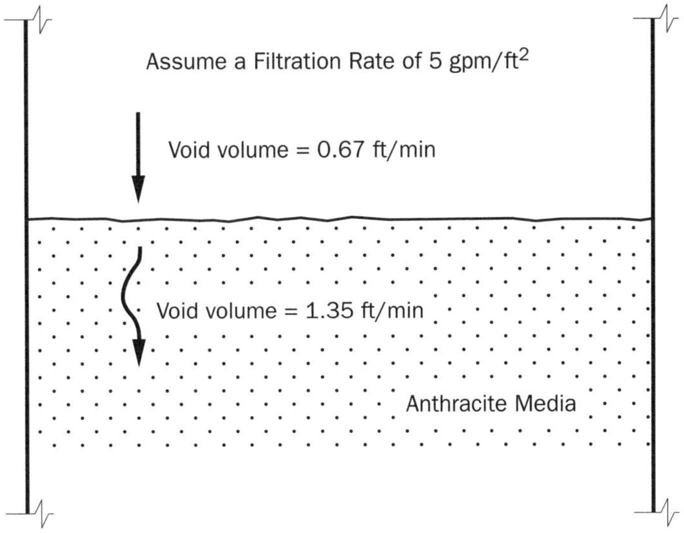

NOTES:

1. The velocity of the incoming water doubles immediately as it settles.
2. Because of the indirect path the water must take around the media grains, the actual velocity will be more than calculated by the void volume.

FIGURE 5-26. Acceleration

apart. Polymers are normally required for this purpose and also to allow the filter to trap and remove very small particles at the increased velocity.

Media Loss

A number of problems can contribute to media loss, including lack of proper freeboard, improper sequencing, normal attrition, high backwash rates, and upset gravel. Comments on each of these are in the following sections.

Media Expansion/Freeboard

Once the filter media has been selected and the amount of expansion due to fluidization has been determined (if any), the amount of media freeboard can be calculated. The media freeboard is the distance from the top of the expanded media during backwash to the bottom of the wash trough.

During backwash, the upward flow of water is passing through the entire plan area of the filter. However, as the water rises, it comes in contact with the wash troughs. The troughs can take up as much as 20 to 25 percent of the plan area of the filter. As a result, the water must accelerate by that amount. The zone of accelerating velocity is underneath the trough where the water coming upward must move around the trough, as shown in Figure 5-27. If any of the expanded media enters the zone of accelerating velocity, it can easily be washed upward along the side, over the top, and

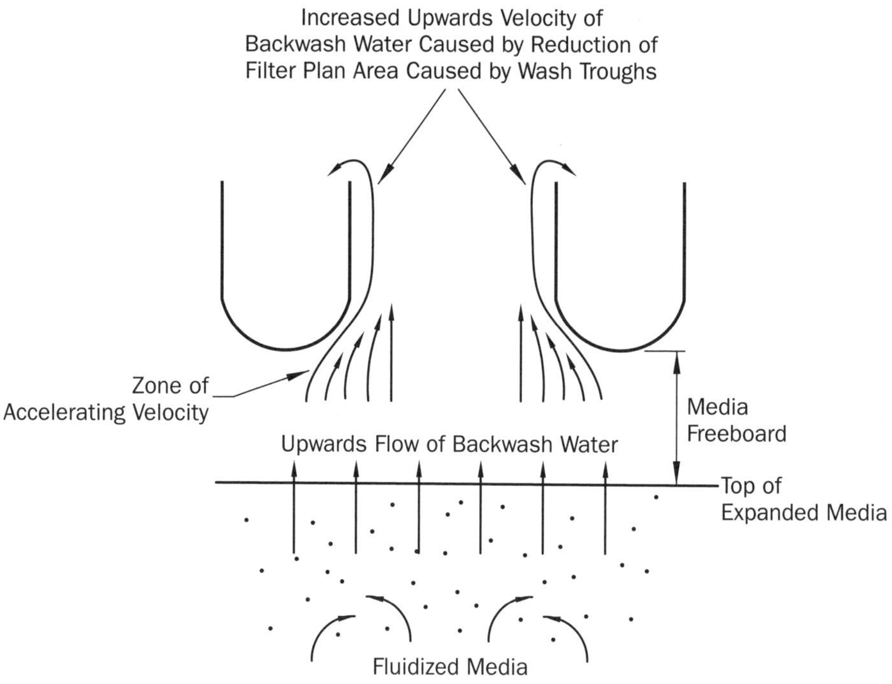

FIGURE 5-27. Accelerated rise rate of backwash water around troughs

into the wash trough, where it will be lost to waste. A significant quantity of media can be lost in this way, unless there is adequate distance between the bottom of the wash trough and the top of the expanded media (freeboard). It is recommended that at least 9 in. to 12 in. of freeboard be designed into the placement of the wash trough.

A 30-in. media bed might experience a 30 percent bed expansion (9 in.) during backwash. The total distance from the top of the media (at rest) to the bottom of the wash trough should then be 18 in. to 21 in.

If there is inadequate freeboard, a shallower media bed would have to be used, or a winged trough should be provided that is capable of disrupting the acceleration of water and shedding fluidized media grains. Simply turning down the backwash rate to avoid media loss is not acceptable. Lower backwash rates would result in inadequate cleaning of the media and fouling.

Sequencing

Improper sequencing of valves and pumps during the backwash cycle can result in the loss of media during surface wash or air scour. High flow rates, high levels of agitation, hydraulic shock, and uncontrolled air can all be caused by improper sequencing of controls. Because of the potential for damage from improper sequencing of the controls, it is recommended that the controls be finalized and approved using an empty filter, *before* installing filter media. If there are problems with an existing plant, the necessary modifications should be made to the controls *before* the media is replaced. The recommended sequencing is discussed in Chapter 7.

Surface Wash

Operating a surface wash system together with the high-rate backwash can result in the loss of media if water is weiring over the wash troughs. Increased agitation from having the surface wash and the high-rate backwash occurring together is adequate to result in the loss of a considerable amount of filter media. It is not uncommon for filters to lose between 6 in. to 12 in. of material from this cause.

In a related issue, the spray nozzles on surface wash sweeps need periodic maintenance. Nozzles can wear out or even come off the sweep arm, which can potentially result in very high surface wash rates and the resulting loss of filter media. Therefore, it is highly recommended that surface wash sweeps be inspected at least once every 3 months.

Air Scour

The most common cause of loss of media caused by air scour is operating with air alone or in a concurrent air/low-rate water wash situation where the water level is at the lip of the trough or weir. Fluidized media grains can easily be lost in water weiring over the troughs. Media can also be lost by cycling the air vent valve (if used) once the air scour cycle is already in operation, which creates an unstable situation causing boiling. In addition, improper support gravel designs can result in localized boiling and loss of media during air scour.

It is recommended that the air scour cycle be adjusted such that the water level is well below the wash trough when the air scour or concurrent cycles are in operation. A recommended air scour backwash sequence is included in Chapter 7.

Normal Attrition

Figure 5-2 shows that filter anthracite contains an amount of fines left over from the manufacturing process. During the initial installation of media, the anthracite is backwashed and then skimmed of fines. Refer to the section on Skimming in this chapter for an in-depth discussion of that subject.

During normal operation, garnet sand and silica sand are relatively durable and will last a long time. On the other hand, filter anthracite breaks down slowly with time. It is common for filters to lose as much as 1 in. or more of media per year because of normal attrition. Therefore, the anthracite must be periodically topped off to the original level and backwashed and skimmed again.

Backwash Rates

If the proper media freeboard is provided, media can still be lost by utilizing excessive backwash rates that result in high expansion, which may encroach too far into the media freeboard. Backwash flowmeters should be calibrated periodically using rise rates in the filters. In addition, the amount of media expansion during backwash

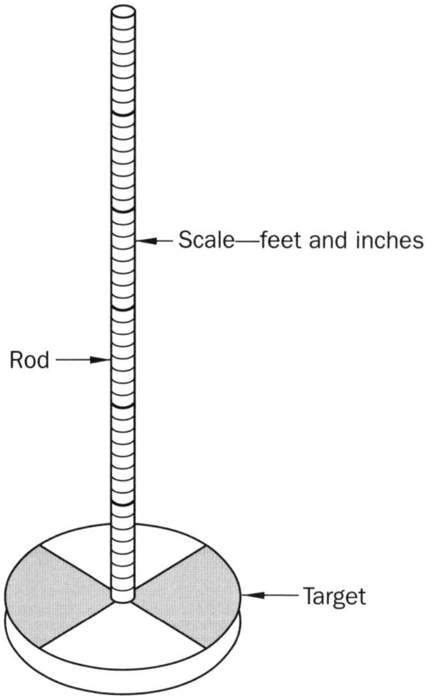

FIGURE 5-28. Expansion gauge

should be measured frequently to verify that the optimum backwash rate is used. If the rate is too low, the media will not be cleaned properly, and there is a risk of fouling. However, a backwash rate that is too high can result in the loss of filter media and significant increases in the loss of water and the use of power.

The amount of expansion can be measured and calculated using a device similar to that shown in Figure 5-28. The author has successfully used a steel tape measure with a broad end. The rod or scale can be pushed down into the water toward the end of the backwash until media grains are observed washing over the top of the target or end of the scale. The distances are measured from that point to some fixed point on the filter, such as the top of the wall. Once backwash is completed, the rod or scale is again inserted down into the media until it rests on top of the media. The difference between the two measurements is then divided by the total depth of media to determine the percentage of expansion. The calculations for media expansion are also shown in Figure 5-22.

Upset Gravel

When support gravel layers become upset, a hole can sometimes be created, as shown in Figure 4-22. There may or may not be an actual hole in the gravel. The layers of gravel may have simply become completely intermixed in that area. Such an area typically becomes a weak spot in the support gravel where unusually high backwash rates may occur. When that happens, an actual hole may be created in the filter media above the gravel. In addition, if higher than normal rates are experienced in the vicinity of the hole, there will be other areas in the filter with lighter than normal backwash rates, and those areas will probably become fouled. It is very common for mounding and fouled media to occur close to a hole in the media. Holes in the media and upset areas in the gravel, as shown in Figure 4-22, are commonly referred to as a blown filter. Chapter 4 has a discussion of identifying and probing for upset or blown gravel.

Installation Procedures

The installation procedures are considerably different for fluidizing and nonfluidizing media.

Fluidizing Media

Installation procedures for fluidizing filter media are very similar to that of gravel, as described in Chapter 4. The major differences are that media does not have to be leveled by hand. In addition, each layer will have to be backwashed and skimmed separately. If support gravel is used, great care must be taken in installing the first one or two layers of filter media so as not to damage the gravel. For example, garnet sand is normally a very thin layer, typically 4 in. deep or less. If hourglass gravel is used, as much as $1^1/_2$ in. of the garnet sand will sink into the gravel, leaving only $2^1/_2$ in. or less on top. Under these conditions, workers in the filter must continue to stand on walking boards, as described in Chapter 4, until at least 12 in. of filter media is in place.

Preparation for Installation

1. Develop a load chart, similar to the one shown in Figure 4-13, which lists the total quantities of filter media and the quantities to be used for any individual filter. Once the load chart is completed, verify the quantities of materials on site. If bulk bags are used, they should be marked with the quantity on each bag. If 1-ft^3 bags are used, all of the bags on each pallet should be counted.
2. Bring the water level in the filter up to the top of the gravel or direct media retention underdrain. Wet sand bridges should stay on top of the supporting material, whereas dry sand might penetrate too far.
3. It should be noted that it is not necessary to backwash the gravel clean at this point. Once all the media has been installed, it can all be backwashed clean at the same time.
4. Provide two or more walking boards for each worker in the filter for the first one or two layers of filter media. The intent is to protect the support gravel from damage or movement. Keep a list of the number of walking boards and tools that are in the filter, to ensure that they are all removed when the work is complete.

Methods of Installation

Bulk Bags

A description of bulk bags is given in Figure 4-8. Bulk bags are typically 4 ft square and between 2 ft and 4 ft tall, depending on the type of material contained. However, when the bags are loaded, they tend to bulge. When a bag is picked up, it tends to sag, which makes it taller. It should be noted that the four lifting loops on a bulk bag must be picked straight up using a lifting bar similar to the one shown in Figure 4-10, or equivalent. The exception is bags that have forklift sleeves instead of lifting loops. Safety instructions regarding the use and handling of bulk bags can be obtained from the supplier.

To install media with bulk bags, a bag should be lowered close to the layer below using a crane, as shown in Figure 4-14. The bag should be lowered to within 2 ft or 3 ft of the layer below and pulled open or the outer tie string cut so the discharge chute will extend downward. The bag should then be lowered so the bottom of the discharge chute is within 6 in. of the layer below. The inner tie string may be pulled open or cut to discharge the media. A very small amount of media will come out of the discharge chute when the inner string is released, forming a small pile. Utilizing this method reduces the force exerted on the gravel or layer below, to minimize disturbances.

Once the discharge chute is open, the bag should be raised slowly with the crane. Media will come out of the bag slowly, forming an increasingly larger pile. Once one bag has been emptied in this manner, the next bag can be discharged directly on top of the pile in a similar manner. The new material will fall on top of the previous pile and will spread out uniformly without disturbing the layer below. In a large filter, it may be desirable to install the media in several piles to spread it out. In that case, follow the previous instructions for installing the first bag at each new location.

One-Cubic-Foot Bags

Installing filter media with 1-ft^3 or 100-lb bags is similar to the bulk bags in that they must be emptied very carefully within 6 in. of the media until an adequate pile is developed. Pallets can be lowered into the filter within reach of the workers but should not be set on the layer below. As an alternative, conveyors could be used. Otherwise, it is strictly a manual operation. The use of 1-ft^3 or 100-lb bags is very labor intensive and should be avoided on large filters.

When using 1-ft^3 or 100-lb bags, all strings, wire ties, or bag debris must be removed continuously from the filter.

Splashboards

As an alternative, super bags or 1-ft^3 bags can be emptied onto a splashboard, as shown in Figure 4-15. The problem with using this method is that the bottom end of the splashboard tends to dig into the layer below, unless a baseboard is provided. If a baseboard is provided, it must be dug out carefully and removed without damaging the layer of gravel below. Because of the problems with protecting the layer below, this method should only be used as a last resort.

Some installers have used a wooden chute or canvas snout to install media, especially when the space between the wash troughs is too narrow for a bulk bag to pass.

Eductor Method

As an alternative, an eductor can be used to install the media, as shown in Figure 4-16. A fire hose is usually required for this method to be effective. Media will discharge from the eductor with great force. Therefore, great care must be taken to prevent damage to the gravel below. The eductor should be discharged against the sidewall of the filter or against a splashboard to dissipate the energy.

An eductor can be an effective method of installing filter media but should only be used in those cases where access to the filter is limited. The eductor method requires the use of a considerable amount of water, which will have to be eliminated from the filter by the filter-to-waste piping or other drains. In addition, the pressure or force involved is such that anthracite may be broken down considerably. As a result, additional skimming may be required on the anthracite. However, the sand and gravel are more durable and may not need any extra skimming.

Wet Method

If the bulk bags will not pass between the wash troughs, the material may be installed by the wet method, as shown in Figure 4-17. The limitations of this method are that only small amounts of media may be placed on the trough because of the strength limitations of the trough and weir boards. Calculations should be made prior to installation to determine the strength of the trough and how much material can be placed on top at any one time. Concrete troughs will probably be more amenable to using this method.

Depending on the strength of the trough, the horizontal splashboard may need to span at least two troughs to carry the load. If the troughs are not sufficiently strong, the media may have to be raked off of the wall of the filter, if possible.

Caution: Material should never be discharged from a bulk bag directly into water over the top of the filter. The free discharge from a bulk bag carries such force that it can create water hammer, which will damage the gravel layering below.

Splitting Bulk Bags

A particular layer of media may only require half a bulk bag to complete that layer. Chapter 4 has instructions on splitting bulk bags.

Media Installation

It is not necessary to hand-level filter media. All walking boards and tools must be removed from the filter and the filter backwashed. The backwash process will level the media and drive the fines to the surface. For this step, it is not necessary to use air scour, even if the plant has that feature. The intent here is to hydraulically classify the media grains for skimming purposes. After 5 to 8 min, terminate the backwash and drain the filter completely. The media layer may now be skimmed in accordance with the recommendations found in the next section.

It should be noted that each layer of media must be skimmed separately, and the layers should be installed in the same order in which they were designed to remain. For example, garnet sand should be first, followed by silica sand, with anthracite last. If the anthracite is installed first and the sand put on top afterward, the sand would backwash down to its position, but the sand fines would be distributed throughout the anthracite. The author has observed filters where this was done, and the sand fines caused short filter runs. The sand fines were scattered through the anthracite so uniformly that normal skimming of the anthracite was not sufficient to remove them.

When the garnet sand and silica sand are installed, it is only necessary to conduct a normal hydraulic backwash for skimming purposes. It is not necessary to backwash the bed clean at that point.

Because the anthracite contains a significant amount of fines, numerous backwashes will be necessary to clean that material. Several of the backwashes may be done as part of the skimming processes. Before placing the filter in operation, backwash until clean using complete cycles, including air scour if available. The process of cleaning the anthracite will also clean the support gravel and sand.

Nonfluidizing Media

When using nonfluidizing media, the material should be installed in a similar manner as previously described. The media should be rough leveled, backwashed until clean, and disinfected (if required).

Disinfection

After all work is completed and before the filter is placed in service, the entire filter should be disinfected (refer to the section on Disinfection in Chapter 11).

Skimming

Both anthracite and sand are produced by a screening operation somewhat similar to that for support gravel. However, there are important differences. For example, support gravel is processed from naturally occurring material that is typically surface-mined near rivers or streams. The rounded nature of the gravel is caused by the action of water and abrasion with other materials in a waterway. On the other hand, anthracite is deep-mined in chunks, which have to be crushed before the material can be screened to the desired sizes. Differences in the screening processes for sand and anthracite are discussed in the following sections.

Wet Production Process

Filter anthracite has been historically screened using a wet process similar to the gravel, as shown in Figure 5-29. The water spray assists in forcing the finer grains through the screens while the vibratory effect carries the coarser particles off the end of the screen.

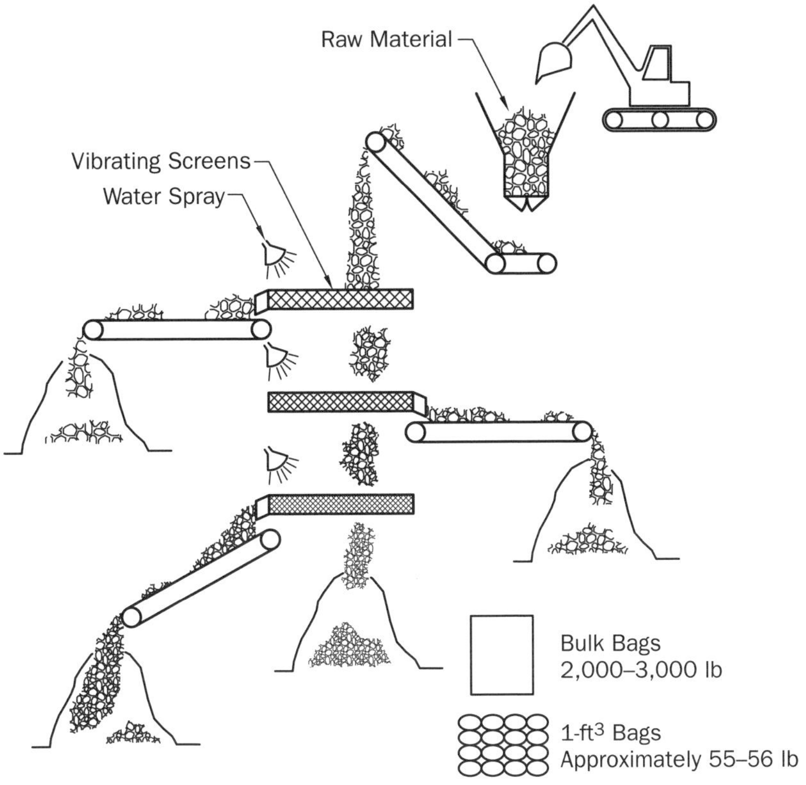

FIGURE 5-29. Anthracite production (wet method)

Water from the spray on the screen is typically recycled and contains a significant amount of fines. As a result, the finished media typically has several pounds of water per cubic foot of material. In addition, the media will contain fines that were recycled in the water, in addition to those fines that pass through the screening process. Using the wet method, anthracite has historically been produced with a UC of 1.4 to 1.7. The UC is the slope of the line between the 60 percent passing (finer) and the 10 percent passing (finer), as shown in Figure 5-2. Slowing down the material as it moves over the screens produces the lower, more desirable UC material, which improves the filtration process.

Dry Production Process

Sand

Filter sand is typically a by-product obtained from the same source as the gravel. However, while filter gravel is screened using a wet process, sand is commonly dried first, because wet sand "bridges." The water spray normally used to force gravel through the screen is not as effective with sand, because of this bridging effect. In addition, the water spray can force grains through the screens that might not otherwise have gone through.

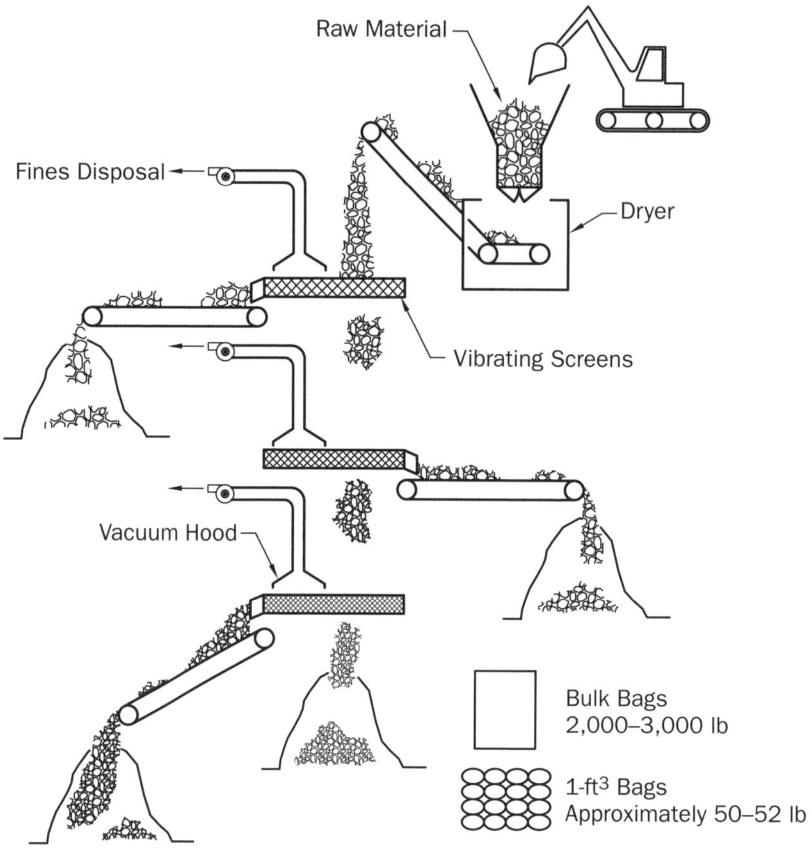

FIGURE 5-30. Anthracite production (dry method)

Anthracite

In recent years, the dry method of screening anthracite has been used to produce media with a UC in the range of 1.2 to 1.3. Using the dry method, shown in Figure 5-30, the raw anthracite material is first dried and then screened. Instead of a water spray that forces grains through the screens, gravity causes the individual grains of anthracite to fall through the screens. In addition, vacuum hoods are placed over the screens to remove some of the dust that occurs in the screening and production processes. The use of gravity and removal of some of the dust results in a lower UC product with less fines. In addition, the dry product weighs less as shown in Figure 5-29 and 5-30.

Anthracite is a brittle material. If the drying process is not controlled properly, it may cause internal stress on the grains of anthracite, which can crack and weaken them. After being installed in a filter, anthracite that has been stressed can break down faster with normal use than might otherwise be expected. Strict quality control measures in the drying and processing phases of manufacturing are important in ensuring a quality product.

Friability is a term used in the water industry to describe the ease with which a material crumbles. The easier it crumbles, the higher the friability. Specifications used to purchase anthracite should include a request for low friability material.

Classification

Even though the dry screening method can result in a lower UC, the material still has fines and some dust which is smaller than the specified range, as shown in Figure 5-2. Above the 10 percent passing (finer) line, a change of slope of the graph is apparent. At about 2 or 3 percent passing (finer), the line changes slope again and passes off of the chart at a very shallow angle. These changes of slope indicate a rapid progression toward finer grains of media.

Filter anthracite is normally shipped to the job site in either 1-ft^3 or 2,000–3,000-lb bulk bags. The individual grains of anthracite vary widely in size, as indicated by the graph in Figure 5-2. These different-sized grains are uniformly mixed together during the manufacturing process and shipment, as shown in Figure 5-16. However, when the media is installed in a filter and backwashed, all of the particles of different sizes within a given layer of media will "classify" with the largest particles on the bottom and the smallest, or finest, on the top, as shown in Figure 5-17.

Surface Filtration Versus Bed Penetration

In a slow sand filter, a *schmutzdecke* (biomass) should be developed, which results in most if not all of the filtration of suspended solids taking place at or near the surface of the media. In a conventional granular media filter, the intent is for the solids to penetrate throughout the media so that the entire bed depth can be used for solid storage. An additional benefit of solids penetration is to prolong filter runs, while producing high-quality water.

If the fines are allowed to remain on the surface of any layer of media in a conventional plant, surface filtration will result on that layer, which will have the effect of shortening the filter runs and requiring higher use of backwash water to keep the media clean. To promote the penetration of solids into the bed and lengthen filter runs, the fines must be removed from the top of each layer separately. Fines removal (skimming) is done *after backwash* when any media layer is first installed.

Quantity of Fines

As shown in Figure 5-2, approximately 5 percent of the anthracite material in that sample should be removed by skimming, which accounts for most of the fines. The size of the remaining material on the surface would be approximately 0.93 mm. Skimming 5 percent of an anthracite layer 24 in. deep would result in removing approximately 1¼ in. of material. A similar approach may be followed to determine the amount of fines in sand or garnet sand.

Skimming (General)

When installing filter media, each layer must be laid, backwashed, and skimmed separately before the next layer is installed. When the sand is installed over the support gravel, there should be several inches of water over the gravel. Sand should always be wet when it is installed. Dry sand could work its way through the gravel or direct media retention underdrain (if used), whereas wet sand will bridge and stay in place.

It is not necessary to manually level filter media because the backwash procedure will level the material in addition to cleaning it and classifying the individual grains. Once the material is backwashed, the filter can be drained completely in preparation for the skimming process.

Because the fines have been forced to the surface of the media, it is important not to walk directly on the media, because the fines would be pushed down into the bed without the opportunity for skimming. The workers should stand on boards. Plywood boards $^3/_4$-in. thick by 1-ft wide and 2-ft long are recommended for this purpose. Three boards per worker are recommended, two for standing on and one for moving ahead.

Skimming should be done with a square-edged shovel or pan made for this purpose (Figure 5-31). Standing on the boards, the worker can skim ahead within reach of the shovel. Once an area is skimmed, the worker can stand on that area directly without the use of boards, as long as there is at least 12 in. of media over the gravel. A minimum of $^1/_2$ in. of material should be removed in any skimming, regardless of the media type or the quantity of fines present. Less is not practical.

Depending on the plant facilities, the skimmings may be either dumped into the backwash trough for disposal in the backwash waste basin or collected in buckets and removed manually from the filter for disposal elsewhere.

Skimming Sand

Garnet Sand

Garnet sand is pink in color. However, when it is backwashed, a layer of brown dirt will typically appear on the surface. Even if only 3 in. of garnet sand is desired, it is

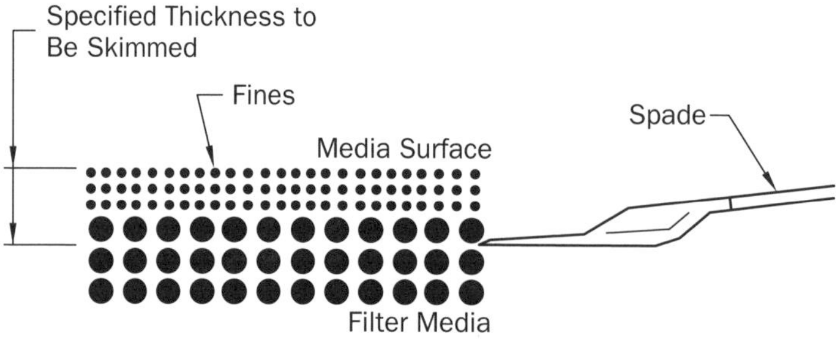

FIGURE 5-31. Skimming procedure

recommended that additional material be provided to allow at least $^{3}/_{8}$ in. to $^{1}/_{2}$ in. to be skimmed off. One skimming is usually adequate.

Especially if installed over hourglass gravel, garnet is a thin layer. As such, great care should be taken to protect the gravel below, including the use of walking boards at all times while skimming.

Silica Sand

Silica sand is normally installed 9 in. to 12 in. deep in conventional plants. For a 10-in. layer, 5 percent would be $^{1}/_{2}$ in. Therefore, at least $^{1}/_{2}$ in. should be skimmed. One skimming is usually adequate, although testing for any remaining fines could be done, as described later in this section.

Garnet is used and the silica sand is at least 9 in. deep, it is usually not necessary to use walking boards in areas that have already been skimmed.

Skimming Anthracite

It is normally sufficient for filter sand to be installed, backwashed, and skimmed once. However, with anthracite, there are several additional requirements.

Soaking

When anthracite is first installed, it is common for small air bubbles to attach themselves to many of the grains. If the anthracite were backwashed immediately, a small amount of the anthracite would be carried out of the filter by these small air bubbles. In addition, anthracite is slightly porous and can absorb 3 to 5 percent of its weight in water, reducing the amount of material washed out if it is allowed to soak. The anthracite can be installed, backwashed, and skimmed immediately. However, it does result in a slightly better filter bed and less loss of material if the anthracite is allowed to sit in water overnight and then backwashed and skimmed on the second day.

Skimming Depth

Skimming anthracite is similar to skimming sand. However, there are typically more fines and dust in anthracite than there are in sand or garnet sand. It may only be necessary to skim the sand once to a depth of $^{1}/_{2}$ in. However, as discussed previously, it may be necessary to skim the anthracite of as much as 1 in. to $1^{1}/_{4}$ in. of fines. The sieve curve for the material will give an indication of how much should be skimmed, as noted in Figure 5-2. When skimming this amount of material, it is desirable to remove the fines in two separate skimmings because of the inefficiencies of the process. If it is anticipated that as much as $1^{1}/_{4}$ in. of material must be removed, the first skimming should be approximately $^{3}/_{4}$ in. The filter should be backwashed and skimmed again. For the second skimming, a visual inspection should be made to determine how much material to remove.

Debris

In addition to fines, there is often some wood debris in new anthracite. Wood debris, from shoring of old mineshafts dug up with the anthracite, may pass through the screening process. Debris of this sort should rise to the surface during backwash, where it can be removed by skimming. If wood debris is present, it should be visible when testing for fines.

Disinfection

Filters should be disinfected after installing new media, skimming, topping off, or other significant work. The exception may be those filters that use granular activated carbon, in which case the manufacturer should be consulted for recommendations. Chapter 11 discusses the recommended disinfection procedure.

Testing for Fines

There is a very simple visual procedure for testing for fines. First, backwash the filter and drain it completely, then climb down into the filter. Referring to Figure 5-32, make a cut straight into the filter media approximately 3 in. to 4 in. deep with a square-edged shovel. Then cut into the media at a low angle with the shovel up to the vertical cut, and raise the triangular shaped section of media up for visual inspection. Looking at the vertical cut, inspect the upper inch or so of the material. If there are fines present, there will be an obvious difference in particle size from the material on top from that which may be found an inch or so into the media. There may be a very fine layer on the surface, followed by an intermediate layer, and the bulk of the material below will appear as more of a grainy uniform texture than the finer material on top. Following this procedure, the

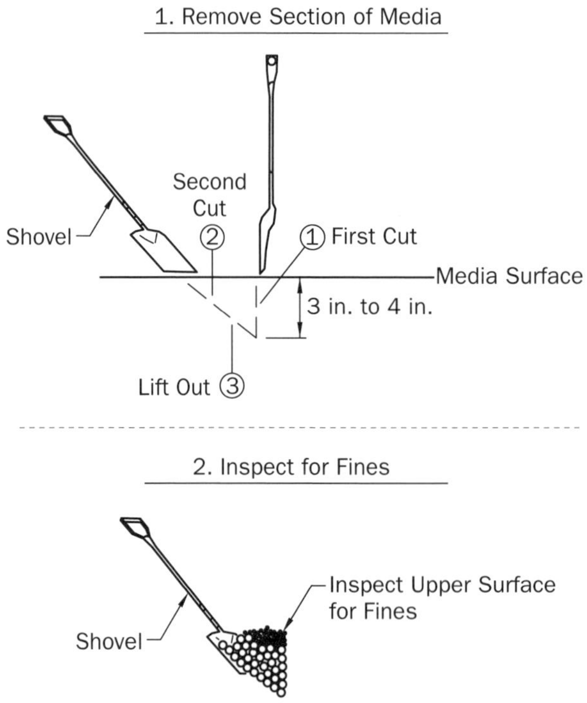

FIGURE 5-32. Fines inspection

depth for the final skimming may be determined. When the skimmings are complete, the surface of the remaining media should be grainy and rough textured. Fines on the surface tend to have a smooth, velvety appearance.

Operational Considerations

Degradation of Material

Filter sand in dual media or multimedia is below the anthracite where it is inaccessible, and it is not possible to skim it after the original installation. However, sand grains are more durable and tend to be worn smooth but do not break down to any significant extent or as rapidly as anthracite. Anthracite is more brittle and tends to break down over time, generating new fines. The amount of fines generated and the frequency with which this occurs varies widely with the quality of the raw anthracite. To the author's knowledge, there is no information that can currently be used with any degree of accuracy to predict how fast anthracite will break down. The friability test is only a rough indicator of how much it might break down. Some anthracite will be quite durable and not break down rapidly, while other material may generate inches of fines in the first few months of operation.

Some anthracite that breaks down will be washed out in the backwash water, while other fines will accumulate on the surface, preventing penetration of solids. If fines are allowed to accumulate, it will shorten the filter run time between backwashes (assuming that the filter equipment and the media are otherwise in good condition). If the filter media becomes upset, turbidity breakthrough of the bed will be the determining factor on filter run time. The accumulation of fines will result in a more rapid buildup of head loss, and the filter runs will terminate on head loss as opposed to turbidity.

Bank coal that has been previously mined and stored in piles on top of the ground may be more subject to breakdown than freshly mined material. Bank coal has been exposed to repeated freeze/thaw cycles, which may cause stress cracks within the grains, which may in turn make it more susceptible to fracturing during normal filter operation. As a result, it is recommended that freshly mined anthracite be specified for filter material. In any case, low friability material should be specified.

Fluidization

For the best skimming efficiency, the media should be completely fluidized to lift the fines to the surface. When using coarse anthracite monomedia, it is not normally possible to provide enough backwash water to fluidize the media. As a result, all of the fines will not be driven to the surface where they can be skimmed off. Over a period of time, additional fines may appear on the surface, which could require additional skimmings. However, when using coarse sand or fine gravel, fines are not usually a problem.

Tests have shown that a rapid shutdown of the high-rate backwash water is best for classification of the media and leaving the fines on the surface. A slow shutdown of the high-rate backwash allows intermixing of the different sizes of grains as the media settles. Because of hydraulic considerations, slow operating valves are normally provided.

Therefore, there may still be fines in the bed after skimming. However, these fines do not normally interfere with proper operation of the filter.

Recommendations

During normal operation, the filter should be checked for fines every 3 to 6 months or until sufficient history is developed on the new filter material. Fines can be inspected visually or with the procedure as shown in Figure 5-32. The fines should be skimmed off when tests show any significant amount on the surface.

Filters in good condition commonly lose as much as 1 in. of depth per year because of normal attrition. When the filter becomes more than 2 in. to 3 in. low, because of attrition and skimming, the media should be backwashed and topped off with at least $^1/_2$ in. more material than is required. The filter can then be allowed to soak, and the new media can be backwashed and skimmed back to the original depth by removing the new fines. As an alternative, the old media can be skimmed first, and the new media added and skimmed separately. If there is any debris on the surface, this will be the preferred method.

At least 5 percent extra material of each type of media should be requested, depending on the job size, to account for potential loss at the job site and skimming. If there is excess anthracite left over, it can be used to top off the filter at a later date.

When media is purchased, it is recommended that freshly deep-mined anthracite with a low friability be specified. Other physical characteristics, such as effective size and UC, should match the specifications for the original material.

Troubleshooting

In this section, technical information is discussed to develop a logical pathway for identifying the condition of the media and any problems that may have occurred.

When troubleshooting filters, it is often helpful to inspect one of the worst filters and compare it with one of the better ones. The first step in the troubleshooting process is to observe a backwash of each.

Observe Backwash

At least one entire backwash cycle should be observed for each filter being inspected. The following items should be evaluated as part of the observation.

Sequencing

Any problems should be identified in the sequence of events during the backwash. The backwash sequence should be close to that described in a previous section as well as in Chapter 7 for either surface wash or air scour.

Backwash Patterns

Watch for any patterns that appear during the backwash, as shown in Figure 5-33. For example, are there areas that clean up faster than others? Are there areas that do

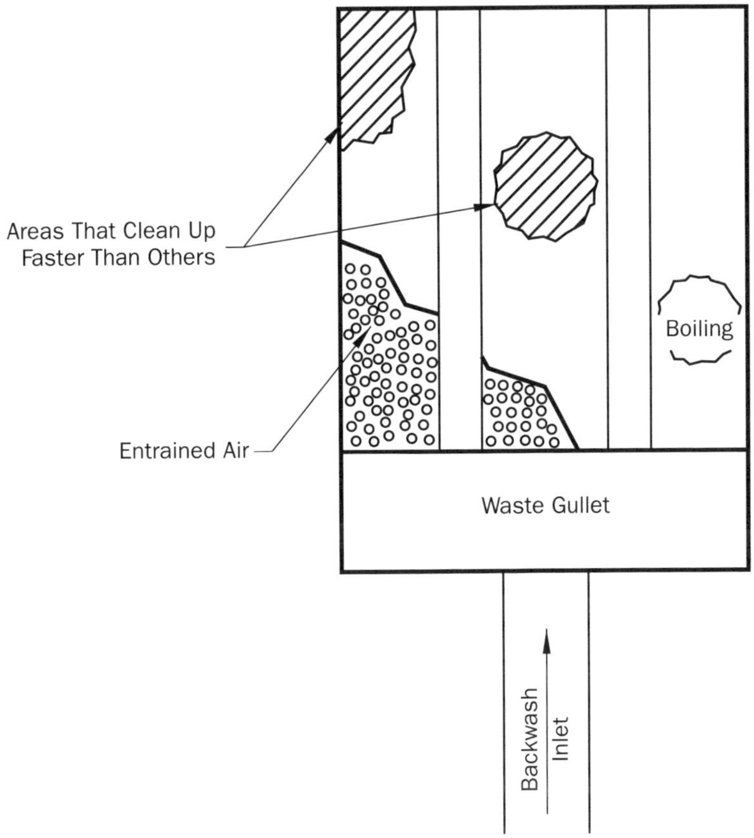

FIGURE 5-33. Backwash patterns (plan view)

not clean up well at all? Areas that clean up faster than others may have a much higher localized backwash rate than areas that do not clean up well. A sketch should be made of the filter to record any unusual observations. Typical diagrams for this purpose are included at the end of this chapter. When probing or excavating the filter, the areas identified should receive special attention.

Entrained Air

Entrained air is characterized by small bubbles spread across the filter. These bubbles are the result of dissolved air or gas in the water that is drawn out of solution by the head loss developed in the media. The bubbles attach themselves to the media grains and are released during backwash. Because this type of air is formed in the media, it typically will not damage support gravel. Entrained air typically will result in shorter filter runs and can cause loss of media from the grains being floated out by the air bubbles. Entrained air can be worse during the winter, because cold water typically holds more dissolved gas than warmer water. The only simple solution for addressing entrained air in an existing filler is to backwash more frequently. If entrained air is anticipated to be an issue, filters should be designed with more positive driving head to help eliminate that problem. Chapter 2 also discusses driving head issues resulting from extrained air and negative drawing head.

Boiling/Uncontrolled Air

Boiling can occur during the air scour or concurrent steps as a result of improper sequencing or compatibility problems between the media and the support gravel (if used). Boiling can also occur during the hydraulic high-rate wash as a result of uncontrolled air in the backwash water. Boiling during the air scour portion of the backwash cycle can occur anywhere, whereas boiling caused by uncontrolled air in the hydraulic backwash water is typically near the flume area.

The location of any boiling should be recorded on the sketch.

Media Loss

If media is lost during backwash, it is usually visible toward the end of the cycle when the wash water is clear. As an alternative, a screen or net on a long pole can be used to capture samples of media that may be washing over the weirs. It is important to determine if media is lost uniformly across the filter or if it is concentrated in certain areas.

All observations should be recorded on the sketch.

Typical diagrams of different filter layouts are included at the end of this chapter for use in recording observations, or a new sketch can be made that more accurately depicts the plant being investigated.

Probe the Gravel

Depending on the previous observations, it is often desirable to probe the gravel, as described in Chapter 4. Particular attention should be paid to probing those areas that exhibited unusual characteristics in the previous observations.

The results of the probing should be recorded on a separate diagram, as shown in Chapter 4.

Core Sampling

Depending on the results of the other observations, media core sampling can be done. It is not 100 percent efficient but can be used to determine the general condition of the media and the relative depth of each media layer. Core sampling is performed on fluidizing media with the device shown in Figure 5-34. The media should first be backwashed so that it is clean. Then the rate should be reduced to approximately 5 gpm/ft^2. The intent is to loosen the media *slightly* so that the sampler can be pushed through the media with minimal force. If the rate is too high, the media will flow around the sampler. If the backwash flow is too low, it will take too much force to insert the sampler, making it difficult to determine when the support gravel is reached (if used).

Using the proper backwash rate, the sampler can be easily pushed into the media. Care must be taken not to punch a hole into the support gravel. Intact support gravel

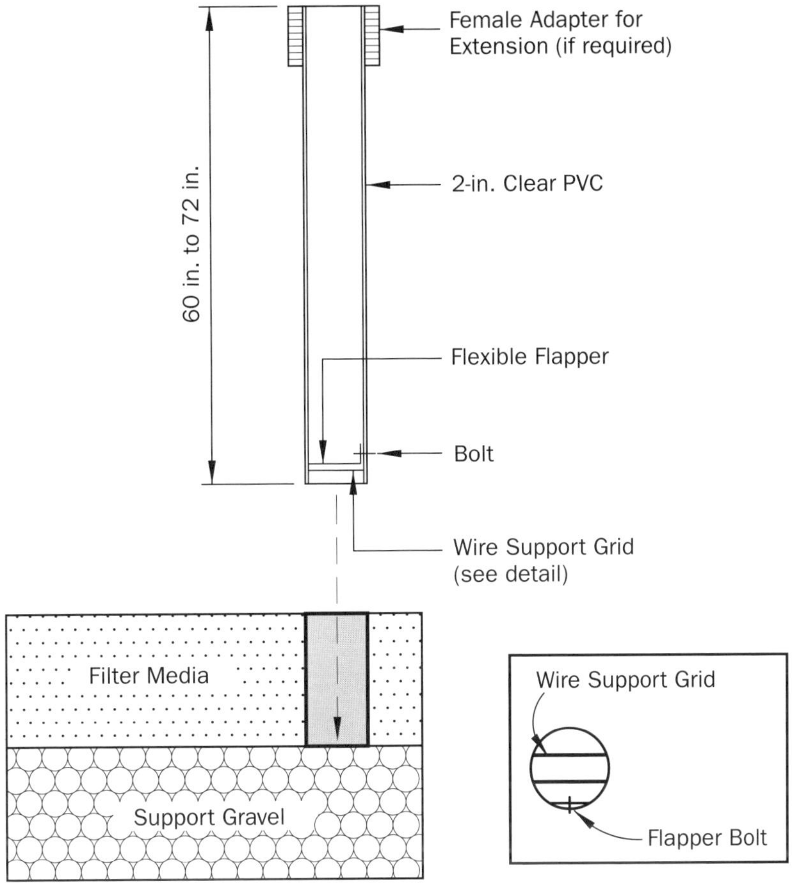

FIGURE 5-34. Core sampler

will feel solid and more "grainy" than the media. The sampler may then be pulled out of the bed with the captured sample. The flapper and support wire in the base will act as a check valve to retain the media.

The sampler and the technique used may not be completely effective in capturing all the media. As a result, the relative depths of various layers may not be exact. Core sampling is a tool to determine the relative depths, condition, and cleanliness of the media. It should be used for general purposes only. For exact depths, manual excavation is recommended.

Backwash Expansion

The backwash expansion can be measured, along with the current height of the media after backwash, as shown in Figure 5-22. These values can be compared to previous measurements taken for that filter. If data are not available, the height of the media can be approximated from the original construction drawings. The backwash rate currently in use should also be recorded and compared with typical or previously established values for that filter.

Media Surface

Observing the media surface can be very instructive in understanding the condition of the media as well as identifying potential problem areas.

After a backwash, the filter should be drained and the water level observed as it breaks through the surface of the media. It is important to know if the surface is level, and if not, where the high and low areas are. It is also important to investigate other surface characteristics, as shown in Figure 5-35. The media surface should be investigated for each of the following situations.

Levelness

After Backwash. Record whether the media is level or not, and the location of high and low areas. Low areas typically are the result of localized high backwash velocity, while high areas indicate the opposite.

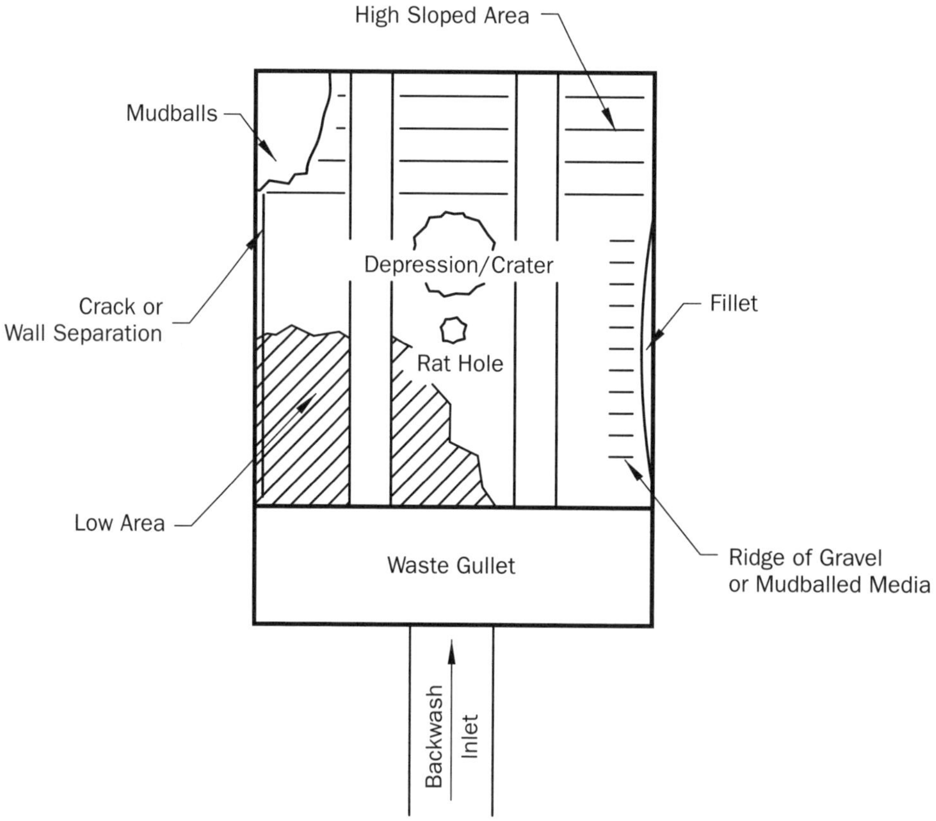

FIGURE 5-35. Media surface (plan view)

In addition to backwash-related issues, the media surface can also be disturbed by incoming water when the filter first comes on line. An *influent condition* can occur if the starting water level is too low or if the incoming water enters the filter too fast.

Low Starting Water Level. When the starting water level is too low, the incoming water will spill over the trough weirs and can gouge a hole in the media, as shown in Figure 5-36. Typically, a long hole will be created on both sides of each trough. The obvious solution to this problem is to ensure that the filter is full before placing it in operation.

Fast Incoming Water. Many filters are constructed so that the incoming water first enters the waste gullet and then flows into the filter through the backwash troughs.

In filter mode, the waste gullet serves to dissipate the momentum of the incoming water. If the inlet pipe is in line with one of the troughs, as shown in Figure 5-37, the momentum of the incoming water may be adequate to gouge a hole in the media and create a mound in the end of the filter.

A similar condition can occur in package plants with a flap gate between the clarifier and filter, as shown in Figure 5-38.

In the case of the inlet pipe lined up with a trough, a baffle is required to dissipate the momentum of the incoming water, as shown in Figure 5-37. For the package

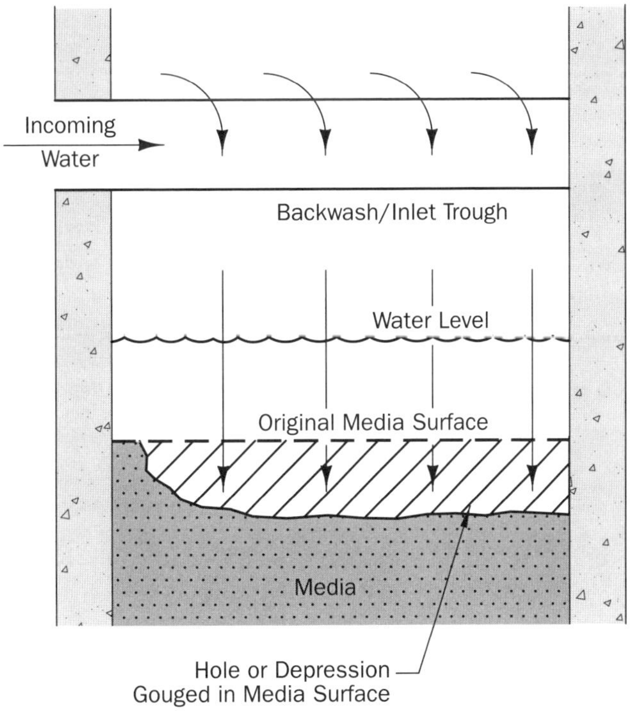

FIGURE 5-36. Low starting water level

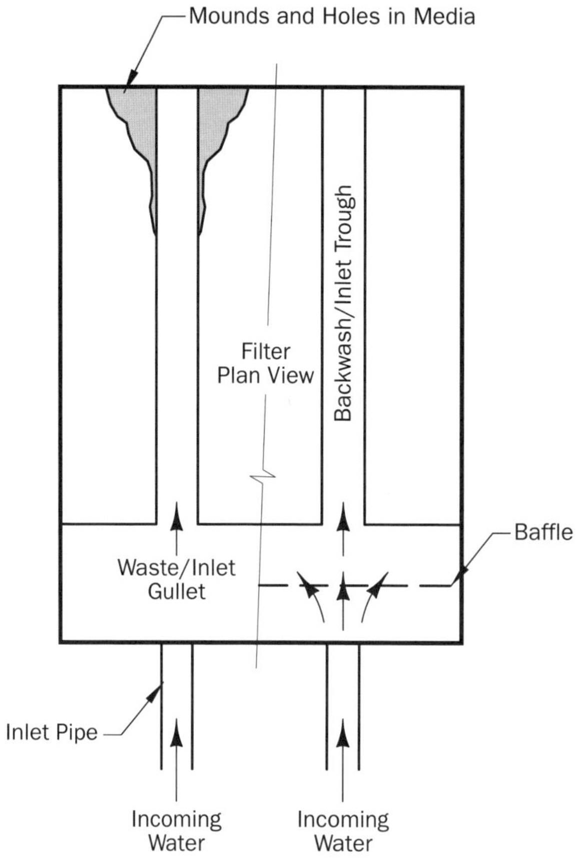

FIGURE 5-37. Fast incoming water

plant, a splash plate or perforated inlet box is required, as shown in Figure 5-38. The intent is to prevent an "incoming water condition" by dissipating the force of the water entering the filter.

Any of the incoming water conditions could result in upsetting the support gravel (if used). However, it should be noted that *if the support gravel is intact, none of these conditions should be apparent after backwash.* They only occur when the filter is placed in service.

Cleanliness

A clean filter should be uniformly black if anthracite is used, with no discolorations. If mudballs are present, record their location and whether they are large or small. Larger areas with a thin coating of dirt or other solids are also of interest. Filters with surface wash will typically have some accumulation of solids in unswept corners, whereas filters with air scour should be uniformly clean.

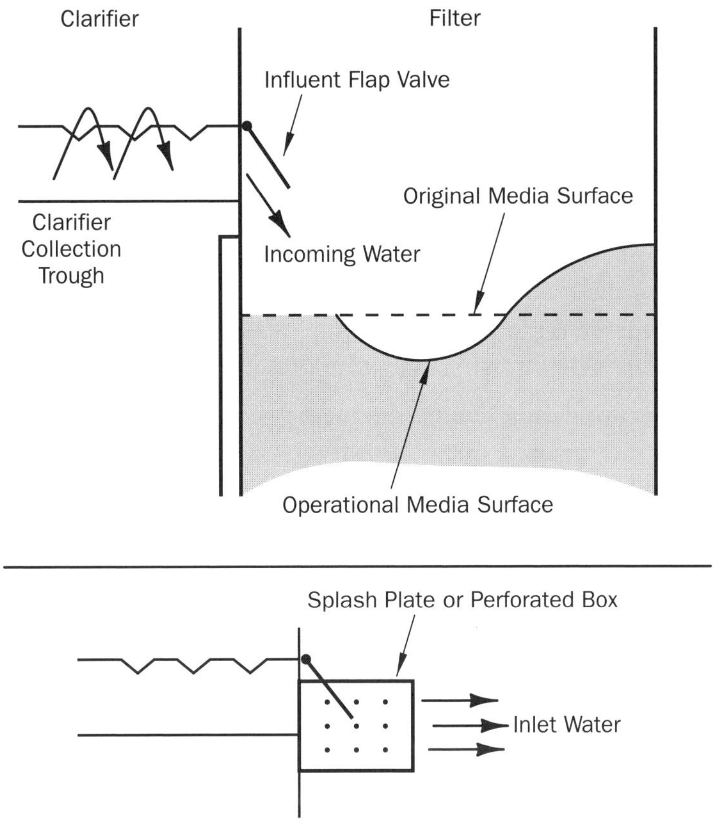

FIGURE 5-38. Package plant incoming water

Craters/Depressions

If there are any depressions or craters in the filter, they should be recorded. Low areas in the filter media can exist in the form of large shallow depressions or as deep craters with steep sides. A large shallow depression may be an indication of failure in progress, while a crater is usually the result of a catastrophic failure. On the other hand, a "rat hole" is usually characterized as a small hole several inches in diameter that may go down through the media and support gravel (if used) to the underdrain.

Cracks

Cracks can exist along the sidewall or in the middle of the filter. Cracks typically represent areas that are not completely mudballed but are not being cleaned properly. Cracks sometimes are characterized by a separation where the media has pulled away from the sidewall of the filter. Cracks and separations are sometimes caused by excessive polymer feed and/or inadequate cleaning of the media.

Sidewall Depression

A sidewall depression extends along one sidewall of the filter and is typically wider and not as deep as a crack. Sidewall depressions are typically the result of a failure of the support gravel. There is probably a high rate of backwash flow in this area, and there is commonly a ridge of mounded gravel or mudballed media adjacent to the depression toward the inside of the filter. A failure of this sort can be the result of hydraulic shock, uncontrolled air, or other gravel problems.

Debris

There are a number of different forms of debris that can appear in a filter. Filters in potable treatment plants that are outside often accumulate pine needles, leaves, and oak seedpods. All of these can eventually become water logged and sink to the bottom of the media and can cause long-term fouling. Wastewater treatment plants may see various forms of plastic debris, which can also result in fouling the filter media.

If the filter is accumulating debris, the media should be skimmed as frequently as necessary to keep the material out of the filter. Steps should be taken to eliminate these items from entering the filter.

Manual Excavation

Depending on the results of the other observations, it may be desirable to excavate one or more areas in the filter to verify the condition of the support gravel or media. Areas that have craters, cracks, depressions, or heavy mudballs are all candidates for excavation.

When digging holes in the media, it is usually necessary to dig a very large hole with a slope of about 45° in order to keep the sides from sloughing in (Figure 5-39). If the filter uses support gravel, care should be taken not to disturb the gravel when digging close to it. Upset gravel will be characterized by mounding or the presence of

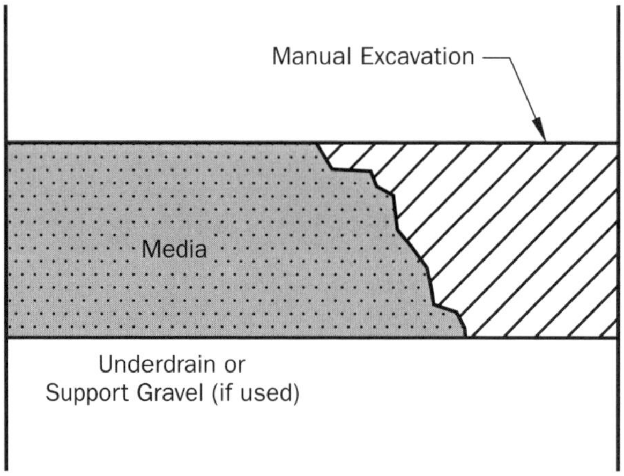

NOTES:

1. A 45° slope may be required to keep sides from sloughing in.
2. Do not disturb support gravel.
3. Observe the face of the excavation and record the profile.

FIGURE 5-39. Manual excavation (plan view)

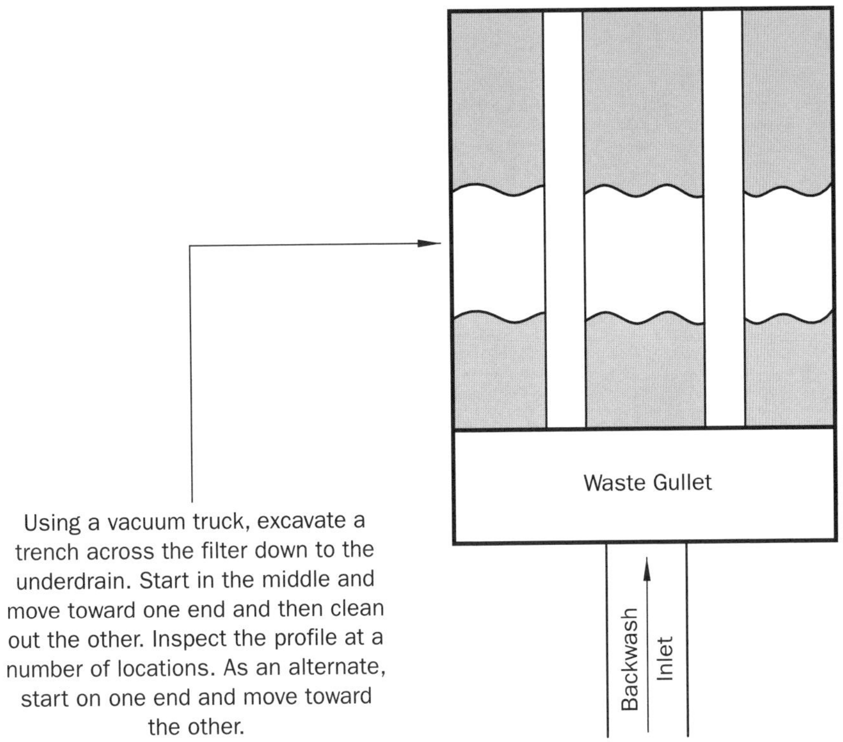

FIGURE 5-40. Complete excavation (plan view)

lower levels of gravel on the surface. Once the excavation is down to the level of the support gravel or the direct media retention underdrain, the face of the excavation should be observed and observations recorded, as shown in Figure 5-41. The observations may include intermixing of two or more layers of gravel, excessive intermixing of anthracite and sand, media layers not being present in necessary depths, improper gravel gradations, craters, blown support gravel, ruptured underdrain, failed grout joints, or the accumulation of debris on top of the support gravel or direct media retention underdrain.

Complete Excavation

If the filter media is to be completely replaced, it is helpful to do a controlled excavation in order to learn as much as possible about the mode of failure, as shown in Figure 5-40. In a small filter, it may be desirable to start the excavation in one corner, proceed across the filter and move toward the other end. During the process, frequent inspections of the excavation profile should be made and the results recorded, as shown in Figure 5-41.

In a large filter, depending on where the problems are visible, it may be desirable to excavate a trench across the middle of the filter, and move in each direction separately. Regardless of how the excavation is done, a profile should be generated, similar to Figure 5-41, in as many locations as possible.

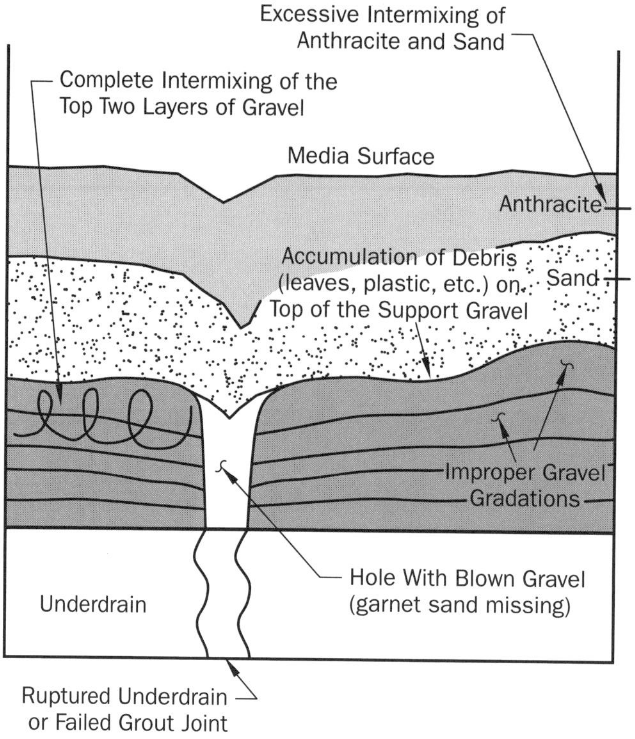

FIGURE 5-41. Excavation profile

Flushing Underdrain

If the media is upset and must be removed and replaced, it is common for the underdrain to be at least partially fouled. It is critical that the underdrain be flushed clean before installing new media. Otherwise, there is a high probability that the new media will rapidly become fouled and upset. The media fouling the underdrain can be pushed upward out of the underdrain blocks during backwash, where it can foul the support gravel (if used) or direct media retention system.

In the case of a pipe lateral underdrain, the laterals should be removed and individually flushed clean, including checking the orifices for plugging. The manifold should also be removed and cleaned if possible. If it is not possible to remove it, the manifold should be flushed clean through the filter-to-waste piping system.

A dual parallel lateral underdrain is much more difficult to clean if it becomes fouled. When the media is removed, the underdrain block should be inspected to determine if there is substantial media fouling. Insert a $^1/_8$-in. rod through the outer orifice holes, as shown in Figure 5-42. By lowering the rod carefully, it will be possible to "feel" media fouling, if present. The media will feel gritty, whereas the bottom of the underdrain block will feel hard and solid if there is no media present. Using this method, it is also possible to estimate the depth of media fouling.

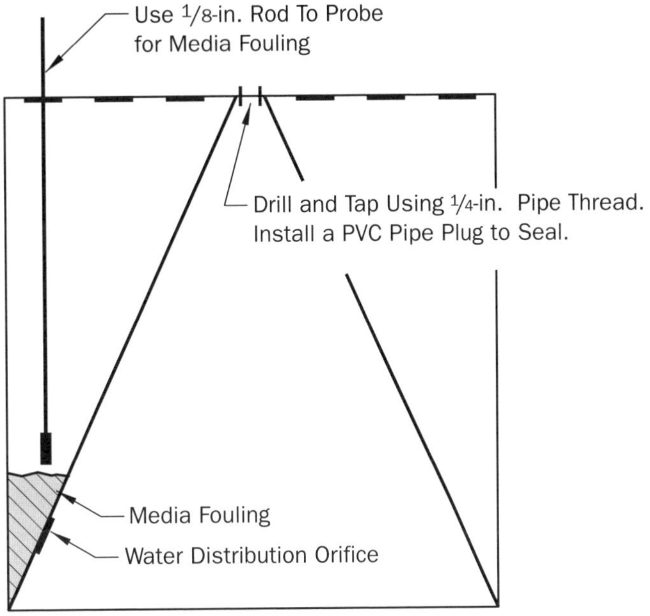

FIGURE 5-42. Probing dual parallel lateral underdrains

Probing by this method will detect media fouling in the secondary chambers of the dual parallel lateral underdrain. The primary chamber should also be investigated. Drill and tap a hole using 1/4-in. pipe thread in the center of the block, as shown in Figure 5-42. The primary chamber may then be probed through this hole. The block over the flume should also be probed to determine if substantial amounts of media are present.

Probing should start on the opposite end of the filter from the flume and then at several intermediate locations. The amount and extent of fouling in each filter should be recorded.

Any significant amount of media fouling the underdrain must be flushed out prior to installing new media. A flushing manifold should be installed, as shown in Figure 5-43 and 5-44. The flushing manifold should extend across the width of the filter, with 1/4-in. copper flushing tubing extending down into each primary and secondary chamber of each underdrain lateral across the filter. The copper tubing should be bent in a curve so that the water jet is directed along the bottom of the underdrain.

The filter should first be filled with water. The flushing water should then be turned on and the media forward-flushed as fast as possible in filter-to-waste mode toward the flume. Depending on the amount of media present, much of it will be retained in the flume, while some may be flushed out in the filter-to-waste piping. It is usually necessary to flush the filter in this manner a number of times to adequately clean the underdrain. Periodic probing will be necessary to determine if the

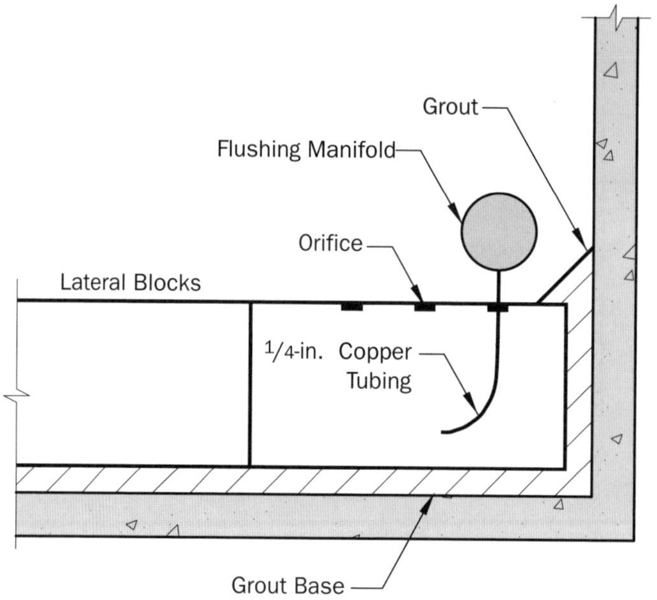

NOTE: Install a flushing manifold on the opposite side of the filter from the flume. Install a 1/4-in. copper flushing tube in each secondary and primary chamber in each underdrain lateral. Fill the filter with water, turn on the flushing water, and forward-flush the media to the flume.

FIGURE 5-43. Flushing dual parallel lateral underdrains (longitudinal cross section)

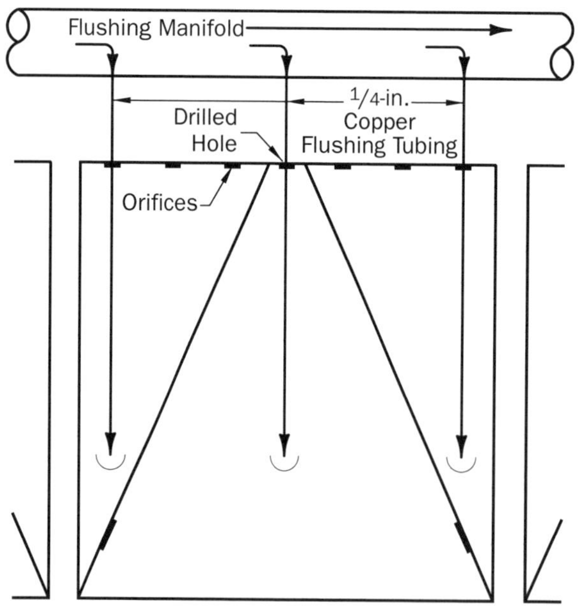

FIGURE 5-44. Flushing dual parallel lateral underdrains (block cross section)

underdrain has been cleaned. If it is not possible to install a flushing manifold all the way across the filter, a few laterals could be flushed at a time.

Once the underdrain is clean, it will be necessary to access the flume to remove the accumulated media. Removal can be accomplished manually or by a vacuum truck, depending on the construction of the filter. Use PVC pipe plugs to seal the holes used for probing in the primary chamber.

Hopefully, problems in filters can be detected early so that they can be resolved before the media becomes badly upset. However, a plant designer should assume that filter maintenance will be required sooner or later. The designer should provide access to the flume through a separate hatch or through the backwash manifold. In addition, the filter-to-waste system should be adequate to completely drain the filter, and the piping should be capable of being taken apart for cleaning. If these issues are adequately addressed by the designer, it will be easier for the operators to maintain the filters.

Lancing Media

In cases where a filter is becoming fouled, plant operators will occasionally attempt to "lance" the media with a water jet or air wand, as shown in Figure 5-45. Lancing is normally done during backwash. The intent is for the lancing to break up or loosen solids in the media so they can be backwashed out of the filter. However, if support gravel is used, lancing is not recommended. If the air or water jet impacts the surface of the gravel, it will cause an upset condition. With a direct media retention underdrain system, either air or water lancing is acceptable. However, care should be taken so that the lance does not damage the underdrain.

The concept of the media needing lancing implies that there is a major problem with the filter. Lancing is normally a stopgap procedure, which should only be used while the real problem is being identified and resolved. One exception is the "edge condition" caused by direct media retention systems, as discussed in Chapters 4 and 6. If this condition is present, the only potential solution is to block off part of the filter or add air scour. If neither of these remedies is implemented, *continuous* lancing is required.

When the filter contains support gravel, and it is desired to lance the media, even though not recommended, a lance of the type shown in Figure 5-46 should be used. Horizontal or upward directed jets reduce but do not eliminate the potential for damage to the support gravel. Without a downward jet, it will be more difficult to push the lance into the media, and it will be more difficult to determine when the support gravel is reached.

Summary

If a filter is in reasonably good condition, topping off and skimming for fines and debris may be the only maintenance tasks necessary. Depending on the various observations that are made, the actual troubleshooting procedure used can be altered. For example, if the filter is badly mudballed or cratered, it is not usually necessary to do an

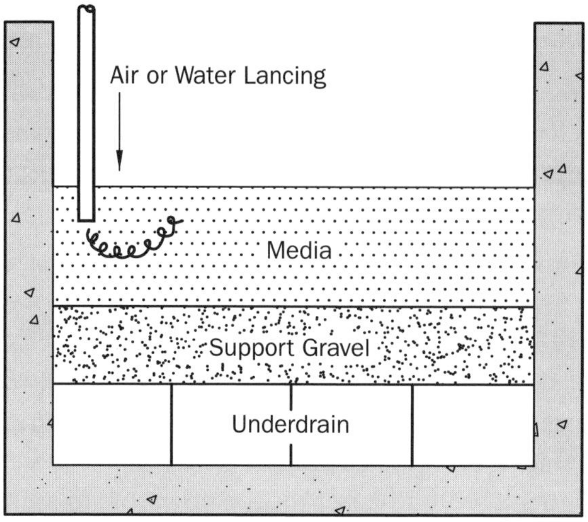

NOTE: Lanching the media is not recommended if support gravel is used.

FIGURE 5-45. Lancing media

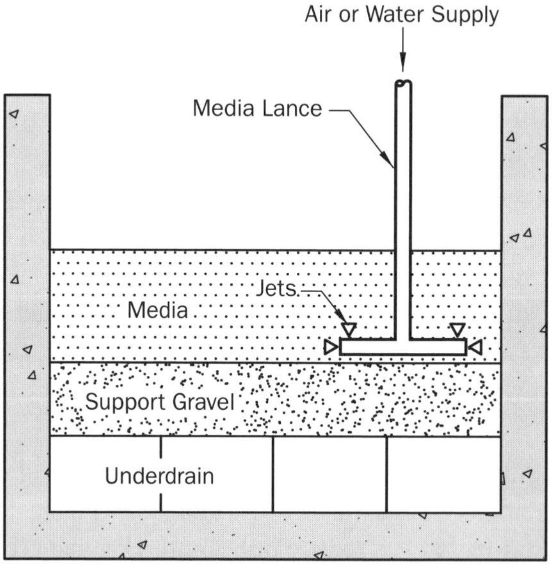

FIGURE 5-46. Recommended lance configuration when support gravel is used

expansion measurement or core sampling. The procedure outlined previously is intended to be a guide to assist in the resolution of whatever problems may be identified. To assist the investigators, diagrams for various types of filters are included in Figures 5-47, 5-48, and 5-49. In addition, a troubleshooting guide is included in Figure 5-50.

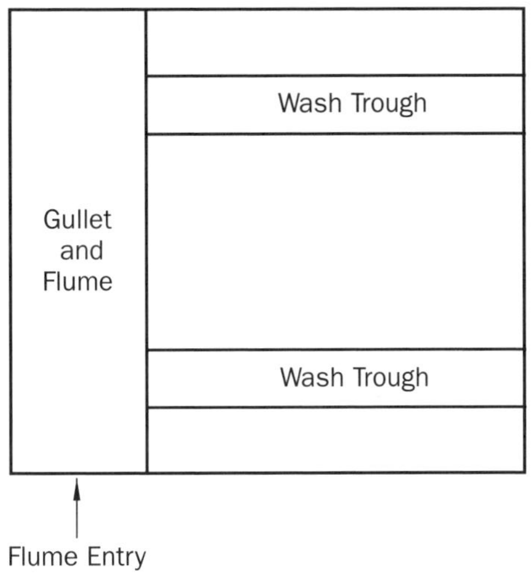

FIGURE 5-47. Side flume with end entry (plan view)

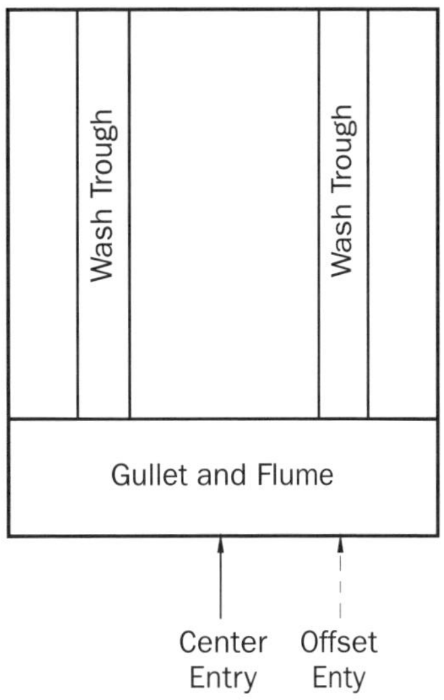

FIGURE 5-48. End flume (plan view)

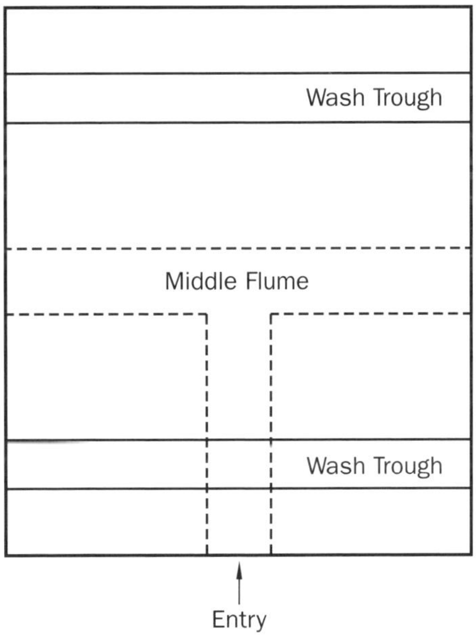

FIGURE 5-49. Middle flume (plan view)

There are a number of issues, such as uncontrolled air and hydraulic shock, that can disrupt support gravel and ruin a media bed in one backwash cycle. Therefore, it is of paramount importance that problems be resolved before the filter media is replaced in an existing filter. In a new installation, the controls should be finalized by operating an empty filter. When the controls are accepted, the media may be installed.

Problem	Probable Cause	Proposed Solution
Excessive intermixing	• High anthracite specific gravity	• Rebed Filter
	• High anthracite UC	• Rebed Filter
	• High anthracite effective size	• Rebed Filter
	• Low sand effective size	• Rebed Filter
	• Low backwash rates	• Increase backwash rate/frequency
Short filter runs (head loss)	• Fines on the surface	• Skim media
	• Did not skim each layer during installation	• Rebed filter
	• Pretreatment problems	• Improve efficiency
	• Polymer feed high	
	• Clarifier inefficiency	
	• Low driving head (refer to Chapter 2)	• Improve if possible
	• Flow rate too high	• Reduce flow rate or add more filters if possible
	• Excessive solids on the surface after backwash	• Increase length of air scour (if used)
		• Top off media to proper level for surface wash (if used)
		• Increase backwash rate or frequency if possible
		• Drain filter and hose down the surface and then backwash again
Short filter runs (turbidity)	• Pretreatment problems	• Improve efficiency
	• Weak floc	
	• Upset gravel/media	• Replace media
	• Fouled media	• Replace media
	• Flow rate too high	• Improve pretreatment or reduce flow rate
High media loss (some media loss is normal)	• High backwash rate	• Optimize backwash
	• Low media freeboard	• Raise troughs if possible
	• Improper backwash/ air scour sequence	• Optimize backwash and air scour
	• Upset gravel	• Replace media
	• High anthracite friability	• Replace media
	• Entrained air	• Shorten filter runs
	• Boiling/uncontrolled air	• Optimize backwash

(Figure continued next page)

FIGURE 5-50. Filter media troubleshooting guide

Problem	Probable Cause	Proposed Solution
Fouled/upset media (craters/depressions, cracks, mounding, rat holes, sloped surface, etc.)	• Accumulated debris on top of gravel	• Resolve problem and rebed filters
	• Low backwash rates	• Resolve problem and rebed filters
	• Support gravel problems (refer to Chapter 4)	• Resolve problem and rebed filters
	• High acid solubility (uncommon)	• Resolve problem and rebed filters
	• Media/support gravel incompatibility	• Resolve problem and rebed filters
	• Installation problems	• Resolve problem and rebed filters
	• Boiling/uncontrolled air	• Resolve problem and rebed filters
	• Underdrain problems (refer to Chapter 6)	• Resolve problem and rebed filters
	• Plenum/flume hydraulic problem (refer to Chapter 3)	• Resolve problem if possible and rebed filters
	• Pretreatment problems	• Resolve problem and rebed filters
	• High polymer feed (uncommon)	• Resolve problem and rebed filters

FIGURE 5-50. Filter media troubleshooting guide (continued)

CHAPTER 6

Underdrain

Introduction

The purpose of an underdrain is to collect filtered effluent that has passed down through the granular media and to evenly distribute the incoming backwash water used to clean the media. Although the purpose of the system is filtration, the reverse flow of the backwash cycle used to clean the media usually governs the design of underdrain systems because of the higher flow rates. During filtration, the head loss of the media and accumulated solids is much greater than that of the underdrain. As a result, the condition of the media controls the flow distribution during filtration, while the underdrain normally controls the backwash flow distribution. The exception would be deep-bed monomedia, utilizing a nonfluidizing backwash. Those types of filters typically use low backwash rates (6 to 12 gpm/ft^2). At rates that low, the head loss through the underdrain may be significantly less than that of the media.

If portions of the media become dirty or fouled in any filter, the water being filtered could be routed through cleaner areas of the filter, which can result in higher local velocities and less efficient solids removal. Therefore, it is of paramount importance to keep the filter media clean, and designing the underdrain and backwash system to accomplish that goal becomes a critical factor.

Air Scour

There are two basic methods in common use for cleaning a filter: hydraulic backwash with surface wash and hydraulic wash with air scour.

As its name implies, surface wash only agitates the media surface and not all of that. On the other hand, air scour agitates the entire bed of media. In addition, it is the author's opinion that air scour combined with hydraulic wash is the most effective method of cleaning a filter. All of the various underdrain types can be provided with air scour either internally, laid on top of the underdrain, or "over the top" of the support gravel (if used).

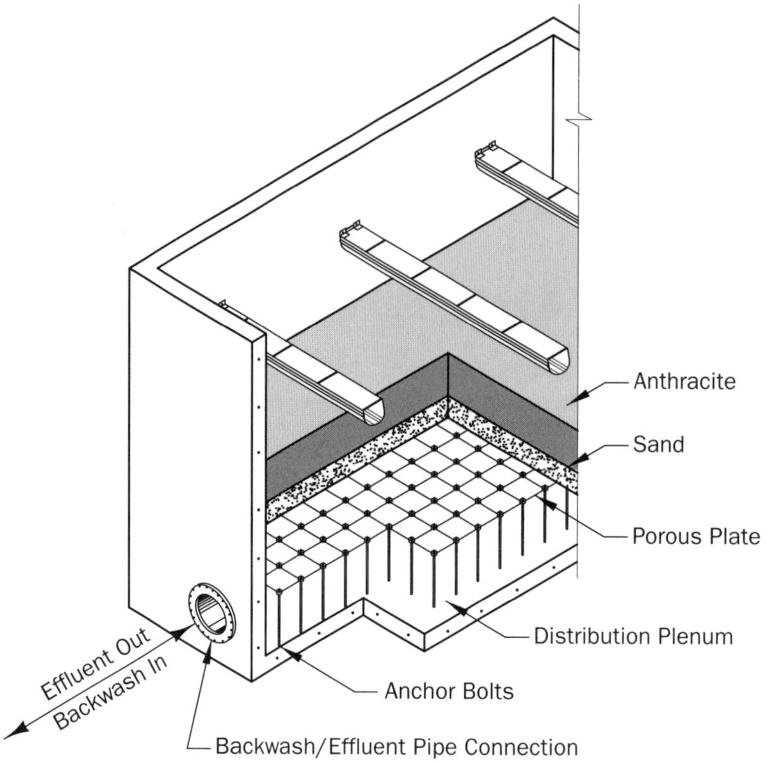

FIGURE 6-1. Typical porous plate filter arrangement

Underdrain Types

There are a number of underdrain types that are commercially available, several of which are illustrated in Figures 6-1 through 6-5. They generally fall into two categories: lateral types or false deck types. The lateral types include the traditional pipe lateral, screen-wrapped laterals, slotted laterals, wedge wire screen, folded plate, and the dual parallel lateral underdrain. Of these lateral types, several can be provided with or without a direct media retention system designed to eliminate support gravel. The false deck underdrain types include a variety of porous plate types as well as nozzle systems, Wheeler pyramids, and others.

A brief discussion is provided on problems that can occur with various underdrain types. However, the emphasis of this chapter will be on the dual parallel lateral underdrain. It is one of the more popular underdrains and is preferred by the author. However, as with any piece of equipment, it is one component of an integrated system in which all of the parts must work together properly for the system to perform as intended. The design requirements for using this type of underdrain, along with problems that may occur and ways to recognize them, will be discussed.

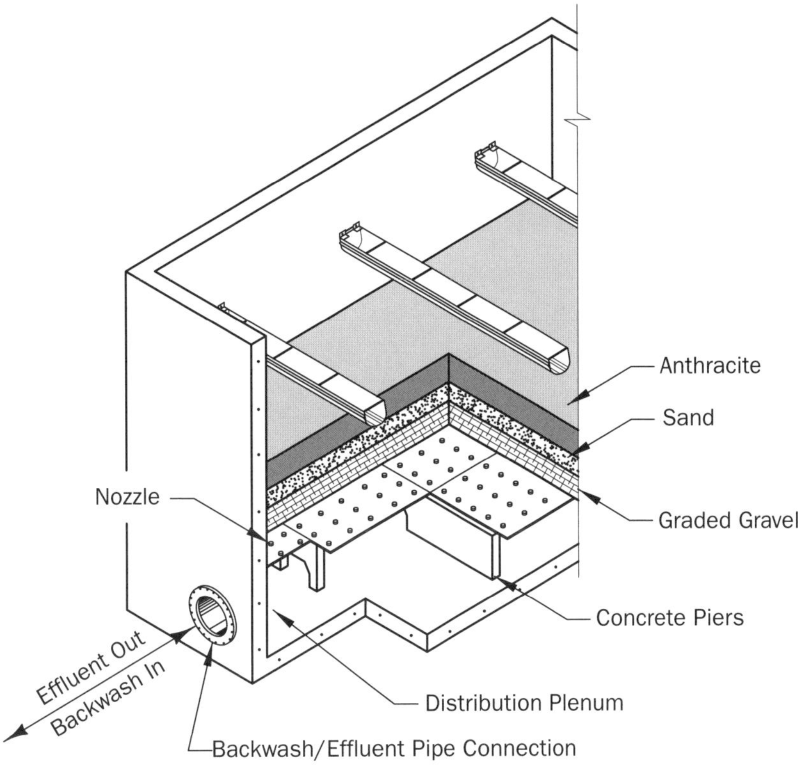

FIGURE 6-2. Typical false deck nozzle filter arrangement

Dual Parallel Lateral Underdrain

The dual parallel lateral underdrain consists of a series of individual *blocks* that are joined together to form a continuous conduit or *lateral*. The individual blocks are constructed of injection-molded high density polyethylene (HDPE). They are between 3 ft and 4 ft long, depending on the manufacturer, and have a bell-and-spigot joint with one or more O-rings for the connection between blocks. The laterals are constructed the length of the filter and are placed side by side to cover the entire filter floor, as shown in Figure 6-5.

Installation

The laterals are set in a bed of grout about 1 in. deep, which is also used to level the filter floor. Grout is also installed in a space left between the rows of laterals. When completed, each lateral occupies a space 12 in. to 18 in. wide, including grout. The difference in width varies with the manufacturer. In addition, the manufacturers have their own individual installation instructions.

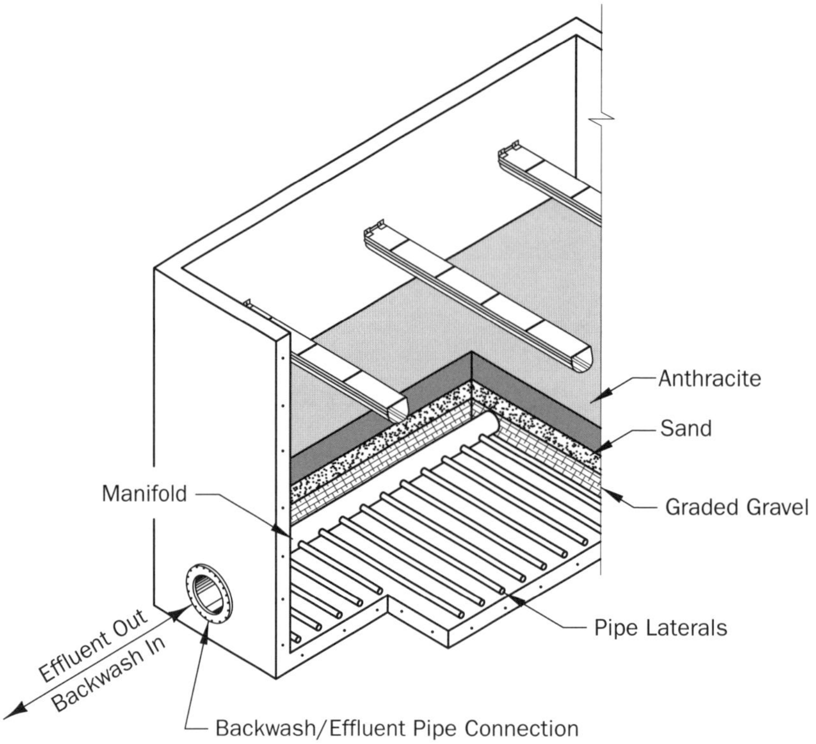

FIGURE 6-3. Typical pipe lateral filter arrangement

Flume

In filter mode, the underdrains collect treated water and route it to a flume located either below the floor or on one side of the filter. For flumes below the floor, cutouts in the bottom of the lateral blocks, over the flume, allow communication from one block to another. When using a side flume, a penetration through the wall and into the block is provided. The arrangement shown in Figure 6-5 depicts a flume underneath the front end of the filter. A pipe from the flume directs the treated water out of the filter and into the clearwell or other treated water storage.

Operation

Two partitions separate the dual parallel lateral block into three compartments: a center primary chamber and a separate secondary chamber on either side. During filtration, the treated water comes down through the secondary orifices on the top deck of the underdrain and into the secondary chamber, down through the water distribution orifice in the bottom to the primary chamber, along the underdrain, and down into the flume.

During backwash, the flow is reversed. The water comes up through the cutout in the bottom of the block from the flume to the primary chamber along the underdrain,

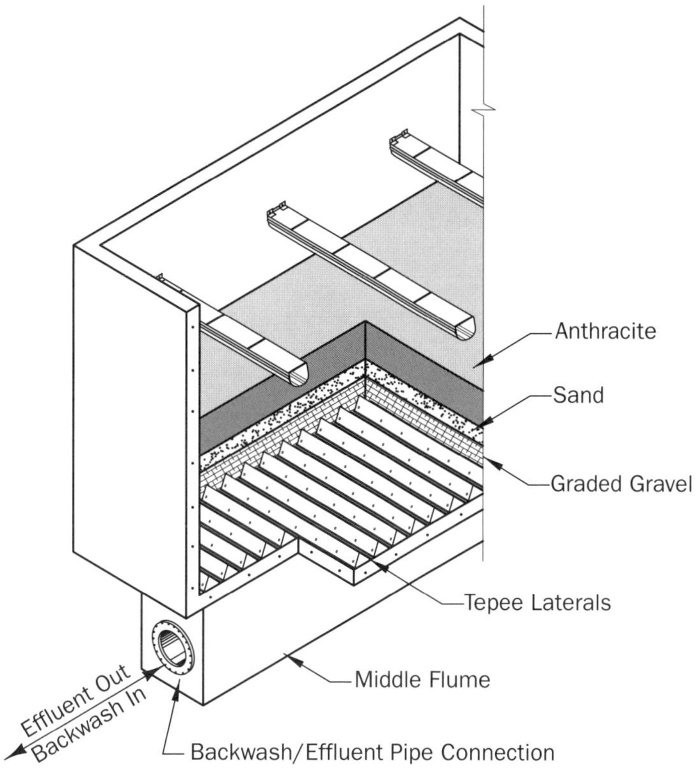

FIGURE 6-4. Typical folded plate lateral filter arrangement

out through the lower water distribution orifices, up through the secondary chamber, and out the top deck orifices.

Air Scour

During air scour, the air comes up through the J-pipes and into the upper triangular space of the primary chamber in the block over the flume in each lateral. An air pocket is formed at that location, as shown in Figure 6-6, and the air is metered into the secondary chamber on each side through a set of air orifices. Another air pocket is formed at the top of the secondary chamber, and the air goes out through the top deck orifices. The internal distribution of air provides for a simple, easy-to-maintain system, with nothing inside the filter box that requires maintenance.

The issues that affect the design of the underdrain system are the backwash cycle and whether or not air scour is used. The remainder of this chapter will discuss problems that can occur with various underdrain systems and design concepts that can be used to eliminate or mitigate those problems.

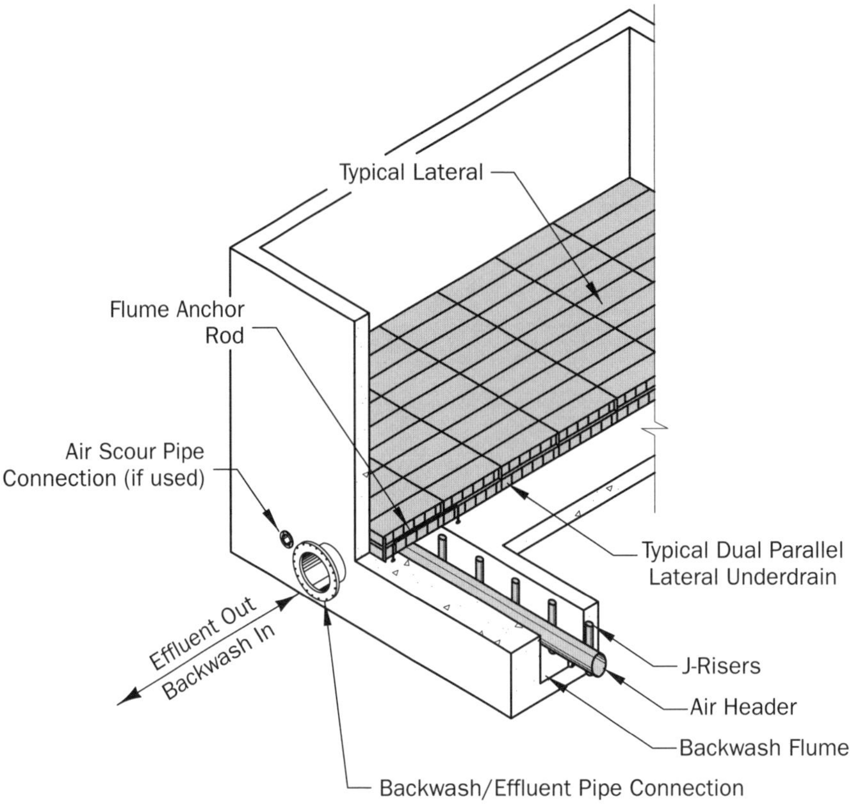

FIGURE 6-5. Typical dual parallel lateral filter arrangement

J-Pipe Design for Below-Grade Flumes

Air System Description

An air scour system is an integral part of the backwash process, which is often used to clean granular media filters. A typical air scour system utilizing the dual parallel lateral underdrain consists of at least one blower with inlet and discharge silencers. In addition, there is normally a pressure relief valve on the discharge piping from the blower. When using positive displacement blowers, there is also a common vent valve that regulates the amount of air delivered to the filters, as shown in Figure 6-7. A system of piping is included to distribute air to each of the filters, and a local automatic air isolation valve is provided for each filter.

Inside each filter with a below-grade flume, there is an air manifold with separate J-pipes that distribute air to each of the laterals (Figures 6-5 and 6-8). Once the air is discharged into the triangular center section of the underdrain, it is channeled from one end of the lateral to the other. The air is metered into the secondary chamber of the underdrain lateral through a series of orifices at different levels, up through the top deck orifices, and into the support gravel or media (Figure 6-6).

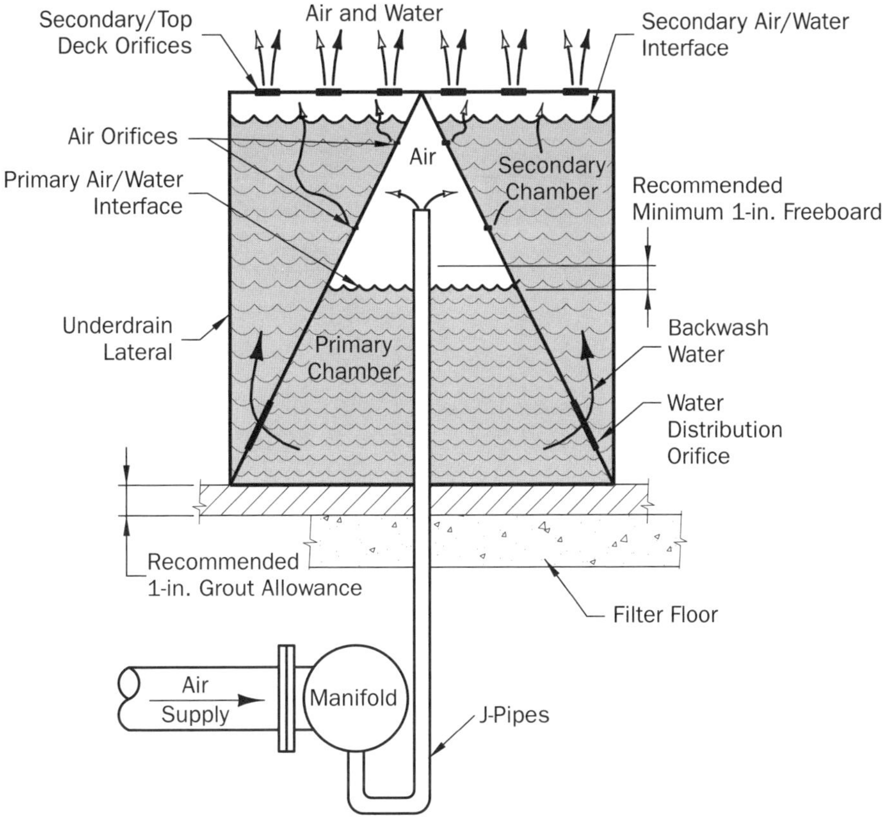

FIGURE 6-6. Dual parallel lateral underdrain

The J-pipe configuration is intended to provide a means for water to be driven out of the air manifold. Piping coming off the manifold vertically, as shown in Figure 6-9, is not adequate for that purpose. Water cannot be efficiently driven directly up out of the manifold. The residual water can move along the manifold in a wave action and cause air distribution problems. With the J-pipe configuration, air is driven down and out of the manifold into the individual J-pipes and up through them into the underdrain (Figure 6-10).

Stages of Air Scour

There are three separate and distinct stages of the air scour operation during backwash. Each stage has its own unique design considerations. First, the water must be purged from the air piping at a low rate. Water in the piping is shown in Figure 6-8. Second, the J-pipe arrangement must deliver air uniformly to the various rows of underdrain during the air scour cycle. Third, a low flow rate of backwash water is added to the air scour to assist in removing solids from the bed. During this concurrent cycle, there must be adequate pressure to maintain the flow of air as the water level rises to its high level shut-off. Each of these cycles and their effect on system design and operation will be discussed in detail in the following sections.

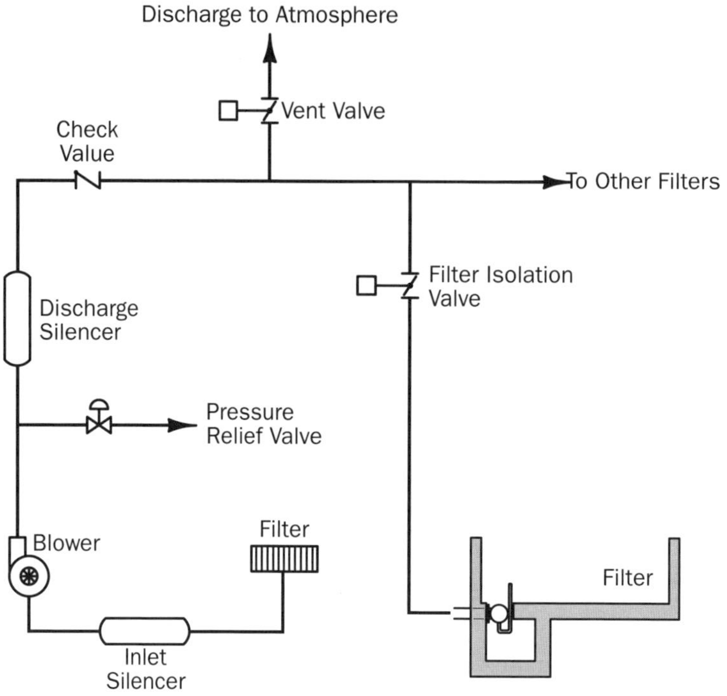

FIGURE 6-7. Air scour system components for a positive displacement blower

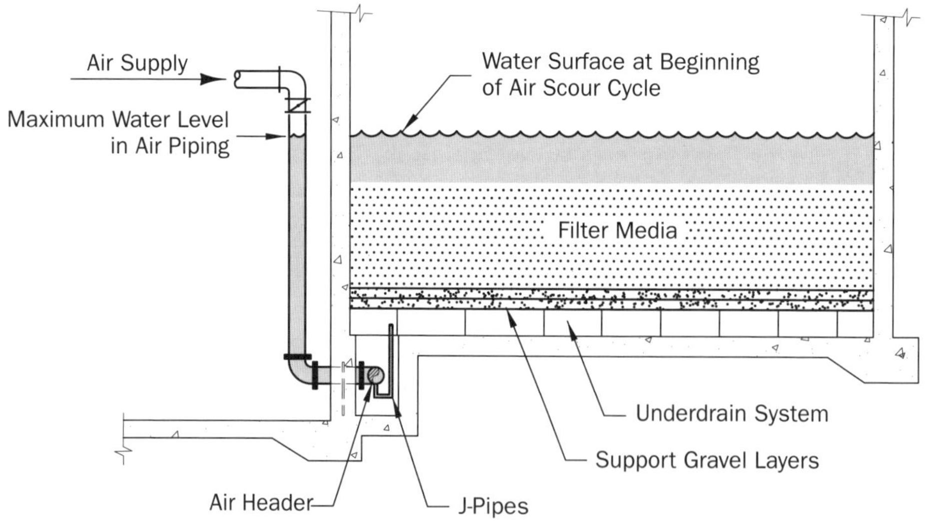

FIGURE 6-8. Typical air piping

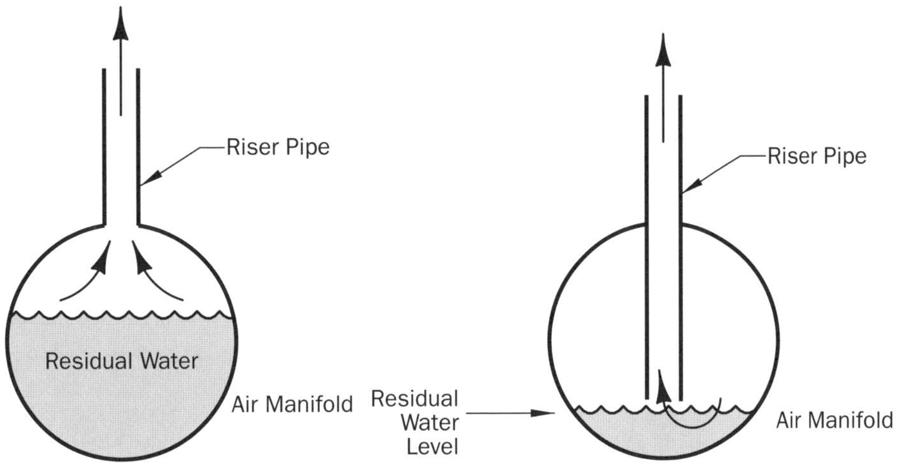

FIGURE 6-9. Vertical riser configurations

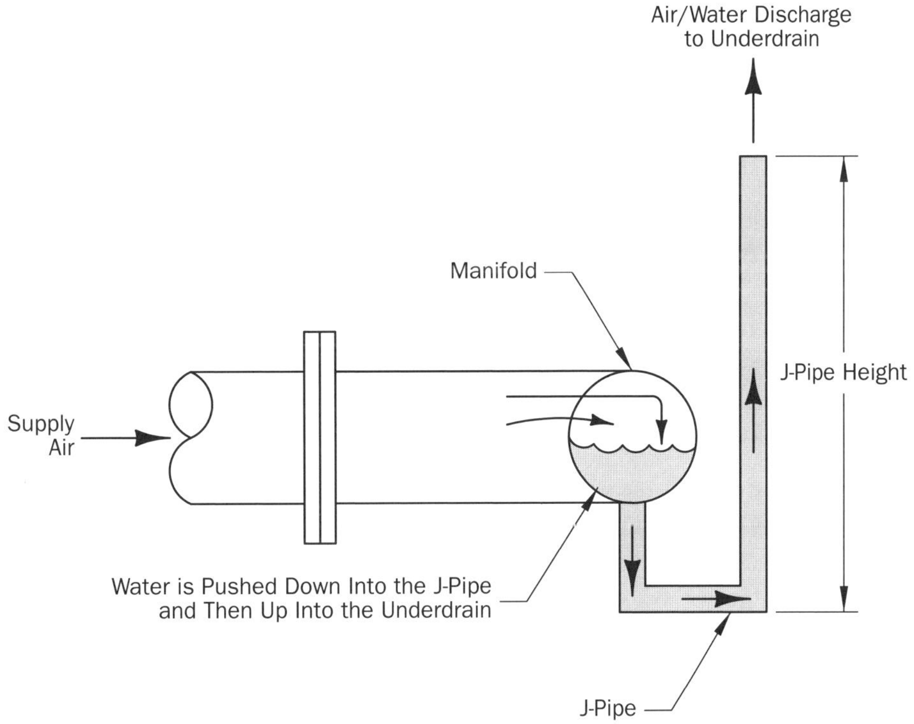

FIGURE 6-10. Air scour system operation (dual parallel lateral underdrain)

Purging the Air Piping. Because of the design of the air system and the way the controls and valving normally work, it is common for water to displace the air and percolate into the air system piping after backwashing is complete. It is not unusual for the air piping to be full of water to the water level of the filter when the air scour begins (Figure 6-8). As a result, the first step in the air system operation is to purge the water out of the air piping system and J-pipes at a low rate. A typical design air rate of 4 standard cubic feet per minute (scfm)/ft^2 is equivalent to a backwash rate of approximately 30 gpm/ft^2, which is very high. Attempting to purge water at that high of a rate may result in overpressuring the air system because of excessive head loss and could result in surging, which could in turn disrupt the support gravel (if used). To minimize the effects of this stage, the air piping should be designed to the extent possible to be above water level in the filter.

Vent valves are typically provided on air scour systems with positive displacement blowers to control the amount of air delivered to the filters (Figure 6-7). Electric valves are recommended for this purpose because they operate smoothly and are easily adaptable to accurate computer control.

The main air piping is normally sized large to reduce friction caused by the high velocity air. However, the vent valve needs to be sized for modulation purposes so that it operates in a range of 30° to 60° open during the purge cycle. Sizing the valve in this manner will usually result in it being one or two pipe diameters smaller than the main air piping. A minimum valve speed of 60 sec to travel 90° is also recommended. If the vent valve is sized properly, it can more easily be modulated to an intermediate position to deliver a small amount of air to the system to purge the water at the beginning of the air scour cycle. Calculations should be made to verify the correct sizing. The actual valve position for purging is usually determined in the field.

Depending on the plant design and the location of the blower, a considerable amount of air piping could be full of water at the beginning of the backwash cycle. The more piping to be purged, the longer it will take. The goal is for a small amount of air to appear in the filter once the water has been purged. Purging the air usually occurs at a low water level, close to the media. While purging, the static water pressure greatly exceeds the friction loss of the piping and J-pipes. The low volume of air used should be adequate to drive the water out of the main air piping but may not drive all the water out of the header in the filter.

Once the air rate and vent valve position have been determined to purge the water, the elapsed time can be set in the controls by observation. Please note that this procedure should be implemented after the underdrain is installed, but before support gravel or media are installed. The intent is for the controls and backwashing procedures to be functioning perfectly to prevent damage to either the gravel or media.

Air Scour Cycle. The second stage of the cycle is air scour. Once the water is purged out of the main air piping, the vent valve is modulated to the closed position (or the position required for the design air rate). The residual water in the air header will be driven out through the J-pipes into the filter and the air will be delivered to the underdrain to begin the air scour. The air scour cycle will continue for a set period of time. During this time, it is important that the incoming air is split equally to the various underdrain laterals through the individual J-pipes. In addition, it is important that all of the water is driven out of the air manifold in the filter. In order for these events to occur, there must be adequate head loss in the J-pipes. The head loss must be

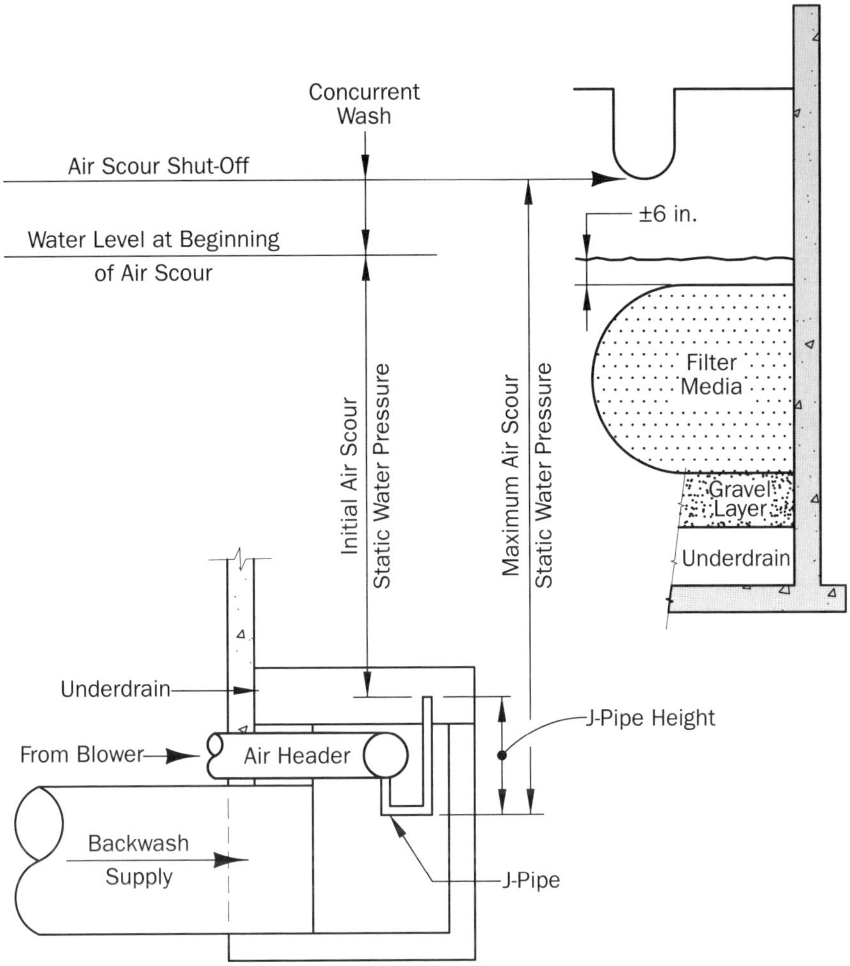

FIGURE 6-11. Air scour system operation (dual parallel lateral underdrain)

equal to or greater than the height of the J-pipe, as shown in Figure 6-11. At the moment the water is driven up through the J-pipe and out into the filter, the column of water above is balanced by the air pressure below. The air pressure must be equal to the static height of water above the J-pipes plus the head loss in the main air piping system, the head loss in the manifold, the height of the J-pipes, and all entrance and exit losses. The design pressure for the blower is calculated by adding all of these factors together and including a contingency.

Concurrent Wash. The third stage of the air scour cycle is concurrent wash. The air scour cycle normally lasts 1 to 2 min. After that time, a low rate of water backwash is added to the air scour. For fluidizing media, the low rate concurrent wash is normally on the order of 5 gpm/ft^2. For coarse monomedia, the wash rate may be 6 to 12 gpm/ft^2, with no follow-on higher rate.

The highest level of agitation for cleaning the media occurs during the concurrent wash step, in the author's opinion. During this stage, the water level will be gradually rising. The effect is to increase the backpressure on the air system and blower. In addition, the increased pressure results in a slightly reduced flow rate because of air compression.

The air is normally turned off when the water level reaches the bottom of the wash trough. Residual air can be purged out of the underdrain by the low-rate wash before the water begins weiring over the trough. Maintaining air scour for a longer period of time can result in the loss of fluidized media when the water begins to weir over the trough. Some coarse monomedia will not fluidize. In that case, there would be no loss. With media that fluidizes, a winged trough could be used to prevent loss if a long concurrent cycle is used.

The blower should maintain an ordinarily constant flow rate during this period of rising water level. The effect on the J-pipes is to slightly reduce the volume and velocity of the air caused by increased pressure, with a corresponding reduction in head loss. As a result, when sizing J-pipes, it is important to consider both the starting and ending water level to ensure that adequate head loss is maintained in the J-pipes at all times. Satisfying this condition may require that the J-pipes generate more head loss than their height in order to ensure proper operation. If the J-pipes do not generate the required head loss, then it may result in fluctuating or uneven air distribution in the filter.

It is important to understand that simply crimping the ends of the J-pipe or installing a discharge orifice is not adequate to drive all of the water out of the system. Crimping or welding on an orifice may generate the necessary head loss at the end of the J-pipe but will not create adequate velocity within the J-pipe to lift and purge the water from the piping.

Operating Rates and Pressures. The choice of the type of blower used for air scour also affects the operation of the system. It is important that the blower maintain a constant air flow rate through the three stages of operation to ensure that the system will operate as designed.

There is a different operating rate for each of the three stages of air scour backwash. A relatively low air rate is used during purging that must overcome the static water pressure in the filter. During the full air scour cycle, the full air rate is delivered against the head loss of the air piping system and J-pipes. The static pressure is highest at the end of the concurrent stage because of the increasing water level. At the shutoff point, the full air rate must be delivered against the highest pressure. The blower must be capable of operating under all of these conditions.

J-Pipe Design

Air Rate. The J-pipes are sized for the nominal rate for air scour. Typical design rates are 3 to 5 scfm/ft^2, with 4 scfm/ft^2 being a typical value, as mentioned previously. Using the length of the lateral, the total air flow for each lateral, and therefore each J-pipe, is easily calculated. However, the actual air rate can be significantly different depending on atmospheric temperature and pressure. The worst-case temperature during the summer is usually used for calculation purposes.

Design for Uniform Distribution. The critical factor in sizing J-pipes is that each one must develop adequate velocity and head loss at the design flow rate to adequately purge the water out of the pipe. In order to accomplish that, the head loss in the J-pipe must be equivalent to, or greater than, its total height, as shown in Figures 6-10 and 6-11 and as previously discussed.

Typically at the beginning of each air scour cycle, the water level is drawn down to within 6 in. of the top of the media before the air scour is started. That is the pressure at which the J-pipe system will operate when purging the water. Because of the compressibility of air and the temperature gain of the blower, calculations should be made at each point within the air system to determine the velocity and the associated head loss. If the J-pipes do not generate adequate head loss, all of the water may not be driven out of the header. When all the water has been driven out of the J-pipe, the volume and velocity of the air must be adequate to balance the column of water it has displaced. That condition occurs when the head loss in the J-pipe exceeds its height.

If all the water is not driven out, water remaining in the air header may move back and forth with a wave motion. The movement of water could result in the laterals alternately being active and going dead as the wave passes by them.

Pipe Sizes and Type. Calculations for sizing J-pipes will rarely come out as an exact pipe size. The results may not even come close to standard pipe sizes. Another important consideration is that engineering calculations are inherently conservative. In this case, however, it is very important that the minimum head loss be achieved. Therefore, the recommended procedure is to use at least the next size *smaller* pipe than the calculated pipe size. If the calculations are close to the next smaller size, a size smaller than that may be desirable. Once a pipe size has been selected, the actual head loss can be recalculated. The new value can be used to calculate the blower system pressure.

Normal pipe sizes and weight schedules provide only limited choices for obtaining the desired inside diameter. However, tubing sizes should also be considered that give the designer additional flexibility in selecting the proper inside diameter.

It should be noted that whether pipe or tubing is used, 304 stainless steel is recommended as the minimum material. Higher quality materials may be required for more corrosive applications.

Maximum Air System Pressure

Once the air scour system is in operation for a set amount of time, a low rate of concurrent backwash water is introduced into the filter in conjunction with the air scour. As a result, the level of water slowly rises to approximately the bottom of the wash trough where the air is typically turned off (Figure 6-11). Using coarse monomedia, the concurrent wash is sometimes extended to the point where water is overflowing into the trough. The point where the air is actually turned off represents the maximum water level during air scour. As mentioned previously, the blower needs to be designed to produce the design rate of air at that maximum pressure. In addition, the increased pressure from the higher water level has the effect of compressing the air slightly, reducing the volume and therefore the head loss in the J-pipes. All these factors should be taken into account in the design.

Ramping the Vent Valve to the Design Rate

When the purging is complete, the vent valve will close to deliver the design air rate to the filter. However, because of the remaining water in the manifold, it is highly desirable that the vent valve move slowly. Otherwise, a hydraulic surge could occur from trying to push the remaining water out too quickly. A minimum vent valve speed of 60 sec to travel 90° is recommended, while some plants use a speed of 120 sec, which is better.

Over-the-Top Air

Over-the-top air can be used with any underdrain type, as shown in Figures 6-12 through 6-14. It can also be used with either support gravel or a direct media retention underdrain.

Support Gravel

When support gravel is used with over-the-top air, the air distributor pipe is usually installed above the support gravel, as shown in Figure 6-12. In that case, the important design criteria is to support the air distribution piping properly to prevent disruption of the support gravel and limit vibration of the air distribution piping. Air

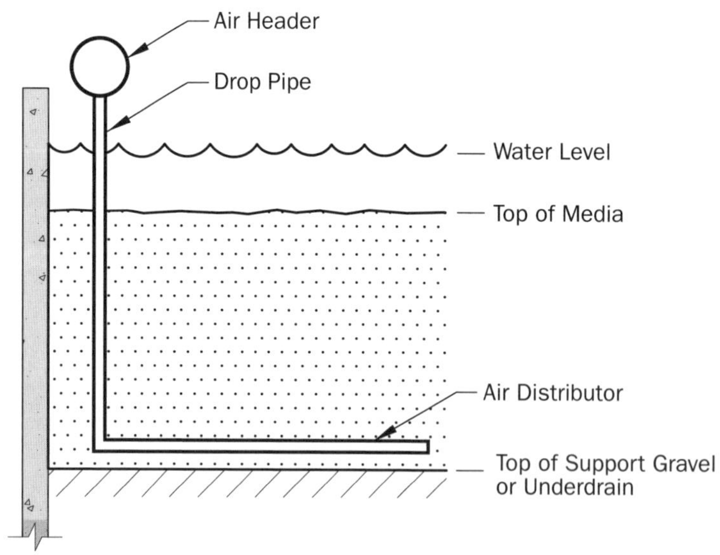

FIGURE 6-12. Over-the-top air

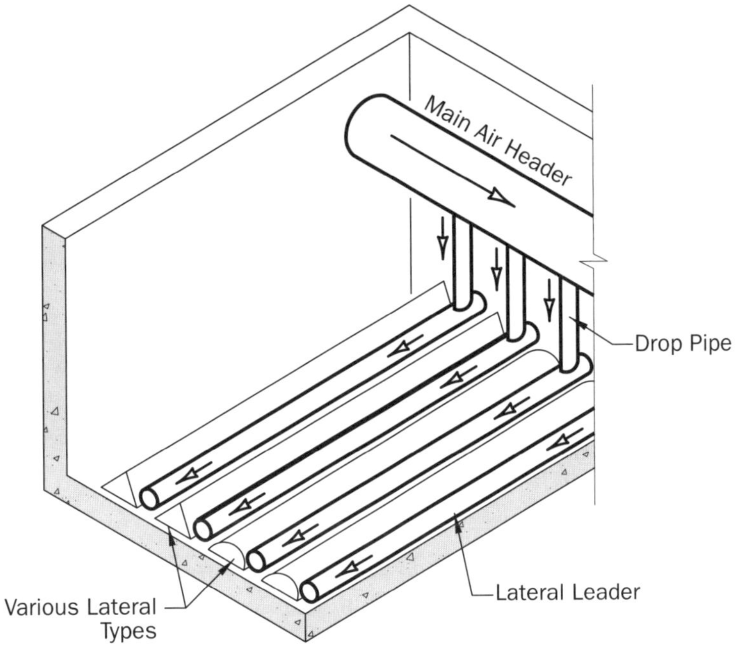

FIGURE 6-13. Air distributor between various lateral types

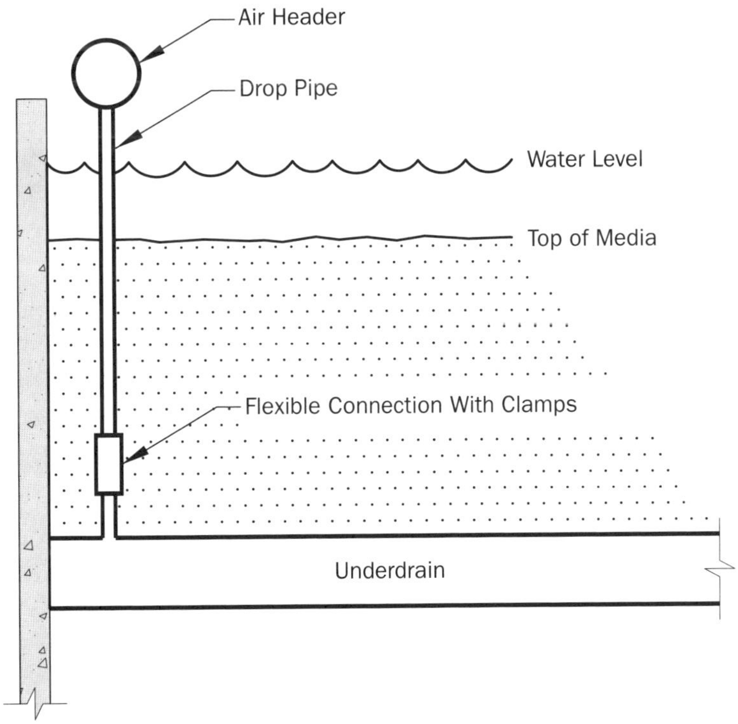

FIGURE 6-14. Air drop pipe connecting directly to the underdrain

distribution piping can also be installed below the support gravel between the laterals, as shown in Figure 6-13. The support gravel (if used) will have to be of the hourglass design in that case.

Direct Media Retention Underdrain

When direct media retention underdrains are used, the drop pipe may connect directly to the underdrain, as shown in Figure 6-14. Connection is typically made to the underdrain by using a flexible sleeve with two hose clamps.

Design Considerations

- Evacuating water from the air header is usually not an issue with this design, because the header is generally above the filter and therefore does not accumulate water. The volume of water in the drop pipes is small and is not normally a problem.
- Stainless-steel drop pipes and air distribution pipes are recommended.
- Over-the-top air is an acceptable alternative for wastewater plants and rehabilitation of existing filters. It is not recommended for new potable water treatment plants. The sidewalls are the weakest point of any filter in good condition. A series of drop pipes would provide another pathway for short-circuiting because of the slick surface of the piping.
- Another important consideration is the maintenance aspects related to the use of over-the-top air scour. In order to be effective, the air distribution pipes should be close together (12 in. or less). Each air distribution pipe will also have a number of nozzles for introducing air into the filter. The total number of nozzles, distribution pipes, and flexible connectors with hose clamps (if used) represent exposed items in the filter that will require maintenance at some point. The preferred air scour system would be one in which there are no exposed components requiring maintenance.

Blower Types

The two types of blowers commonly used are positive displacement and centrifugal. The blower air flow rate is determined by multiplying the desired unit rate by the filter plan area. The unit rate is usually specific to the underdrain to be used. Typical rates are 3 to 5 scfm/ft^2, with 4 scfm/ft^2 being commonly used. It may be possible to use less than the given range, but higher rates are not recommended.

The type of blower used may vary depending on whether over-the-top air is used or a header with J-pipes is installed in the flume to send air up through the underdrain.

Over-the-Top Air

A positive displacement or a centrifugal type blower may be used.

J-Pipes in the Flume

Once an air rate is selected, the J-pipes are designed accordingly. The underdrain can handle a range in air flow, but the J-pipes may not. Completed filters should not

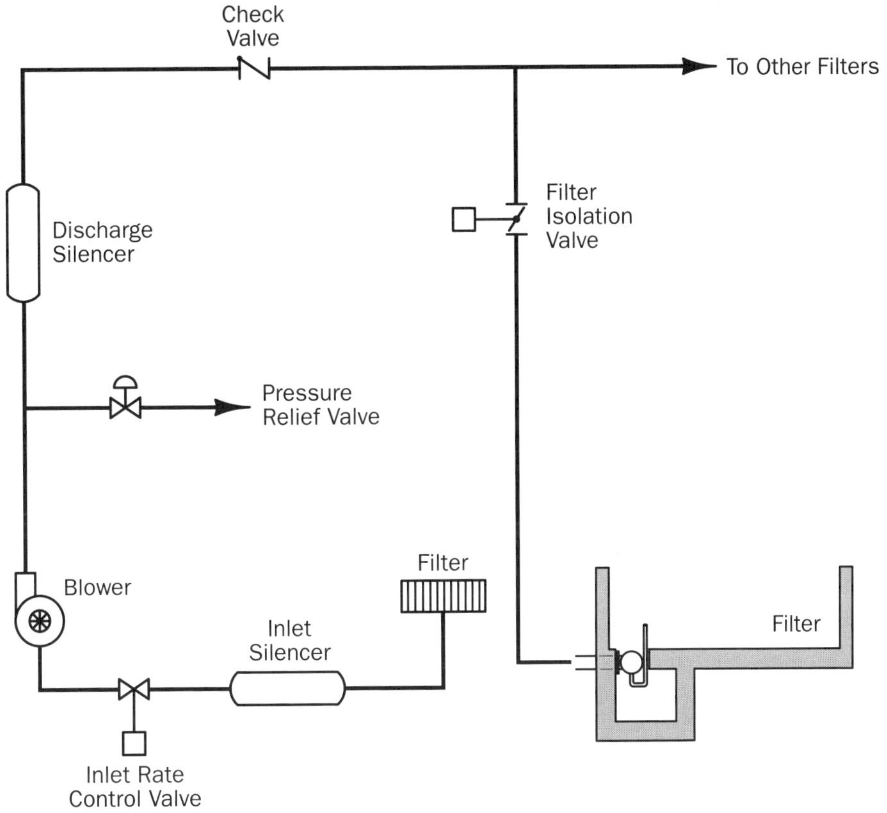

FIGURE 6-15. Air scour system components for a centrifugal blower

be run with less than the design air flow. With less than the design air rate, the velocity in the J-pipes may not be adequate to purge the water or prevent any air from reentering the piping. The maximum blower pressure is determined by calculating the friction loss in the piping, valves, and fittings, and the total height of water from the bottom of the J-pipe to the designed air shut-off point.

Centrifugal blowers typically operate on a curve similar to that for a pump. The higher the flow rate, the lower the pressure and vice versa. The components of an air scour system using a centrifugal blower are shown in Figure 6-15. The principal difference from a positive displacement blower is that the flow through a centrifugal blower is modulated by an inlet flow control valve instead of a vent valve. However, an inlet flow control valve must be sized similarly to a vent valve. The air purge rate will depend on the size of the distribution piping and to some extent, the amount of piping expected to have water. In either case, the purge rate will be relatively low. If the designer intends to use a centrifugal blower, care must be taken to ensure that it will operate successfully in each of the three stages of air scour.

Because of the wide range of requirements of air scour, a positive displacement blower is preferred for air scour backwash systems because the air rate is more consistent when the backpressure changes. When combined with a vent valve, positive displacement blowers can easily meet the requirements of the three stages of air scour. Positive displacement blowers do have some slippage, so there is some minor difference in flow rate as the pressure changes. However, the difference is less noticeable than with the centrifugal type.

Miscellaneous Design Details

Air Piping Height

The air piping is normally routed through the pipe gallery to the blower(s). The recommended procedure is for most of the piping to be located above the backwash water level in the filter, whenever possible. As with many other issues, it is not always possible to locate the air piping properly, especially in plant rehabilitations. As noted previously, water will commonly enter the air piping to the backwash water level. The proper controls sequence can be used to purge the water at the beginning of the air scour cycle. More submerged piping will result in a longer purge cycle. However, the important concern is to prevent water from damaging the blower or its silencers.

Check Valves

A check valve is recommended to protect the blower and silencers from water flowing backward from the filter. It is recommended that the check valve for that purpose be installed on horizontal piping downstream of the silencers.

Pressure Relief Valve

A relief valve is also recommended to protect the blower from operating against a closed valve. The blowers normally operate at low pressure (6–10 psi). The relief valve should be sized for the pressure and air flow of the blower and will normally be several sizes smaller than the air piping. For example, a relief valve on a 6-in. pipe might only be a 3-in. unit.

It is important that the pressure relief valve *not* be used to control the pressure. It does not result in a steady-state operation. When a pressure relief valve is activated, it should be designed to vent enough air to reduce the pressure to an acceptable level. However, at some reduced pressure level, the relief valve will reset and will actuate again when the high set point is reached. As a result, it is common for a relief valve to activate and reset numerous times when the pressure is high. Those surges can be damaging to support gravel (if used) if the maximum set-point pressure is reached.

Flowmeter

Although not absolutely necessary, a flowmeter is recommended to verify the blower operation and assist in setting the air vent valve position. Flowmeters can be as simple as a pitot-tube type or a more complex mass density meter. If the simpler meter is used, a thermometer is also required to enable actual or inlet conditions to be converted to standard cubic feet per minute.

Air Filters

An inlet filter is required on the air supply to protect the blower and the flowmeter and to protect the treated water in the underdrain from contamination.

Air Piping

Rust from steel pipe, strands of fiberglass, or decomposition of other pipe material types could potentially result in fouling of the underdrain, support gravel, or direct media retention system (if used). As a result, the use of thin-walled Schedule 5 stainless steel is recommended for all air piping because of its ability to handle high temperatures and resist corrosion. The higher cost of material is offset, at least in part, by the lower weight of Schedule 5 material and ease of handling.

Sequence of Operation

The normal sequence of operation for a dual parallel lateral underdrain using an air scour system is as follows:

1. Open the vent valve when the blower is off and the air scour system is not in use.
2. Start the blower with the vent valve open.
3. Open the filter air isolation valve. At this point additional water may enter the air piping system to the water surface elevation.
4. Close the vent valve to the desired purge position.
5. Maintain the vent valve in that position until a small amount of air appears in the filter.
6. Close the vent valve to the desired position to deliver the desired amount of air to the filter for the air scour cycle.
7. Open the backwash valve to a low-rate position for concurrent wash.

Air scour should be terminated as follows:

1. Open the vent valve.
2. Turn off the blower.
3. Close the local air isolation valve.

Chapter 7 has a more detailed discussion of backwash procedures.

Lateral Length

Long Laterals

When a front flume is used, the J-pipe from the air header enters the lateral near one end. All of the air for that lateral enters at this location and moves down the length of the lateral. As laterals become longer, a point is reached where the air rate becomes so high that the velocity of the air in the primary chamber could create high and low pressure areas in the underdrain, as shown in Figure 6-16. If this occurs, air distribution in those areas will not be as good as in the center of the filter. The author has no actual data to support this theory. However, filters with long laterals (over 40 ft from the flume) do exhibit these characteristics.

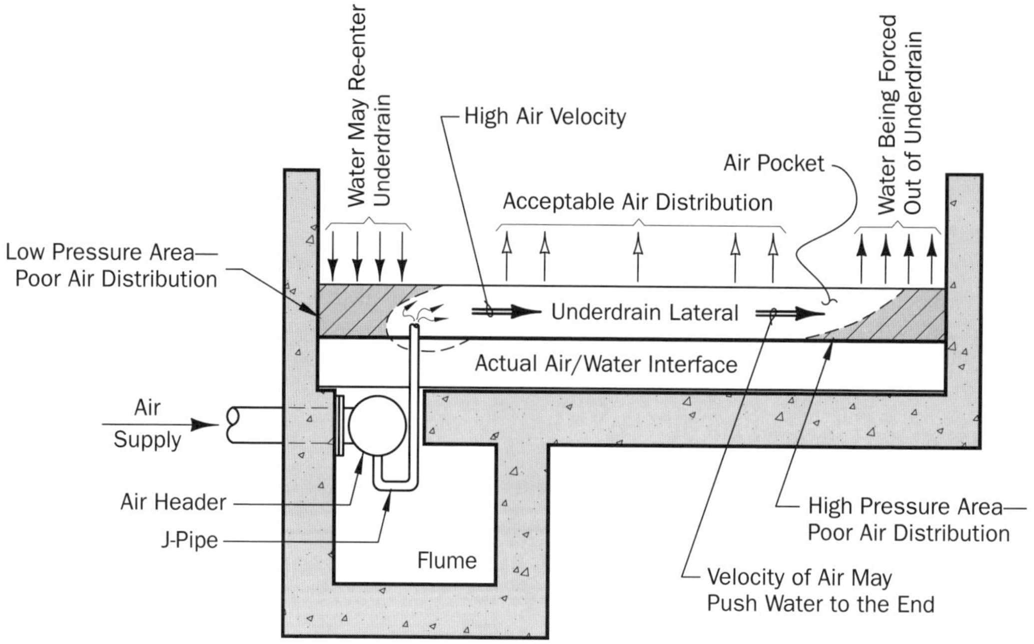

FIGURE 6-16. Long lateral with end flume

The author witnessed air distribution tests in a plant with a long lateral, as shown in Figure 6-17. When the air was initially turned on, the air distribution was reasonably good across the filter. However, within 30 to 45 sec, the air pattern demonstrated in Figures 6-16 and 6-17 was established. Once the concurrent wash began, the additional pressure from the incoming water was sufficient to force a more uniform air distribution. For a plant with long laterials, it is recommended that the air-only cycle be limited to approximately 45 sec before starting the concurrent wash. The goal of any wash cycle is to clean the media, and the highest level of agitation occurs during the concurrent wash cycle. Therefore, by following the recommended procedure, it should be possible to adequately clean the media.

Even though the filter with the long lateral previously described could work, it is best to keep the laterals shorter. It is recommended that laterals not exceed 36 ft if possible.

A number of factors influence the magnitude of possible problems with long end fed laterals including the manufacturers' orifice design, underdrain block geometry, the method used to introduce air into the underdrain laterals (over the top or with J-pipes), and the configuration of over the top air (if used). As a result, some underdrain manufacturers may have better success than others using long laterals. If long laterals are desired, full-scale laboratory testing should be conducted.

If it is desired to have a longer filter, a middle flume is recommended, as shown in Figure 6-18. With a middle flume, the air is fed in the middle of the filter and travels in

FIGURE 6-17. Uneven air distribution in a long lateral

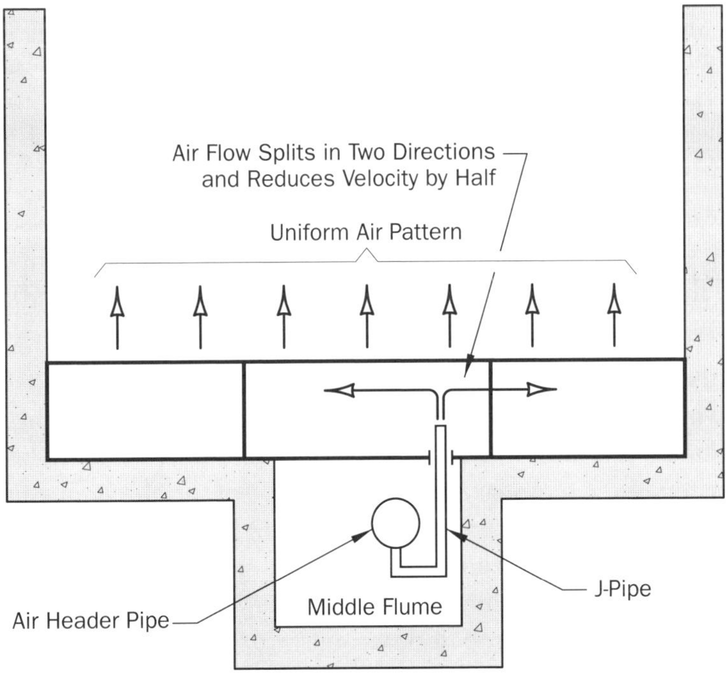

FIGURE 6-18. Long lateral with middle flume

both directions equally. As a result, the effective length of the lateral is reduced by half. It should be noted that reducing the effective length of the lateral also reduces the lateral velocity (V_L). As a result of the lower lateral velocity, the center flume design will become larger than a corresponding end flume. However, it should be an acceptable trade-off for having longer laterals. Refer to Chapter 3 for a more extensive discussion of plenum and flume design.

Short Laterals

When the head loss of the underdrain during backwash is expected to be low with respect to the media, short laterals are preferred. Distribution efficiency is dependent, in large part, on the condition of the media when the underdrain head loss is low.

Low head loss can be caused by the design of the underdrain or can be the result of low backwash flow. An example is the low backwash rates (typically 6 to 12 gpm/ft^2) used with nonfluidizing coarse monomedia, because underdrains are designed specifically for that application.

The flume can be used to control distribution to the various laterals when the underdrain head loss is low. In this case, large flumes are recommended with an orifice to create head loss at the entry to each lateral. As long as short laterals are used, the maldistribution will usually be acceptable. Distribution efficiency in the underdrain is less important when short laterals (less than 15 ft long) are used. Chapter 3 has a more extensive discussion of plenum and flume design.

Recirculation

Before a backwash cycle including air scour begins, the underdrain laterals are full of water. As the air begins to come into the lateral blocks, some of the water is displaced. An air space is created within the primary chamber, with another at the top of each secondary chamber, as shown in Figure 6-6. The air space at the top of the secondary chambers may be as much as 1½ in. deep.

Once the water is displaced and the air spaces are created, a steady-state condition will exist, with air coming into the underdrain blocks and percolating upward through the block and the various orifices and up into the support gravel (if used) and the filter media.

As part of the steady-state air scour condition, a phenomenon called *recirculation* also exists. As shown in Figure 6-19, air is metered out of the primary chamber through several levels of orifices into the secondary chamber. From there, the air bubbles rise to the air space at the top of the secondary chamber. Because of the air coming up through the water, conditions in the air space are very turbulent. The bubbles coming up through the water lift columns of water with enough force to splash against the underside of the top deck of the underdrain, as shown in Figure 6-19. Some of these columns of water will impinge on the back of the top deck/secondary orifices. When this occurs, some of the water is carried up through the top deck/secondary orifices and out into the filter. As a result, a considerable amount of water is driven out of the underdrain lateral. In order

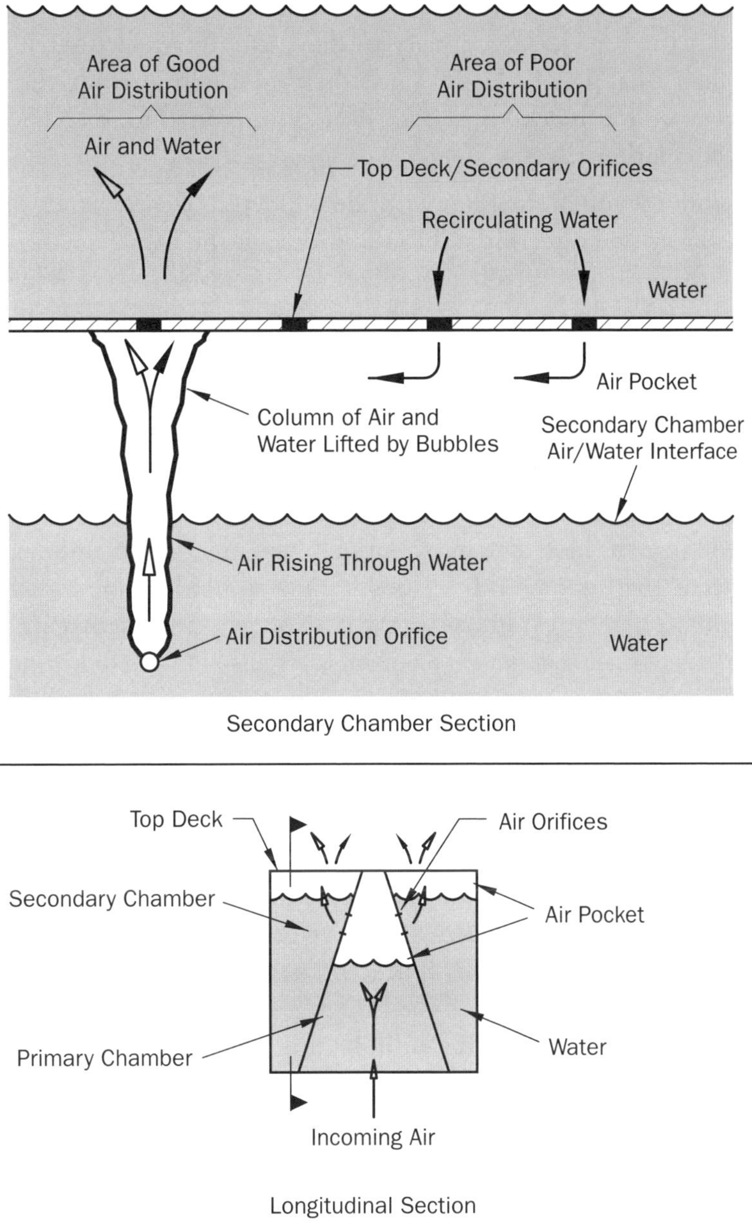

FIGURE 6-19. Recirculation

to maintain a steady-state condition, an equal amount of water must return into the underdrain in another place.

Within each underdrain lateral block, it is possible for a recirculation phenomenon to be established wherein water is driven out of the block in one end and recirculated back into the block in the other end. The recirculation phenomenon is usually confined to each individual 3-ft to 4-ft block, although the phenomena could exist in a longer section of the lateral. The result is good air distribution in the area where both air and water are emerging, and relatively poor distribution where the water is recirculating back into the top deck/secondary orifices. One manufacturer has provided a feature with their dual parallel lateral underdrain blocks to provide the water with a pathway to reenter the block with less effect on air distribution. The intent is to reduce this recirculation effect.

If viewed from the top of the filter with the full depth of media and water, this phenomenon would not be discernible. However, during the initial air distribution tests with 4 in. to 6 in. of water over a bare underdrain, it may be visible. When conducting tests in shallow water, the goal is to achieve relative uniformity of air distribution. Once a concurrent wash is added to air scour, the additional pressure of the incoming water is sufficient to even out the outward flow of water and essentially eliminate the recirculation phenomena.

The recirculation phenomenon will not affect the overall cleaning of the filter media. The highest level of agitation occurs during the concurrent wash when the air and water distribution will be at its best. The recirculation phenomenon is mentioned here to alleviate any concern that may arise from observing a somewhat uneven air distribution during the initial tests with 6 in. of water. A stronger, more damaging recirculation phenomenon, called the *air shutdown problem*, is discussed in the following section.

Air Shutdown Problem

Two basic procedures may be used during an air scour backwash: sequential wash and concurrent wash.

Sequential wash involves performing an air scour, turning off the air, and then beginning a hydraulic wash. Each of the two events is performed separately. In a concurrent wash, an air scour is provided, and a low rate of water wash is added to the air to create the concurrent condition. Once the concurrent wash has been operating until the water rises to the bottom of the trough, the air is turned off and the low-rate water wash is increased to the high-rate flow until the media is clean, and the backwash cycle is ended. Chapter 7 has a more detailed description of the concurrent wash cycle.

If a sequential wash cycle is performed, a recirculation phenomenon may exist that is sufficiently strong to disrupt support gravel (if used).

During air scour, a steady-state condition is created within the underdrain lateral. If the air is shut off without using a concurrent wash, water will immediately attempt to reenter the block through the top deck/secondary orifices to displace the air. At the same time, the air is attempting to escape upward, as shown in Figure 6-20. The opposition of the air and water may continue for several seconds without any movement. Afterward, the air in the secondary chamber of the underdrain will begin to move along the lateral until it reaches a point where a significant release of air will occur in a small area. While this is occurring, there is a horizontal movement of water above the underdrain in the opposite direction of the air bubble. (Figure 6-21.) When the air is released on one end of the lateral, water will reenter through the other end.

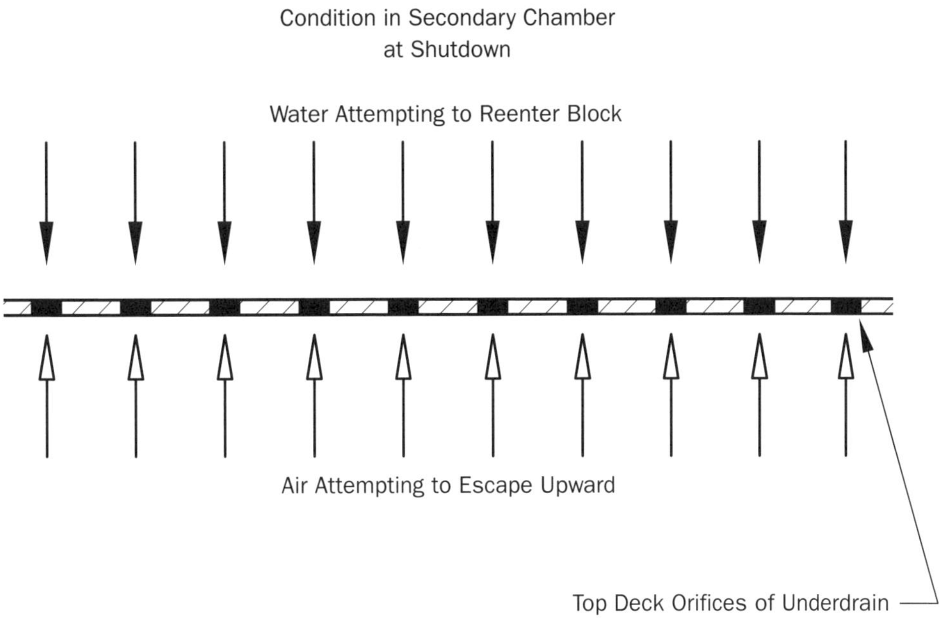

FIGURE 6-20. Air shutdown phenomenon

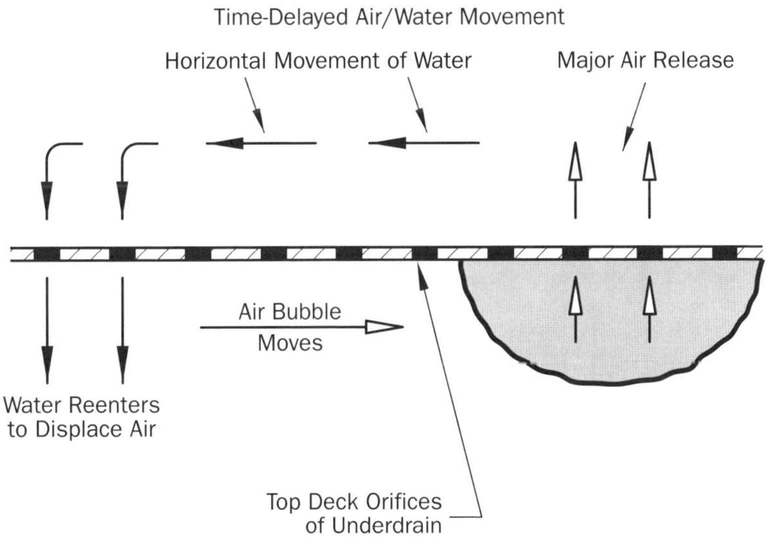

FIGURE 6-21. Air shutdown phenomenon

The horizontal movement of water above the underdrain can be strong enough to peel off the top layers of support gravel (if used). Therefore, sequential air scour should never be used with a dual parallel lateral underdrain when support gravel is provided.

One manufacturer has made provision within their underdrain block to mitigate or eliminate the movement of air. However, because of the potential liabilities, it is recommended that sequential air scour never be practiced when support gravel is used. If a direct media retention system is used to eliminate gravel, there is no danger of disruption, and sequential air scour may be practiced.

When a concurrent air/water wash is used, adding the low rate of water to air scour creates enough backpressure within the block to force the air out uniformly without causing damage. The incoming flow of low-rate water displaces the air without having water from above reenter the block. Stable conditions then exist that will not disrupt the support gravel.

Air Piping

It is common for one blower to provide air for the air scour backwash to a number of filters. In some larger plants, one blower may serve a group of filters, with another blower for a different group. An air supply pipe is provided from the blower to each of the individual filters it serves. The elevation of this pipe can have a major impact on the operation of the system. In Figure 6-22, the location of the wash trough is such that the backwash water level is normally below the operational water level.

When the air is turned off, water will reenter the J-pipes and fill the air header pipe with water to the backwash water level. It is important to keep the air piping as high as possible to limit the amount of water that must be purged. At a minimum, the piping should be above the wash trough, although higher is preferred. If it is not possible to keep the piping high, a high loop may be added to the piping, as shown in Figure 6-22, to protect the blower.

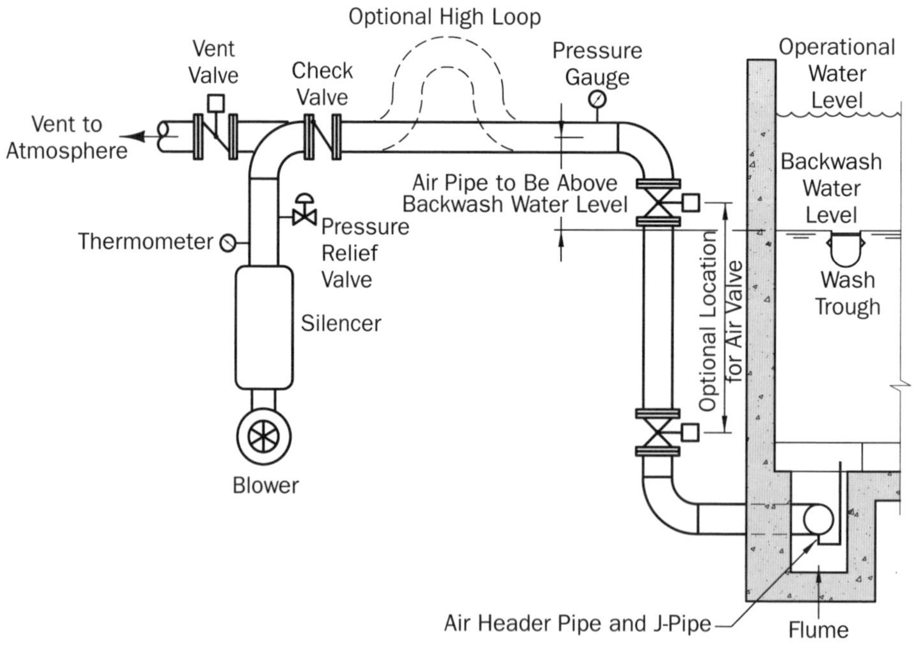

FIGURE 6-22. Air piping

Direct Media Retention System

Numerous direct media retention systems are in use to eliminate the need for support gravel. These systems include screen-wrapped laterals, wedge wire screen or nozzles, perforated plates, or porous plate caps. Several of these types are provided in lateral form or can be fitted to laterals.

Support gravel relies on depth and the fact that wet sand bridges. The porous plate cap material operates in a similar manner. The porous plate cap is designed to fit on top of the dual parallel lateral underdrain and will be the focus of this section. It is made of HDPE beads, which are sintered or heat-welded together to form a sheet up to 1 in. thick. The length and width of the resulting porous plate is designed to fit on top of each underdrain block separately. The porosity and head loss of the porous plate is similar to support gravel. Caulk is used to seal the porous plate to the underdrain block, and a series of stainless-steel screws is used to attach them together, as shown in Figure 6-23.

The benefit of using a porous plate cap instead of support gravel is that less height is required in the filter, and it is not as prone to upsets. On the other hand, it may be

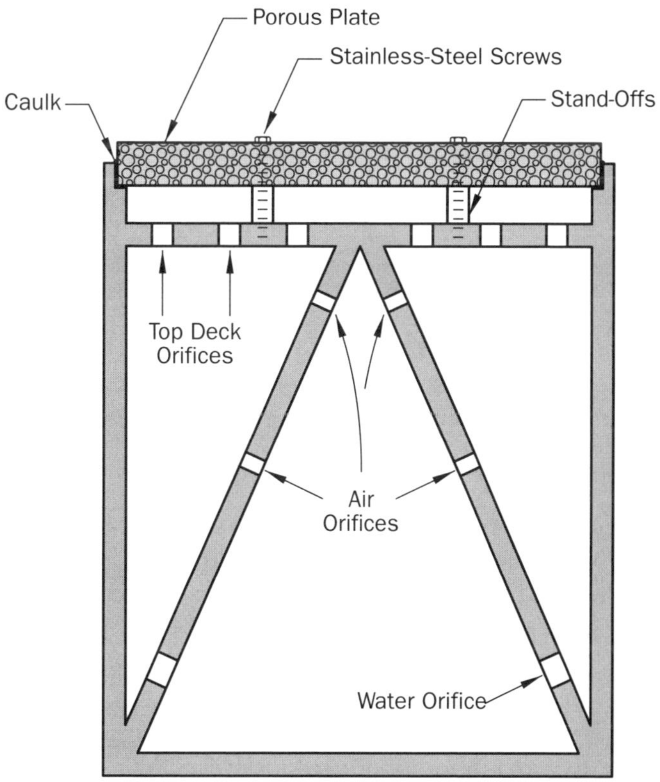

FIGURE 6-23. Porous plate cap media support

subject to the occurrence of an *edge condition*, which is discussed in Chapter 4 and in the following sections of this chapter.

Another important factor to consider is that the selection of porous plate must be matched to the media to be supported. For example, some porous plates will support garnet sand, while others may not.

The best use for porous plate is in the rehabilitation of older filters where height is at a premium.

Underdrain Failures

Dual Parallel Lateral Underdrain

Structural failures of the HDPE dual parallel lateral underdrain are relatively few. The HDPE material and the triangular internal structure are such that the block is strong as well as resilient. Filter failures are usually caused by J-pipe design, air flow or pressure control, hydraulic shock, sequencing of controls, valve sizing, backwash rates, flow control, or gravel and media design. Failures of this type caused by plant equipment and components are not related to the actual underdrain block itself. Failures that occur in the block are usually the result of joint problems during installation, expansion and contraction, grouting problems, severe hydraulic shock, or uncontrolled air.

Grout Failures

Failures in the grout can result in the disruption of support gravel (if used) and the loss of media. If a grout failure occurs over or near the flume, backwash water can emerge through the grout in an uncontrolled manner at higher than normal velocities and cause upsets in support gravel. Whether support gravel is used or not, filter media can be lost because of failures in the grout, as shown in Figure 6-24 and Figure 6-25. If this occurs, media can accumulate in the underdrain or the flume to the point where the underdrain is fouled, which would result in a major failure. Therefore, it is very important for the grout to be installed properly.

Quality Control for Grout

- When mixing grout using cement in bags, be sure that there are no chunks where the cement has begun to set up in the bag.
- The proper amount of time should be used to ensure complete mixing of the grout. Improperly mixed grout could result in an area being softer and weaker than intended. Mixed grout should be very homogeneous.
- Care should be taken to ensure that the proper proportions of sand, cement, and water are used. Refer to the manufacturer's instructions.
- Nonshrink grout is not recommended because it may expand.
- A backing strip is required under the blocks that span the flume. The backing material should be provided by the underdrain manufacturer. If the backing material is missing, it will result in an excessive amount of grout running down into the flume and could also result in a weak area that could fail.

FIGURE 6-24. Grout failure in clay tile (next to flume)

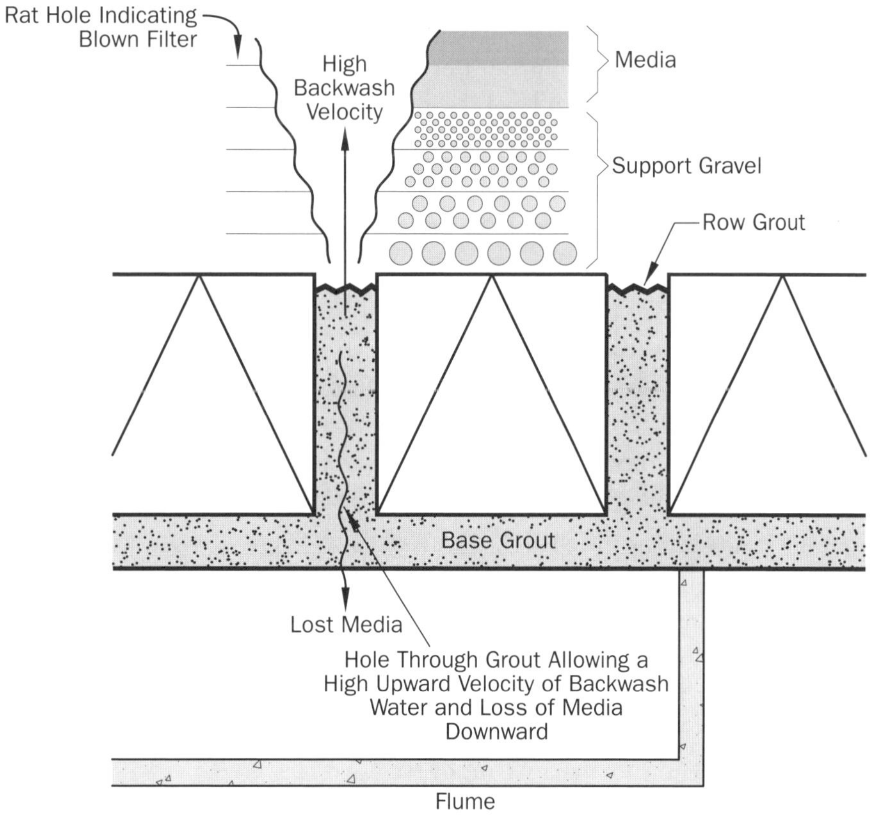

FIGURE 6-25. Grout failure

- The grout as installed should be relatively loose so that it flows evenly over and around the rebar over the flume and the lugs on the side of the underdrain block. Row grout should be loose enough to flow evenly and fill all the voids. However, if it is too loose, the grout may shrink and crack excessively. On the other hand, if the grout is too stiff, it may leave air pockets, which would result in a weak section. It is also recommended that the grout be installed in complete rows without leaving a cold joint.
- Installing grout during very hot weather can create several problems. The grout may set up rapidly, leaving voids, and there may be problems with contraction when it cools. If the underdrain and the grout are to be installed during hot or cold weather, refer to the manufacturer's instructions for special procedures to be followed.
- If the underdrain and grout are installed when the weather is either very hot or very cold, it may be desirable to cover the underdrain with water once the grout has set up. Water will help prevent excessive temperature differentials on the underdrain and grout. The underdrain manufacturer's recommendations should be followed in this case.
- Refer to the manufacturer's information for complete grouting instructions.

Base Grout

As shown in Figure 6-6, the dual parallel lateral underdrain should be set in a 1-in.-thick bed of base grout. The base grout performs two functions: it provides a base for the underdrain and is also used to level the filter floor.

The air manifold and J-pipes are usually anchored to the sidewalls of the flume or, in some cases, embedded in the concrete. The height of the J-pipes is calculated, in part, on the basis of the anticipated 1-in. bed of grout.

When leveling the floor, more than 1 in. of base grout is sometimes required. Depending on the location of the high and low spots on the floor, there could be substantially more than 1 in. of base grout adjacent to the flume. In this case, the J-pipe could be lower in the primary chamber of the underdrain than intended. The filter floor should be checked for levelness before installing underdrain or base grout. If it appears that the floor is more than 1 in. out of level, construction should be stopped until the impact on the height of the J-pipes can be determined.

During air scour, the air pocket in the primary chamber occupies approximately half of the block height. The location of the J-pipe within the air pocket is important. It should be well up in the air pocket without restricting the discharge of air. High volumes of air will be discharged at high velocities, especially in the case of long laterals. Therefore, the J-pipe cannot be too high in the primary chamber, or the flow of air would be restricted. On the other hand, if the J-pipes are too low, there is the risk of their being submerged if more than 1 in. of base grout is used.

Submerged J-pipes could cause a wave action to be created in the water, which could result in unstable air distribution within each lateral. As a result, it is recommended

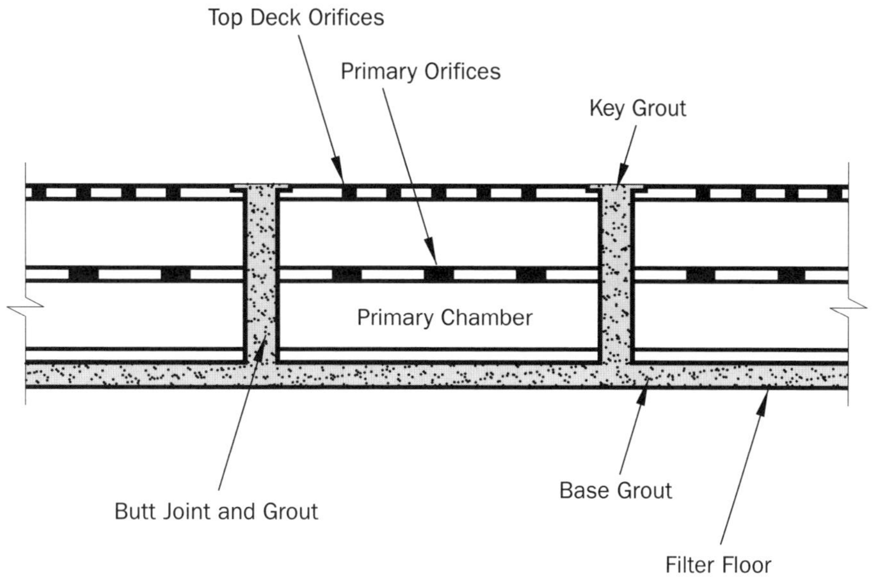

FIGURE 6-26. Clay tile dual parallel lateral underdrain joint

that the J-pipes extend several inches above the air/water interface. The J-pipe should be well up in the primary chamber but not so high that it would restrict the flow of air.

There are several manufacturers of the dual parallel lateral underdrain, and the exact height of the air/water interface may vary accordingly. The J-pipe height will vary accordingly. It is recommended that the plant designer work with the manufacturer to establish the height of the J-pipes within the discussed parameters.

The air scour feature requires the underdrain to be laid within a specific tolerance, which may vary depending on the manufacturer. If the underdrain is not within that tolerance, it could result in uneven distribution of air.

Joint Problems

Clay Tile Laterals. A form of grout intrusion can occur with the older clay tile version of the dual parallel lateral underdrain. A clay tile block is manufactured by extrusion and is cut into blocks and baked. As a result of the process, the joints are not always square. The blocks are set in the filters in rows on a thick layer of base grout, as shown on Figure 6-26. When installed, the ends of each block first are "buttered" with joint grout and set end to end. The joint grout forms the seal between adjacent blocks. An example of joint grout failure is shown in Figure 6-27 and Figure 6-28. Such a failure could result in upsetting the support gravel and a loss of media.

Once the joint grout has set up, additional grout is poured between the rows of underdrain to complete the installation. If the joints are not square and if the joint grout is not properly installed or cured, loose row grout can enter one of the blocks, as shown in Figure 6-29. The flow of water through the rest of the lateral during normal operation

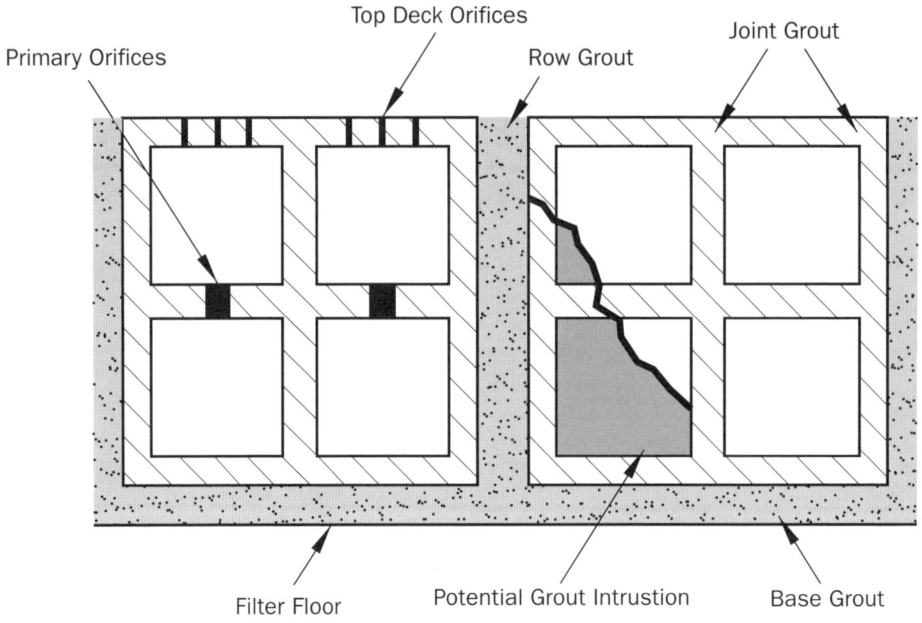

FIGURE 6-27. Clay tile dual parallel lateral underdrain—potential grout intrusion

FIGURE 6-28. Grout failure in clay tile (butt joint)

FIGURE 6-29. Grout intrusion in clay tile

can be greatly retarded or stopped completely, which can result in the filter media not being cleaned properly. The media over that part of the lateral would be dead.

Plastic Laterals. The plastic version of the dual parallel lateral underdrain is not normally subject to internal grout intrusion. Those lateral blocks have a snap-lock joint and an O-ring seal that prevents grout intrusion if the block is properly assembled. However, row grout failures can still occur.

When installing the plastic dual parallel lateral underdrain, the individual blocks are joined together to make a continuous lateral the length of the filter. Knife-edge joints on the outside edges and on the triangular inner partitions are intended to provide a seal to the next block. One or two O-rings are typically used in the bell-and-spigot joint to ensure a watertight seal. If the blocks are not assembled properly, the triangular inner partitions can be misaligned, as shown in Figure 6-30. If this occurs, it will result in the leakage of air and water, and the resulting operation of the lateral will be unsatisfactory. Great care should be taken during installation to prevent joint failure by inspecting each lateral when it is complete and before it is installed.

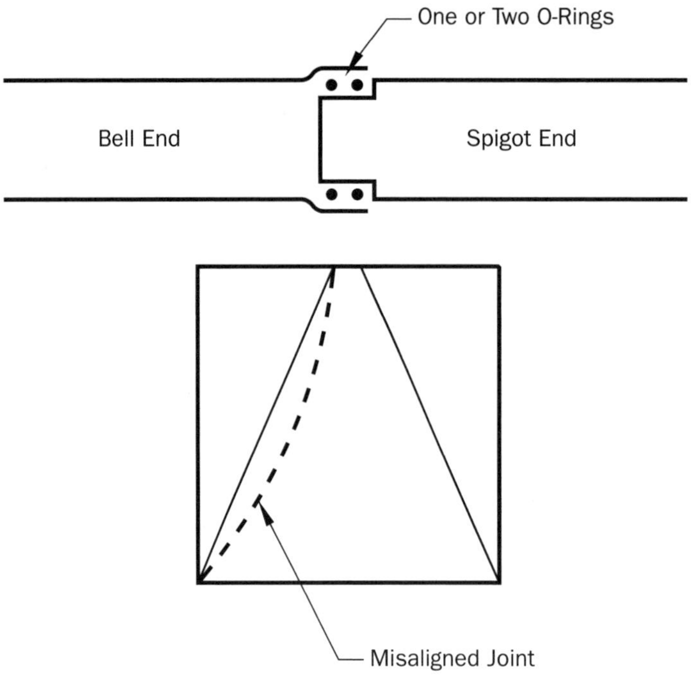

FIGURE 6-30. Misalignment of joint

Cleaning. Because of the potential liability, a significant quality control effort is needed during installation of the clay tile underdrain to ensure that the grout is properly installed. Once media percolates into a clay tile underdrain in substantial amounts, it is extremely difficult to remove. Cleaning the plastic dual parallel lateral underdrain is less difficult but still requires a significant effort. For new installations, the plastic version of the clay tile underdrain is preferred, even for hydraulic backwash applications. The snap-lock joint in the plastic block makes a positive seal between blocks and eliminates the grout intrusion problem.

Expansion and Contraction

The bell-and-spigot joint of the underdrain blocks is similar to that shown in Figure 6-31 and has a snap-lock connection.

HDPE is a plastic material that has a very high coefficient of expansion. If the underdrain is installed on a very hot day, the underdrain will be expanded significantly. If the underdrain is allowed to sit without cover until the weather changes and air temperatures fall to freezing, the block will contract and could cause severe damage, as also shown in Figure 6-31. If severe contraction occurs, the snap lug could break the end of the bell and expose the O-ring. If the O-ring is exposed, it may not be functional and could cause the release of water and air, which would be very disruptive to the support gravel, if used, and the filter media. Problems may also occur if the underdrain is installed on a very cold day and later operated in warmer conditions.

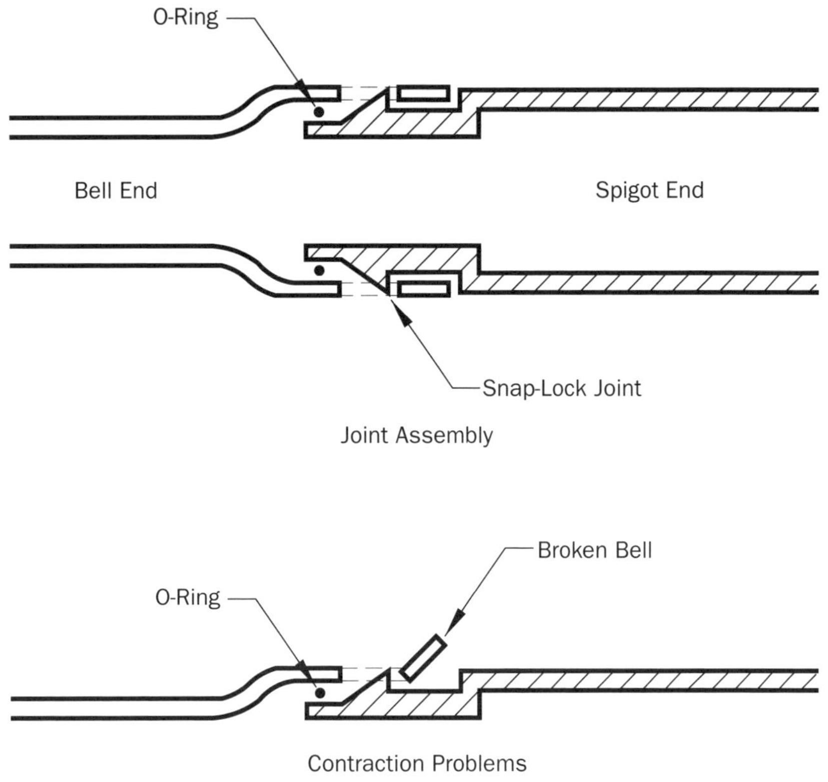

FIGURE 6-31. Expansion/contraction

If the underdrain is installed in either very hot or very cold weather, the manufacturer's specific temperature-related installation instructions should be followed.

Direct Media Retention Underdrain

If a porous plate cap is used on top of dual parallel lateral underdrain to eliminate gravel, great care should be taken in the design and installation of the system to achieve the maximum coverage of the filter floor. Direct retention underdrains are prone to edge fouling with hydraulic only backwash, as discussed in Chapter 4. In order to eliminate or mitigate that potential problem, the maximum coverage of the filter floor is recommended. If a seal angle is used on one end or a grout fillet on the other to seal the ends of the underdrain, these items should be kept to a minimum in order to retain the maximum floor coverage. The underdrain should also come as close as possible to the sidewalls. The use of air scour backwash may also mitigate or eliminate the edge condition problem. Chapter 4 discusses additional recommendations on this issue.

Pipe Lateral Underdrain Failures

In addition to problems caused by hydraulic shock or uncontrolled air, pipe lateral failures can be caused by inadequate lateral support, lateral breakage, connection failures, or large orifices.

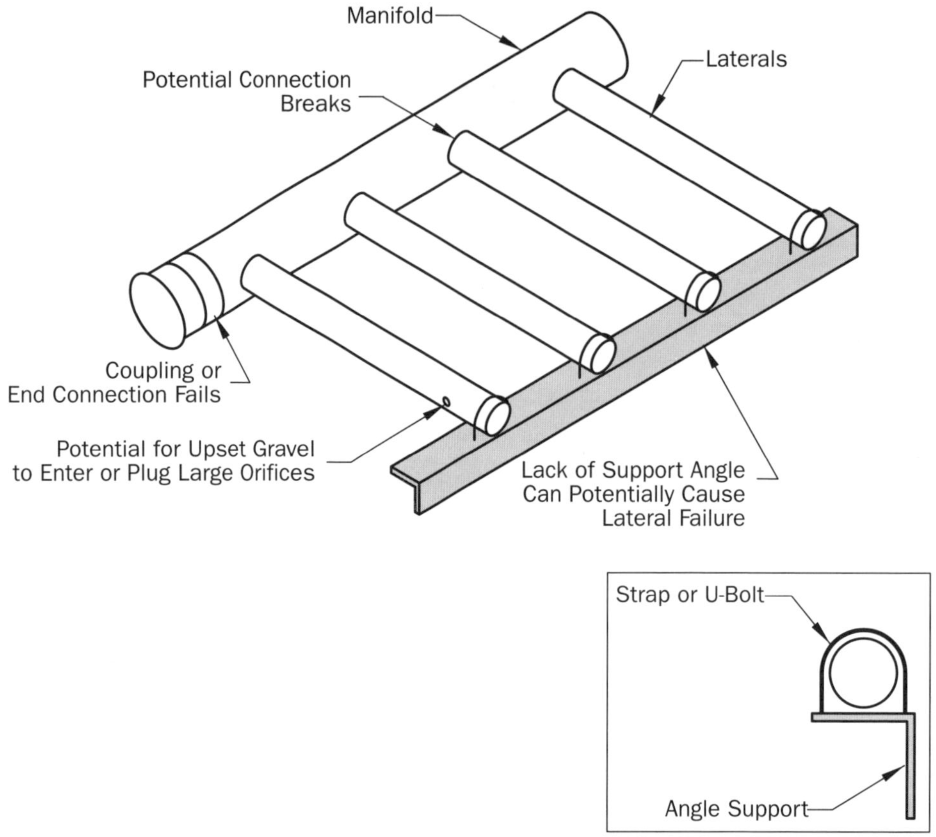

FIGURE 6-32. Typical pipe lateral underdrain failures

Lateral Support

Laterals are commonly made of PVC. Supports are required near the end of the laterals, as shown in Figure 6-32. Without support, laterals can be broken by the installation of, or downward pressure of, the support gravel and media. The supports should be anchored to the floor and/or endwalls. The hold-down strap or U-bolt holding the lateral to the supports should be of corrosion-resistant materials.

Improper installation of the support gravel can also cause excessive downward pressure on the laterals. When installed, the base layer of gravel must be carefully worked in underneath the laterals to ensure proper support.

Connections

The connection of the lateral to the manifold is a potentially weak point. Some manufacturers simply drill and tap the manifold and screw in the lateral. However, it is a much stronger connection to first weld a half-coupling to the manifold, whether using a PVC or steel manifold.

There may be a connection or coupling where the manifold penetrates the filter wall. Rigid connections are preferred. Flexible or sleeve-type connections are more subject to failure, which can result in an upset filter.

Orifice Size

When the orifices are sized, they must satisfy the hydraulic requirements of Chapter 3, as well as be compatible with the support gravel. It should be noted that orifice holes drilled in PVC laterals will contract slightly after removal of the drill. The support gravel covering the laterals should be three to four times the size of the orifice. Gravel that large will not enter or plug the orifice holes. For example, a $^{3}/_{8}$-in. orifice would require the gravel to be 1 in. to $1^{1}/_{2}$ in. in diameter.

False Deck Underdrains

There are a number of false deck type underdrains that will have similar problems, including structural failure of the deck, high backwash inlet velocity, high flume velocity, and broken or fouled tailpipes.

Structural Failure

The false deck structure can be constructed of steel or aluminum in package plants or of concrete in larger filters. Failures can occur because of repeated flexure of the false deck during backwash or excessive pressure differential caused by plugging of the underdrain. Grout failures in concrete filters are also a possibility. In either case, a pressure relief vent is recommended as an early warning of plugging, as shown in Figure 6-33.

Inspections

Periodic inspections of the plenum space are recommended.

Records

Historical records of changes in clean bed head loss are also helpful in predicting problems. A slow increase over time may indicate a potential failure.

High Inlet Velocity

High inlet velocity of either air or water could result in unequal cleaning of the media and potential failures. Multiple inlets should be provided, if required, to reduce the velocity. Refer also to hydraulic discussions in Chapter 3.

High Flume Velocity

A plenum space that is too shallow will result in high water velocities near the entry point. Plenums need to be deep enough to slow down the water to ensure equal distribution. As discussed in Chapter 3, the entry velocity in the tailpipes should be two or three times greater than the plenum velocity to ensure adequate distribution. The choice of nozzle or strainer (if used) and the backwash rate will determine the maximum plenum velocity desired.

Broken or Fouled Tailpipes

During installation, it is relatively easy to crack a nozzle head or break a tailpipe. Leftover construction debris in the flume could also plug the tailpipes in backwash mode.

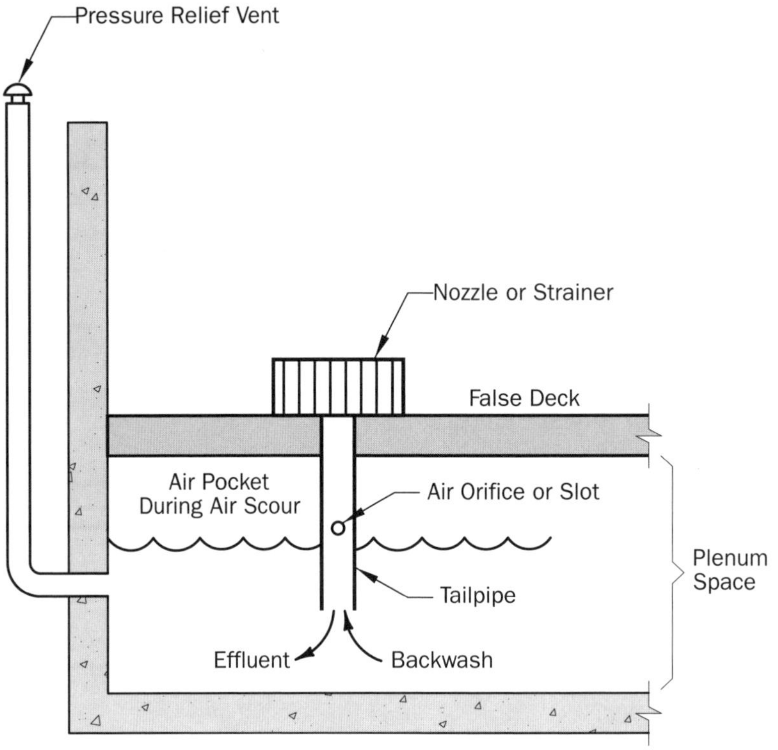

FIGURE 6-33. Typical false deck design issues

Quality Control

Implementing a comprehensive quality control program during construction and periodic inspections after the plant is in service are both important for maintaining a filter in good operating condition.

Lubricants

The nozzles or strainers are screwed into a fitting cast or welded into the false deck. Lubricants are sometimes used to make that operation easier. Some lubricants are not compatible with the materials of construction or the equipment. The manufacturer should verify the proper materials and installation procedures to be used.

Summary

An air scour backwash is an effective way of cleaning filter media and keeping it loosened. However, problems with the air system can also cause damage to the filters, especially if support gravel is used. Therefore, it is essential that all components of the air system be properly designed and operated.

An underdrain troubleshooting guide is shown in Figure 6-34.

Problem	Probable Cause	Proposed Solution
Poor air distribution	• J-pipes sized improperly	• Test before installing underdrain. Replace if necessary. • Possibly increase air flow.
	• Laterals too long	• Shorten air-only portion of backwash cycle • Design for middle flume
	• Submerged J-pipe	• Replace if necessary*
Air pressure relief valve activating	• J-pipes too small	• Replace J-pipes or reset relief valve
	• Relief valve set too low	• Reset relief valve
	• Water in air piping	• Program in a low-rate purge cycle
Blower damaged	• Water in piping and blower	• Install check valve • Replace blower
Top layer(s) of gravel disturbed (if used)	• Air shutdown problem	• Check cycle for sequential operation. Go to concurrent wash.
Grout failure	• Quality control	• Reconstruct*
Joint problem	• Quality control	• Replace underdrain*
False deck failures	• Flexure	• Reconstruct deck*
	• Lateral support	• Reconstruct deck*
	• Broken tailpipes	• Replace nozzle*
	• High inlet/flume velocities	• Install baffles to slow down the backwash water
Pipe lateral failures	• Connections	• Reconstruct filter*
	• Lateral support	• Reconstruct filter*
	• Large orifices	• Reconstruct filter*

*These items are very expensive to correct after the underdrain is installed. Care should be taken during design and installation to correct these issues before they become a problem.

FIGURE 6-34. Underdrain troubleshooting guide

CHAPTER 7

Optimizing Backwash

Introduction

The goal for any backwash cycle is to clean the media and restore it to its original condition in preparation for another filter run. It is important to optimize the backwash cycle to maximize efficiency and minimize the cost of operation. Backwashing a filter involves generation of wastewater with an associated cost of power and loss of net production. In addition, most of the mechanical problems that arise in a filter occur during the backwash process. Therefore, it is of primary importance for all the equipment and controls associated with the backwash process to operate properly.

Many issues are involved in optimizing backwash, including the provision of an adequate backwash flow rate to properly expand the media, the use of sufficient volume of water to clean the media properly, and the use of a highly efficient underdrain to distribute the water. As defined herein, optimizing backwash also involves ensuring the proper operation of each component of the backwash system.

Optimum Expansion

An extended discussion of media expansion and measurement is given in Chapter 5. For the purposes of this discussion, it is assumed that the optimum expansion for filter media occurs when the D_{90} level of media grains are agitated. The percent of expansion is the amount of expansion divided by the media depth times 100, as shown in Figure 5-22. The optimum expansion is best determined by observing the media in a pilot column. By incrementally increasing the backwash rate, the percent expansion achieved for D_{90} agitation can be visually determined. Those results can be translated to the operation of the full-scale plant.

Fluidization

Fluidizing Media

The optimum backwash rate for complete D_{90} fluidization of dual media or multimedia is normally dependent on the anthracite specifications, including the effective size (ES), uniformity coefficient (UC), specific gravity (SG), and perhaps grain shape. In addition, water temperatures have a large effect on fluidization. For example, higher temperatures require higher backwash flow rates.

The dynamic viscosity of water is shown on the graph in Figure 7-1, with the units omitted for clarity. On this graph, it can be seen that the relative velocity changes from about 3.4 units at 40°F (4°C) to approximately 2 units at 75°F (24°C), which is almost a 2:1 difference. As a result, the required backwash flow rate will vary dramatically with water temperature. Because of this variation, the backwash flow rate used should be seasonally adjusted as the water temperature changes, in order to maintain optimum conditions. A typical backwash rate for dual media or multimedia is shown in Figure 7-2. It should be noted that this curve is only effective for media loads using the specified

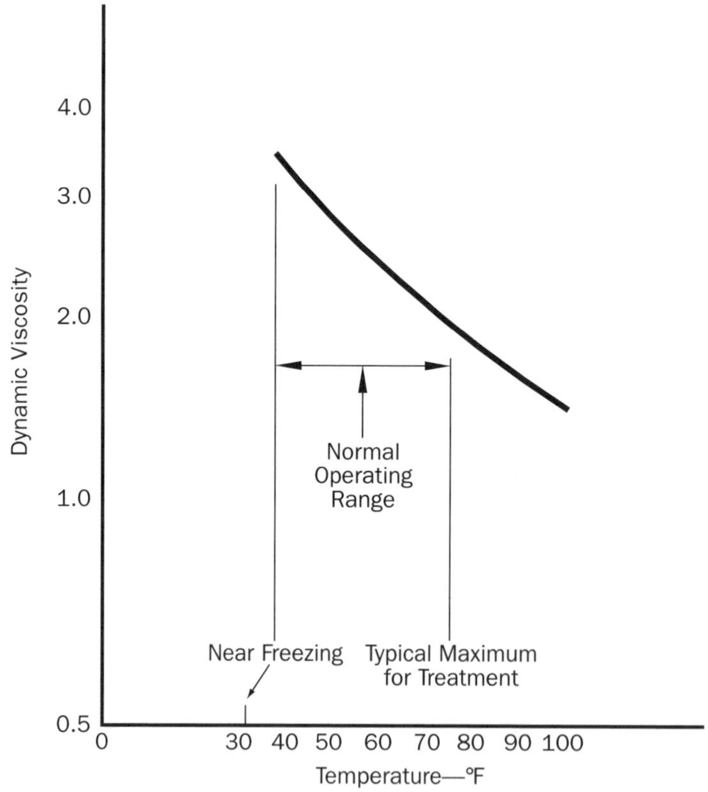

FIGURE 7-1. Viscosity of water

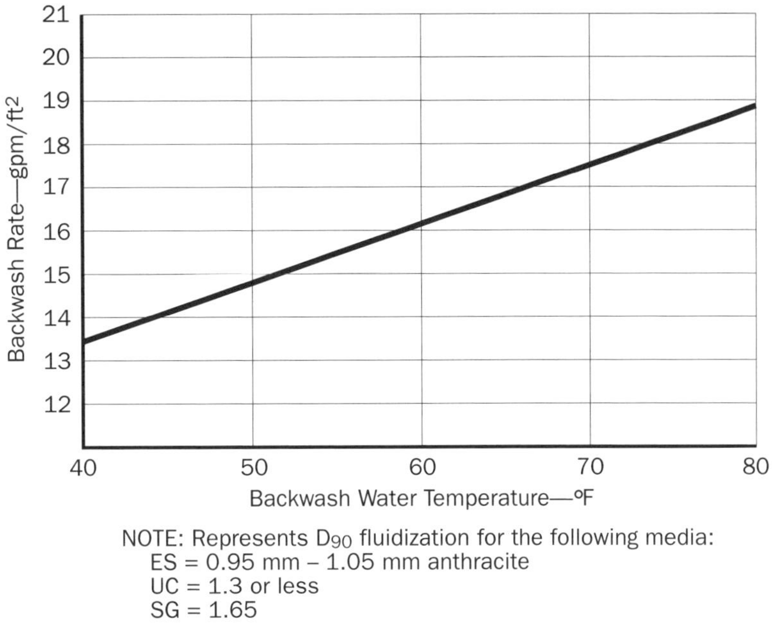

FIGURE 7-2. Typical backwash curve for dual media or multimedia

anthracite. For media with other characteristics, pilot studies will need to be performed to determine the amount of flow and expansion required.

Tests conducted by the author indicate that an increase in uniformity coefficient from 1.25 to 1.65 can require an increase of approximately 40 percent in backwash flow rate for the same expansion. For the same effective size, lower uniformity coefficients reduce the amount of coarse particles in the bottom of the bed and therefore, reduce the backwash rate necessary to fluidize the media down to the D_{90} level. As a result, it is highly desirable to obtain anthracite with the lowest uniformity coefficient commercially available. Lower backwash rates correspond directly with lower operating costs.

Nonfluidizing Media

Nonfluidizing monomedia typically consists of a single layer of 1.5 mm to 3.0 mm anthracite, activated carbon, or coarse sand. The depth of this type of media may be 4 ft to 6 ft. When using coarse monomedia, the backwash rates necessary for fluidization are impractical. The typical backwash rates actually used are in the range of 6 to 12 gpm/ft^2. The head loss of the media during backwash is greater than that of the underdrain in this case. The media will control distribution of the backwash water. As a result, the use of air scour is considered by the author to be mandatory to clean the media.

The term *nonfluidizing* means that the bulk of the media is not fluidized. However, the surface grains can be lifted during air scour, to some extent, if the media is on the small end of the size range listed previously. When using smaller-sized monomedia, grains of media can be lost if the wastewater is weiring over the troughs during the concurrent backwash stage. A substantial amount of media can be lost over a period of time by this method.

Chapter 5 has a more detailed discussion of this issue.

Over/Underexpansion

Many filter plants operate at a fixed backwash rate year-round. When an average rate is used, the media is typically underexpanded and may not be cleaned properly during the summer. Prolonged operation with an inadequate backwash rate could result in the creation of mudballs. In the winter, it is overexpanded when the water viscosity is higher. When overexpansion occurs, there is a much higher potential of media loss. Potential media loss is another reason to adjust the backwash rate seasonally according to water temperature.

The physical characteristics of media are discussed in greater detail in Chapter 5.

Backwash Sequence

Some filters are run for long periods of time between backwashing. High raw water quality and good pretreatment may allow long filter runs. However, it is recommended that filters be backwashed at least every 48 hr to prevent possible compaction of the media.

Different backwash sequences are used, depending on whether surface wash or air scour is used.

Surface Wash

The recommended backwash procedure when using surface wash is shown in Figure 7-3. Subsurface wash systems are less common and may operate differently. The manufacturer should be consulted for information on that equipment.

Step 1

Drain the filter to a beginning level. At many plants, the water level is simply drained to the top of the wash-water trough. However, it is recommended that the water level be taken to at least 6 in. below that point to reduce media loss.

Step 2

Activate the surface wash for an adjustable time period. As the name implies, the intent of this step is to break up the encrustation of solids that typically occurs at the surface of the media. Several minutes should be allowed for this step.

Step 3

After an elapsed time, the high-rate backwash can begin by starting the backwash pump (if used), eliminating air, and opening the rate control valve to the proper position.

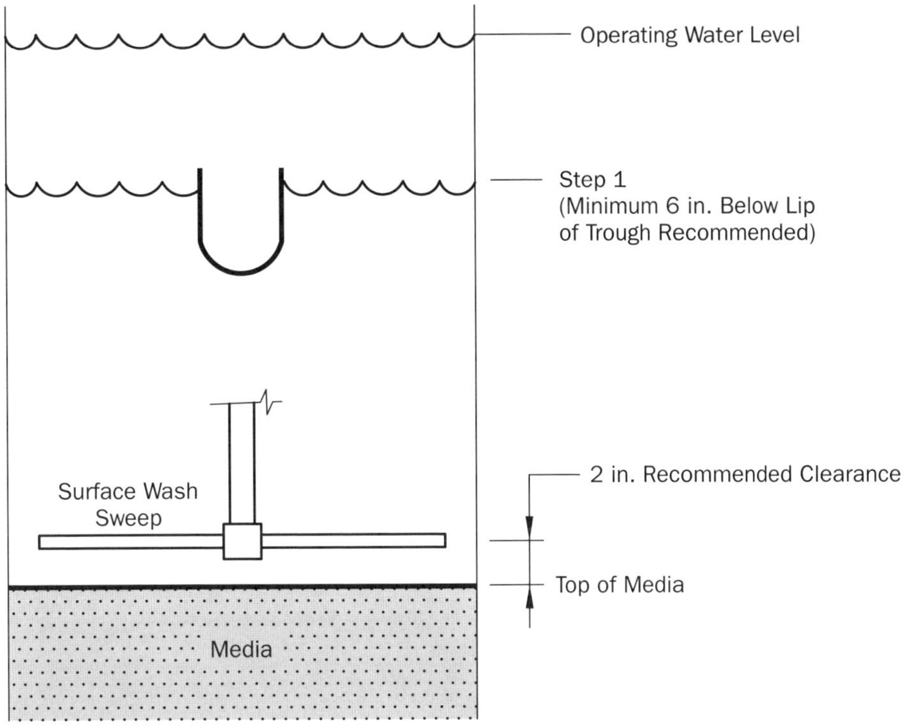

Recommended surface wash sequence:

1. Drain to 6 in. minimum below trough.
2. Activate surface wash for an adjustable time period.
3. Begin high-rate backwash.
4. Turn off surface wash within 1 min of achieving the full backwash rate.

FIGURE 7-3. Surface wash sequence

Step 4

The surface wash should be turned off within 1 min of achieving the full backwash rate. Drawing the water level below the lip of the backwash trough will allow some time for the surface wash sweep to be on before water begins weiring over the trough. If the surface wash is on for any length of time in combination with the high-rate wash, media could be lost. The agitation from the surface wash sweep in combination with the upflow of high-rate backwash is often enough to agitate the media sufficiently to wash some media grains out in the wastewater. When surface wash and high-rate wash are operated together over a long period of time, a significant amount of anthracite can be lost until an equilibrium point is reached. It is not unusual for filters to lose 6 in. to 8 in. of media in that manner.

Step 5

Terminate backwash by closing the backwash and waste valves and turning off the pump (if used). The backwash cycle can be terminated by reaching a low-level turbidity set point or by elapsed time as determined visually.

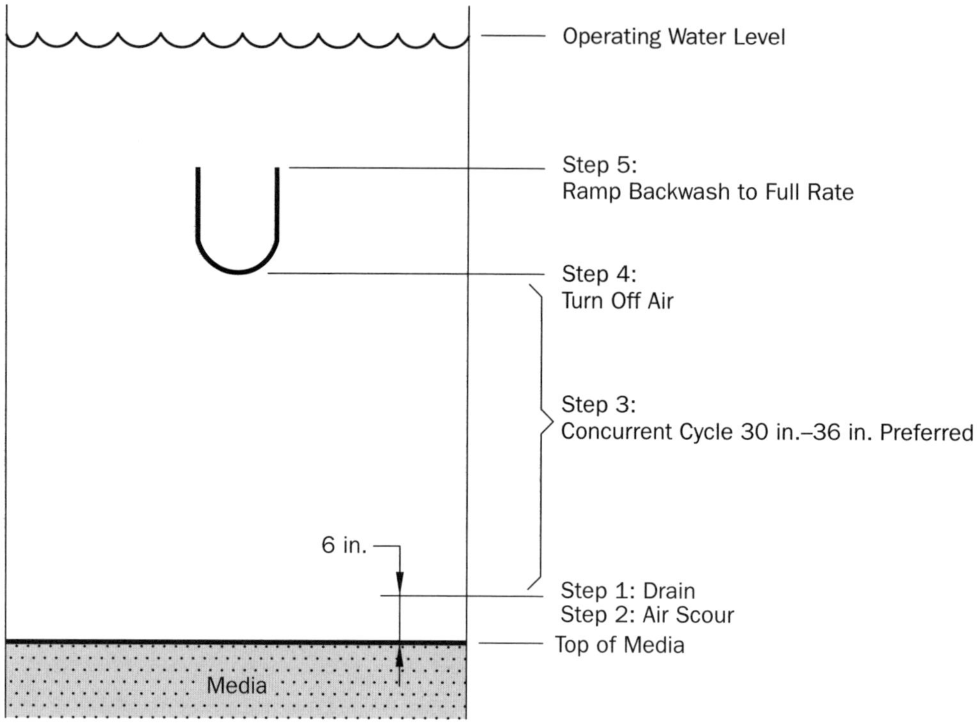

Recommended air scour sequence:

1. Drain the filter to within 6 in. of the media in filter mode, then open the waste valve.
2. Begin air scour (1-min duration preferred).
3. Begin low-rate water wash (concurrent).
4. Turn off air (air purge cycle).
5. Ramp backwash to full rate (purge solids and reclassify media).
6. Terminate backwash.
7. Return to service.

FIGURE 7-4. Air scour sequence

Step 6

Filter to waste and return to service by opening the filter inlet valve and the filter-to-waste valve. Continue in that mode until a low turbidity set point is reached or by elapsed time.

Air Scour Sequence

The recommended air scour backwash sequence is shown in Figure 7-4. For the purpose of this discussion, a positive displacement blower is assumed.

Step 1

Drain the water level in the filter in filter mode to within 6 in. of the media surface in filter mode, after closing the influent valve.

Step 2

Begin the air scour cycle by purging the air piping of water. Open the filter air valve, turn on the blower, and modulate the vent valve (rate control valve) to the purge position. After a set time period (established visually), ramp the vent valve to the full flow rate position. A 1-min duration at the full position is preferred, although controls should be provided to allow for a longer cycle, if desired.

Step 3

Begin the low-rate water wash in addition to the air scour (concurrent wash). Open the local filter backwash valve, turn on the backwash pump (if used), eliminate air, and then modulate the rate control valve, as required. A rate of approximately 5 gpm/ft^2 is recommended for fluidizing media, while 6 to 12 gpm/ft^2 is recommended for nonfluidizing monomedia. Even though these are nonfluidizing rates, the agitation will be greatly increased using a concurrent wash.

Note that the media will become completely mixed during Steps 2 and 3.

Step 4

Turn off the air scour portion of the concurrent wash. Open the vent valve, close the local filter valve, and turn off the blower. The low-rate wash is continued to purge air from the media and underdrain. It is the author's experience that the water coming out of the media at this time should be clean. When using coarse monomedia, the concurrent wash may be continued until both the air and water are turned off.

Step 5

Ramp up the backwash to the full rate for a set period of time. The normal time duration is 8 to 10 min, although controls should be provided for a longer period of time. The intent of this step is to purge loosened solids already suspended in the water above the media surface and to reclassify the media.

Step 6

Terminate backwash by closing the backwash valves, and then turn off the pump (if used).

Step 7

Run filter to waste cycle and return to service as previously described.

Notes:

1. The procedure outlined above is typical for fluidizing media. A distance of 36 in. to 42 in. from the bottom of the wash trough to the top of the media is preferred for this purpose. A shorter distance is workable, but there must be adequate time for a short duration of concurrent wash before the air scour is turned off. Chapter 6 has an additional discussion of this issue.
2. Nonfluidizing monomedia does not need as much freeboard as fluidizing media. However, coarse sand or fine gravel in the approximate size of 1.5 mm

will be lost during the concurrent step if water is weiring over the trough. Chapter 5 has an extended discussion of this issue.

3. In the process applications where it is desired to run continuous concurrent wash with coarse monomedia, material with an effective size closer to 2.5 mm to 3 mm should be used if it meets the removal requirements. The alternative would be to accept media loss or use a winged trough design, which is available from several manufacturers for continuous concurrent applications.

Enhancing Backwash

Using normal backwash procedures, there is usually a turbidity spike when the filter is returned to service. A filter-to-waste or rewash cycle is normally provided to waste the first few minutes of effluent water to eliminate the production of poor quality water.

An experimental process called *enhanced terminal subfluidized wash* (ETSW),* is being tested in some plants to minimize the "ripening period" and reduce the solids passing through the filter during that time period. Pilot studies are recommended to determine if this procedure is effective for the plant in question.

Return to Service

When a filter has been out of service for backwash or maintenance, the starting water level for filter mode should be at the lip of the wash trough as a minimum. When a filter has been down for maintenance, it should always be refilled slowly with backwash water to drive out air.

In either case, the filter should be left full so that the incoming raw water does not overflow directly onto the media and create holes or depressions. Depressions, as shown in Figure 7-5, could result in reduced filter efficiency. In the worst case, the support gravel (if used) could also be upset.

Backwash Water Volume

The volume of water used during backwash is also an important factor in cleaning the media. An amount of water is required to loosen the solids from the media, and another amount is required to lift those solids to the wash troughs. In addition, the amount of water required varies significantly, depending on whether air scour or surface wash are used. Example calculations are given in the following section.

Backwash Volume Using Surface Wash

Surface wash will be effective only if the proper flow of water is used and if the media surface is close to the sweep arm, as shown in Figure 7-3.

---

* Amburgey, Amirtharajah, Brouckaert, and Spivey, An Enhanced Backwashing Technique for Improving Filter Ripening, *Journal AWWA*, December 2003.

FIGURE 7-5. Overflow onto media

The recommended sequence of events in a typical backwash cycle was described previously. The anticipated volume of water necessary is calculated in Table 7-1. Table 7-1 makes the same assumptions as discussed in the recommended sequence: the surface wash is actuated for 2 min prior to backwash and continues for approximately 1 min into high-rate backwash.

Backwash Volume Using Air Scour

At the beginning of the air scour backwash cycle, it is assumed that the water level is drained to approximately 6 in. above the media surface. Once the air is turned on, the water level may rise as much as 6 in. to 8 in. as a result of the space occupied by air. The concurrent wash cycle starts at that elevation and continues until the air/water surface rises to the bottom of the wash trough. The air is turned off and the low-rate wash is continued. The intent is to purge the residual air and water from the underdrain, support gravel (if used), and the filter media before the water begins weiring over the wash trough.

Once the air is purged, the backwash flow can be ramped up to the established high-rate value. Typically, an increase in backwash rate will release additional air from the underdrain or media. For that reason, it is desirable to increase the rate before water is weiring over into the wash troughs. If air is present when the water is weiring over, some media will typically be lost.

TABLE 7-1: Backwash volume using surface wash

Assumptions		
Backwash rate	=	18 gpm/ft^2
Backwash time	=	8 min
Surface wash rate	=	$1^1/_2$ gpm/ft^2
Surface wash time	=	3 min
Calculations		
3 min × $1^1/_2$ gpm/ft^2	=	$4^1/_2$ gal/ft^2
8 min × 18 gpm/ft^2	=	144 gal/ft^2
	Total:	$148^1/_2$ gal/ft^2

A calculation is given in Figure 7-6 for the recommended height and time interval for the low-rate wash. As previously noted, there will be a rise in the water level when the air is first introduced. When the air is turned off at the bottom of the wash trough, the water surface will settle. The actual purge height available (H_T) in Figure 7-6 is, therefore, increased by that amount. If the air is purged adequately by the low-rate wash, the backwash flow should be ramped up to the high-rate position at least 6 in. below the top of the wash trough. Any residual air can be purged at the high rate, before the water weirs over the wash trough. If additional high-rate purge time is needed, the low-rate wash time should be reduced to make up the difference.

In the calculations in Table 7-2, the time intervals are based on 30 in. of filter media. The time intervals are also dependent on the height of the backwash trough above the media and the nature and volume of the entrapped solids. The previous calculations should only be used for comparative purposes. Actual values should be field determined. Following the procedures outlined herein, filters should be backwashed until clean or until no further gain is visible in the clarity of the wastewater.

Another variable in the amount of backwash water used is the cycle time of the valve actuators that are used to control the process. Cycle times of 60 sec are common. In addition, some computer systems use a feedback loop to verify that the previous valve is closed or the previous step is completed before proceeding to the next step. Also, any delays in communications will add significantly to the backwash time. The controls need to be optimized to the extent possible so that the backwash cycle can be terminated efficiently and as quickly as possible. Otherwise, it could result in an excessive use of backwash water and greatly lengthen the time the filter is out of service. The longer the filter is out of service, the greater the impact on the total net production of the plant. In addition, the cost of treatment will be higher.

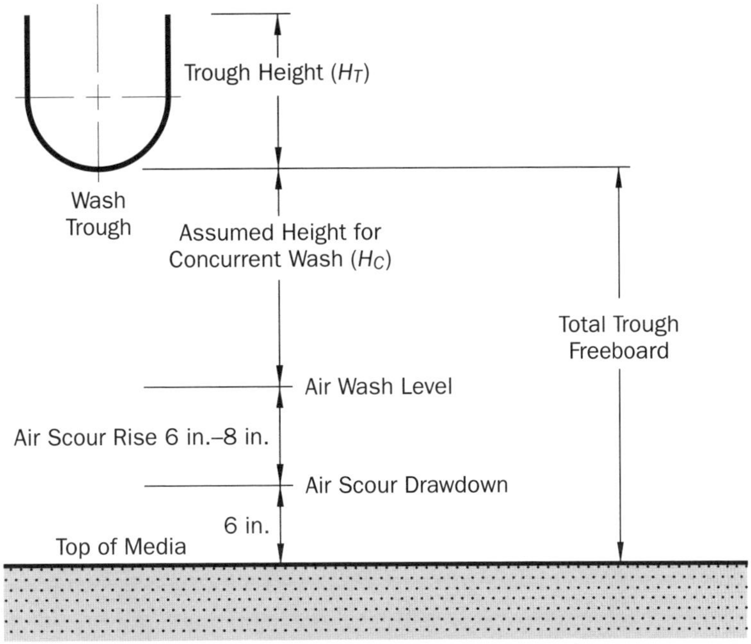

Concurrent height H_C (recommended)	= 30 in. to 36 in.
Purge height H_T (assumed)	= 16 in.
Low-rate wash height (H_{LR})	= $H_C + H_T$
H_{LR} = 24 in. + 16 in.	= 40 in. = 3.33 ft
Assumed low-rate wash	= 5 gpm/ft^2
	= 0.67 ft/min

$$\text{Low-rate time} = \frac{H_{LR}}{\text{Wash rate}} = \frac{3.33 \text{ ft}}{0.67 \text{ ft/min}} = 4.97 \cong 5 \text{ min}$$

FIGURE 7-6. Suggested air scour height

TABLE 7-2: Backwash volume using air scour

Assumptions		
Low-rate wash	=	5 gpm/ft^2 for 5 min
High-rate wash	=	18 gpm/ft^2 for 5 min
Calculations		
5 min × 5 gpm/ft^2	=	25 gal/ft^2
5 min × 18 gpm/ft^2	=	90 gal/ft^2
	Total:	115 gal/ft^2

Underdrain Distribution Efficiency

The goal of any underdrain is to distribute the backwash water as uniformly as possible. Uniform distribution of backwash water will result in a more uniform cleaning of the filter media in all parts of the filter. Maldistribution, or the deviation from the average, has been the traditional yardstick to measure distribution efficiency. For many years, underdrains with a maldistribution efficiency of ±5 percent have been available. It should be noted that this value should include the component of maldistribution contributed by the flume or collection manifold. Underdrains with lower maldistribution capability are available. It is recommended that underdrains be specified with the lowest possible maldistribution.

In addition to distribution efficiency, the underdrain must have higher head loss than the filter media during backwash, if fluidizing media is used. High head loss will ensure that the underdrain controls the distribution against the backpressure created by the filter media. Coarse monomedia designs use low backwash rates and rely on air scour to keep the media loosened and properly cleaned. Chapter 5 has a more detailed discussion of this issue.

Terminate Backwash on Turbidity

It is possible to provide instrumentation and controls capable of terminating backwash when a certain low turbidity is achieved in the wastewater. A turbidimeter can be put on the main backwash waste header, and special models of turbidimeters may be available for installation directly in wash troughs, if desired.

If this capability is provided, it is recommended that it be an option available to the operator and not mandatory for use at all times. When seasonal water quality changes occur, they will result in changes to the water chemistry and the chemical feed systems used. Therefore, it is recommended that a visual check be made on the filters during backwash to make sure they are cleaned without excessive use of backwash water. Even though this can be automated, it is recommended that the operator spend time observing the filters directly, to better understand their operation.

Terminate Filter-to-Waste on Time or Turbidity

The filter-to-waste cycle at the end of backwash can also be terminated on time or turbidity. Depending on the placement of the effluent turbidimeter, the same unit may be used to terminate filter-to-waste. Simple additions to the programming of computerized controls should be all that is necessary to achieve this function. Again, both time and turbidity options should be available to the operator for terminating filter-to-waste. Please note that any settings for terminating filter-to-waste should be reevaluated when the temperature or raw water quality changes.

Polymer Addition

It normally takes a filter-to-waste cycle length of at least 3 to 5 min or more to recondition and recompact the filter media to achieve the desired removal efficiency.

At some plants, a dilute amount of filter-aid polymer is added during the last minute or two of the backwash cycle to precondition the media. The intent is to shorten the filter-to-waste time so that the filter can begin producing water more quickly.

It is important to understand that most of the coagulated solids are intended to be removed in the filter media before the water reaches the support gravel (if used), or the underdrain.

If support gravel is used, there should be no problem in adding a very dilute solution of polymer at the end of the backwash cycle. However, if a porous plate cap on top of a dual parallel lateral underdrain or another direct media retention underdrain system is used, caution is recommended in the addition of polymer in the backwash water. Especially with the porous plate cap, the pore spacing is so small that too much polymer in the backwash water could potentially plug the underdrain over time.

Uncontrolled Air

Uncontrolled air is one of the most common and destructive backwash problems. Sources of air include vertical turbine pumps, combination air/vacuum release valves, leaking valves or fittings, entrained air, or exceeding the minimum pump submergence.

Description and Causes

Uncontrolled air is the accumulation of a large volume of air in the backwash piping, which can result in water hammer during the backwash process, when that air is pumped to the filter. A large air bubble moving in a pipeline is illustrated in Figure 7-7. As the bubble approaches the filter, the pressure in the pipeline is reduced as a result of passing through valves, flowmeters, and pipe fittings. The reduced pressure may result in an elongation of the air bubble, which may cause the leading water to accelerate slightly. When the air bubble reaches the filter, it is suddenly dissipated into the large volume of the filter, and the leading water slows. When the air bubble collapses, the trailing water hits the slower water in the filter at full speed, creating a shock wave that can damage the underdrain and support gravel (if used) and result in loss of media. Therefore, it is highly desirable to eliminate all air in the backwash water. Each of the common sources will be discussed in the following sections.

Vertical Turbine Pumps

Vertical turbine pumps are commonly used to supply backwash water for filters. A typical vertical turbine pump installation is illustrated in Figure 7-8. The pump motor and discharge head are typically above floor level, with the pump column extending downward below the clearwell water level.

Fittings

The fittings downstream of a backwash pump typically include an air release valve, a check valve, a rate control valve, individual filter valves, and other miscellaneous fittings. It is typical for air to leak into the system through the air release valve when the pump is off-line. As a result, it is common for the pump column to be full of air from

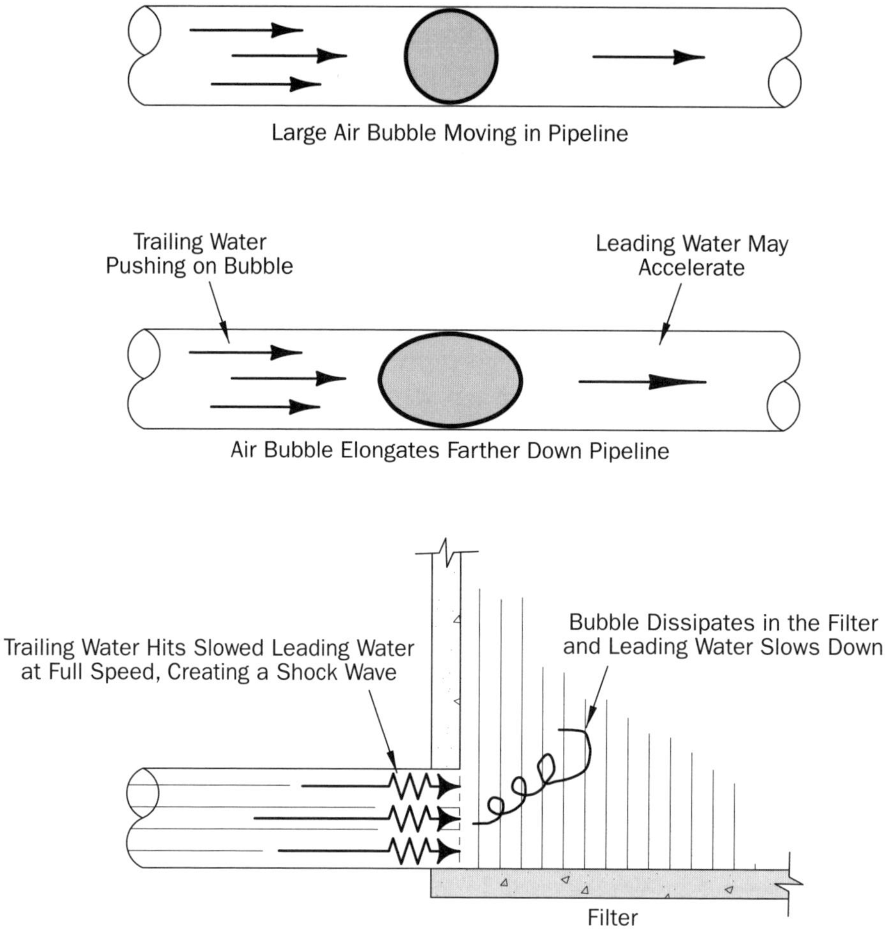

FIGURE 7-7. Water hammer caused by air

the water surface in the clearwell to the check valve. The piping beyond may also be partially full of air if the check valve leaks.

When the pump is activated, the water coming up the pipe column pushes the air ahead of it and out of the pump discharge head into the piping system. It is very difficult to prevent air from entering the pump column when the pump is out of service. Therefore, it is of paramount importance to ensure that the air is eliminated before it reaches the filter. Eliminating the air is accomplished through proper design of the air release valve and the correct sequence of events in the beginning of the backwash cycle.

Water Rotation

Immediately downstream of a pump, the water may be rotating around the perimeter of the pipe as it moves forward, as shown in Figure 7-9. In this case, an air bubble in the water may be in the center of the pipe. Therefore, all of the air may not

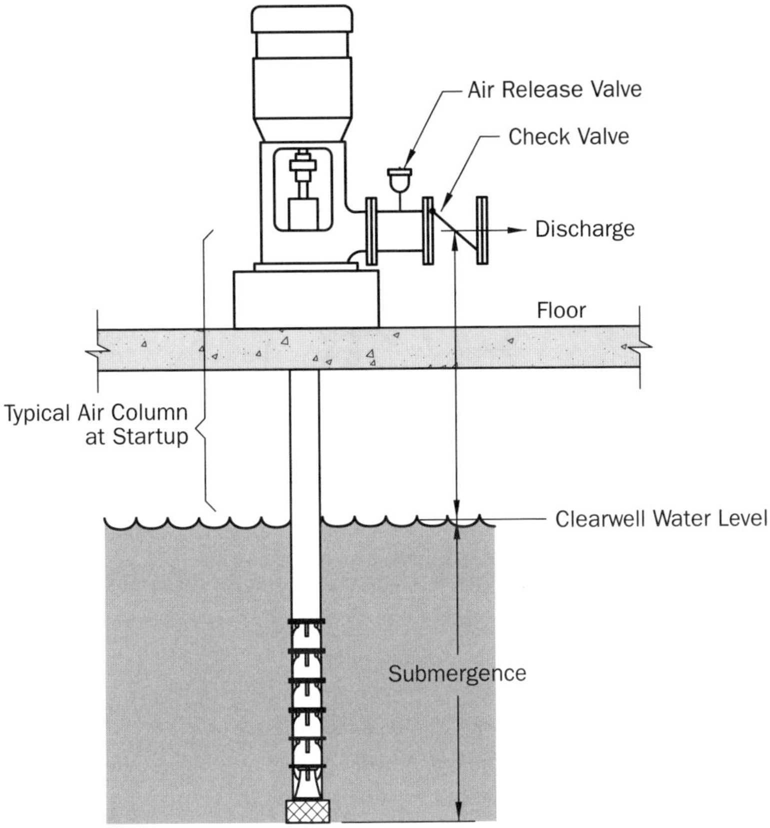

FIGURE 7-8. Vertical turbine pump

be eliminated by an air release valve, even if it is close to the discharge of the pump. As a result, it is very important to eliminate the air from the entire backwash system, with the pump operating against a closed valve. Once the air has been eliminated, the process valves can be opened to begin the backwash cycle. The length of time required should be determined in the field by observing the operation of the air release valves at the beginning of the backwash cycle.

High Loops in Backwash Piping

The backwash piping seldom, if ever, goes straight from the backwash pump to the filter. There is usually at least one high loop in the piping, as shown in Figure 7-10, for clearance or access purposes. High loops provide a space in the piping where air can accumulate. Even a small leak could result in the accumulation of a large air bubble in a high loop, which can cause or aggravate hydraulic shock in the filter during the backwash cycle. As a result, an air release valve should be installed on every high loop in the backwash piping. If the high loop is long, an air release valve should be installed on both ends, to ensure that all the air is driven out.

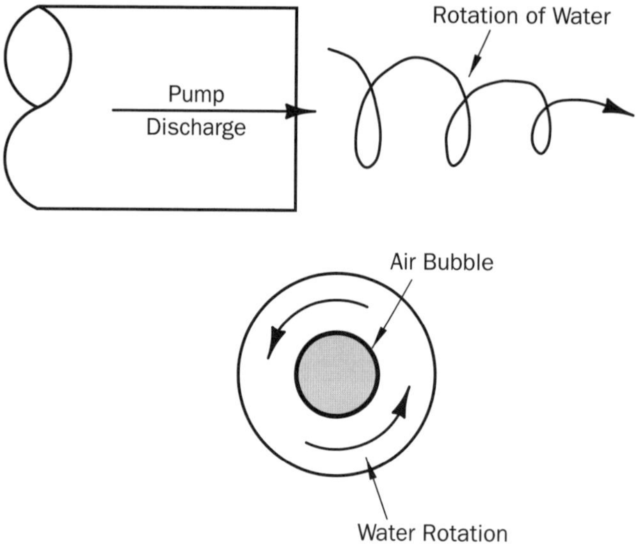

FIGURE 7-9. Water rotation

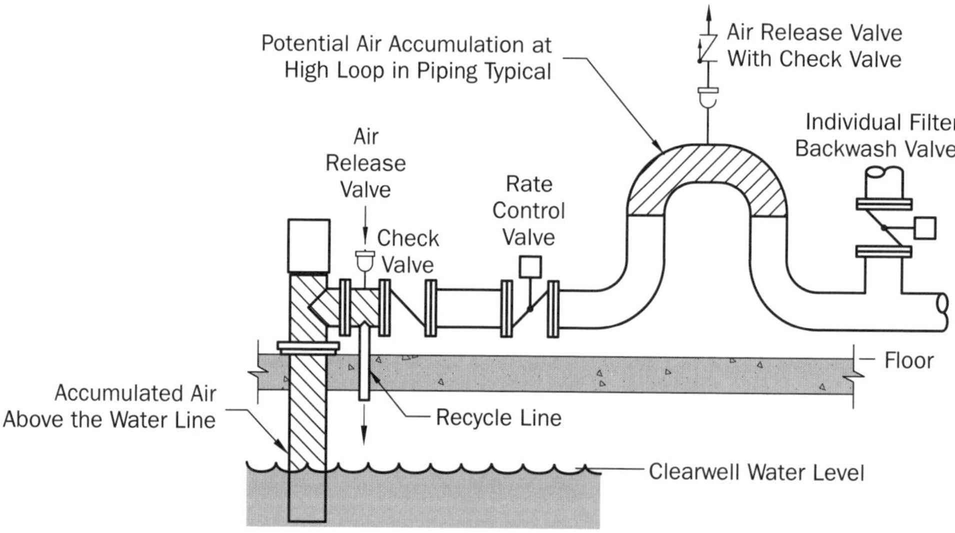

FIGURE 7-10. High loops in backwash piping

Water Velocity

When a low-rate/high-rate combination backwash is used, it may be difficult to purge all the air from the piping. The piping is typically sized for the velocities that occur during high-rate condition. Therefore, when beginning the backwash cycle in the low-rate condition, it may not be possible to eliminate all of the air because of low pipe velocities.

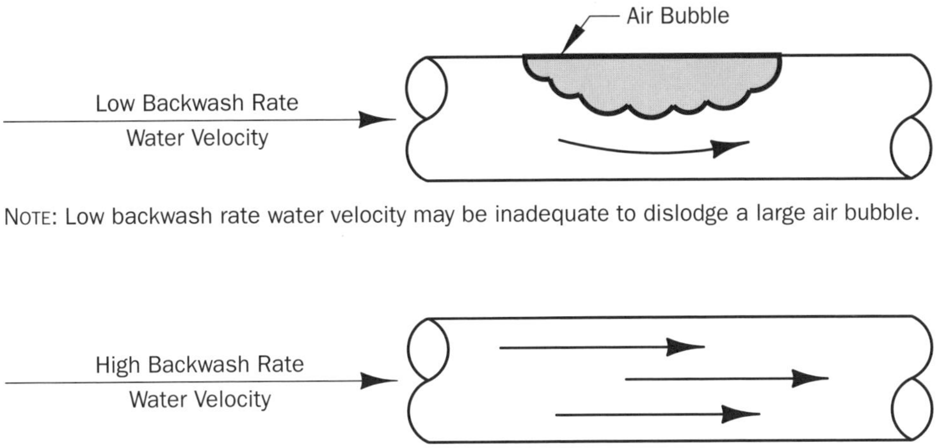

NOTE: Low backwash rate water velocity may be inadequate to dislodge a large air bubble.

NOTE: Ramping up to high-rate water velocity is usually adequate to dislodge remaining air bubbles.

FIGURE 7-11. Residual air in backwash piping

On long horizontal runs of piping, large air bubbles can sometimes "hang up" in tees, fittings, or pipe joints, which are difficult to dislodge at low-rate wash velocities, as shown in Figure 7-11. When initially pressurizing the piping, these air bubbles may not be completely eliminated. When ramping up to the high backwash flow rate, the velocity is usually adequate to dislodge the remaining air bubbles. As a result, it is common for an additional release of uncontrolled air when the backwash rate is ramped up to the maximum level. If the amount of air released is substantial, an investigation should be made to locate where the bubble is hanging up, so that air release valves can be installed at that location.

Gravity Backwash Tank

The backwash supply for some filters comes from reservoirs in the distribution system, while others have dedicated gravity backwash tanks. In either case, if the reservoir is small, it is possible to introduce air into the backwash water. If the level is drained too low in backwash, there is the potential for generating a vortex, as shown in Figure 7-12. A whirlpool effect can be generated that would allow air to come in with the water. Other than providing a larger tank, one common solution is to provide a vortex baffle over the discharge piping in the storage reservoir or backwash tank, which is also shown in Figure 7-12.

Whether a vortex baffle is provided or not, level controls should be provided to prevent a backwash from starting if the level is too low and to terminate the backwash if the level is drained to a point where a vortex could be created.

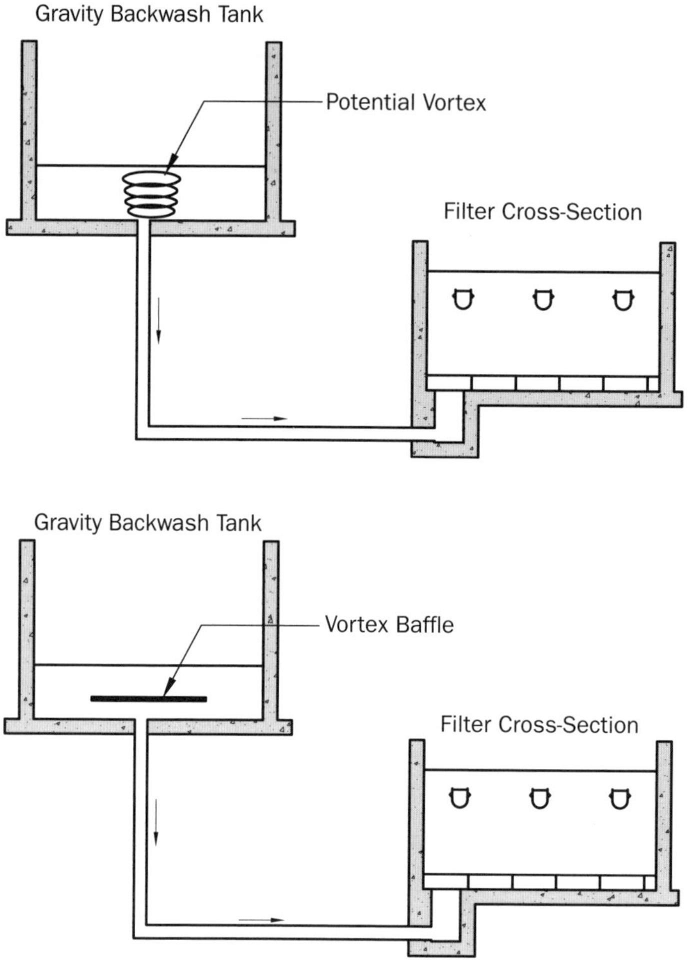

FIGURE 7-12. Uncontrolled air in gravity backwash

Air Release Valves

A typical air release valve is shown in Figure 7-13. Air enters the inlet connection on the bottom of the air release valve and creates an air pocket inside, which causes the metal float to fall. A linkage connected to the float opens a seat when the float falls, allowing air to be discharged upward and out of the system. When the air is driven out, the water level inside the valve rises, causing the float to rise and seat the seal, which prevents water from being lost.

There are two common types of air release valves: a combination air/vacuum release valve and an air-only release valve. Combination valves allow for air to reenter the valve and the piping system in order to prevent a vacuum from occurring. However, in a backwash supply pipe, a vacuum is usually of no concern. It is more desirable to prevent air from entering the system. Therefore, air-only release valves are

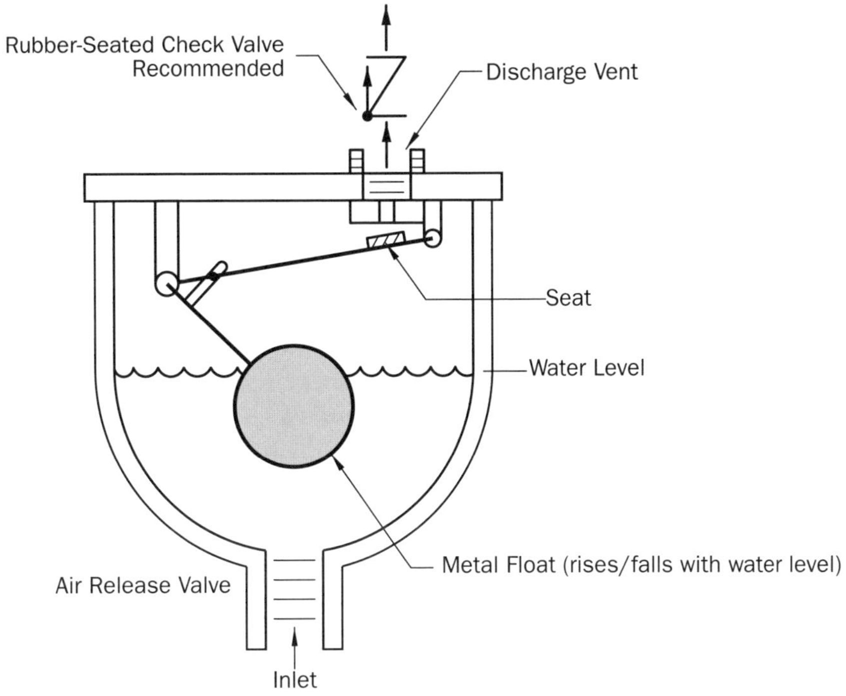

NOTES:
1. When all the air has been eliminated, the water pressure raises the float and seals the seat to prevent loss of water.
2. The float may not seat properly at low pressure.

FIGURE 7-13. Typical air release valve

recommended for this service. Combination air/vacuum release valves should not be used on backwash piping.

The principle of operation of air release valves requires a minimum system pressure to properly seat and seal the valve. The manufacturer's literature should be consulted to determine the minimum pressure required. However, a pressure of 10 psi is generally adequate.

At lower pressures, the air release valve may not function as desired. In fact, if the velocity of the backwash water is relatively high, a siphon can be created whereby an air-only release valve can allow air to reenter the system. Therefore, it is highly recommended that all air-only release valves have a rubber-seated check valve on the discharge vent. The purpose of the check valve is to prevent air from reentering the system during low-pressure conditions. Low-flow conditions can also result in low pressure, especially if the air release valve is downstream of the rate control valve.

Backwash Piping

Air release valves are needed in the backwash piping to provide a means for eliminating air from the system. They should be installed at each high loop or high point in the backwash piping. Long headers should have an air release valve on each end as a minimum, and intermediate points are recommended. Great care should be taken in the placement of the air release valves. If there is a question regarding placement of the air valves, more is better than fewer. Air release valves are not very expensive, and it is better to err on the side of caution.

Air Release Valves for Vertical Turbines

It is common to install an air release valve close to the discharge of vertical turbine pumps. If the pump is allowed to start at full speed, there is usually a sudden surge in pressure at the discharge. Pressure surges of that magnitude can be adequate to slam the float against the seat, although there is still air inside the valve body. The floats used in air release valves are made of hollow stainless steel or copper. If the float is slammed against the seat under high pressure, the float can become damaged, which can prevent it from seating properly. In order to mitigate this condition, some manufacturers make an air release valve with a perforated cage surrounding the float on the inside of the valve, as shown in Figure 7-14. The purpose of the cage is to break up the inrush of pressure so that the float can function normally.

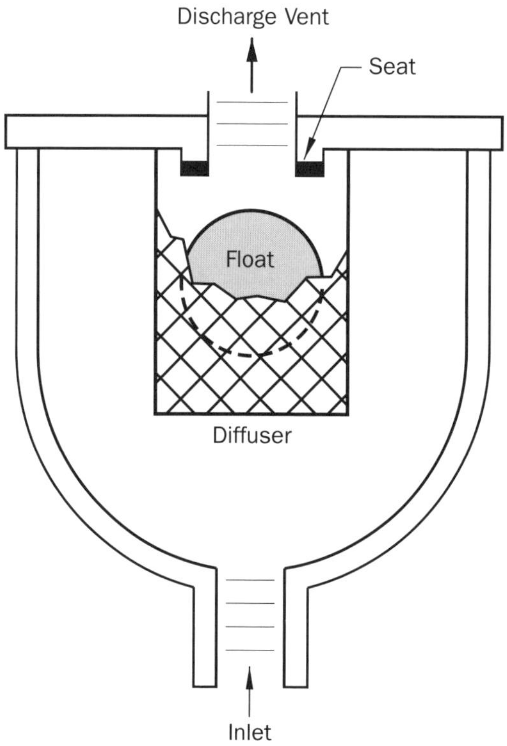

NOTE: Diffuser prevents sudden rush of water from damaging float when pump starts.

FIGURE 7-14. Air release valve for vertical turbine pumps

The potential for a sudden inrush of pressure is another reason why air should be driven out of the system at the beginning of the backwash cycle, with the pump operating against closed process valves. The air can be gradually driven out of all the air release valves, and the low-rate backwash can be started.

Leaking air-release valves are an indication of a damaged float or a worn seal. In either case, the valve should be repaired or replaced as soon as possible.

Maintenance

Some periodic maintenance is required on air release valves. If the air valve continues to leak water, it may be necessary to install a repair kit, including seats and seals. While making these repairs, the float should also be inspected to make sure it has not been damaged. Air release valves are an important part of the backwash system for a filter. As such, they must be maintained like all other equipment in order to ensure proper operation.

Sizing Backwash Pumps

The backwash flow is easily obtained by multiplying the unit rate for the media selected by the filter surface area. The head pressure is the sum of the water lift, the piping and fittings head loss, the underdrain head loss, and the media head loss as shown on Figure 7-15. For most filters, these values typically amount to 20 ft to 25 ft of pressure required by the high-rate backwash flow. However, there is the possibility

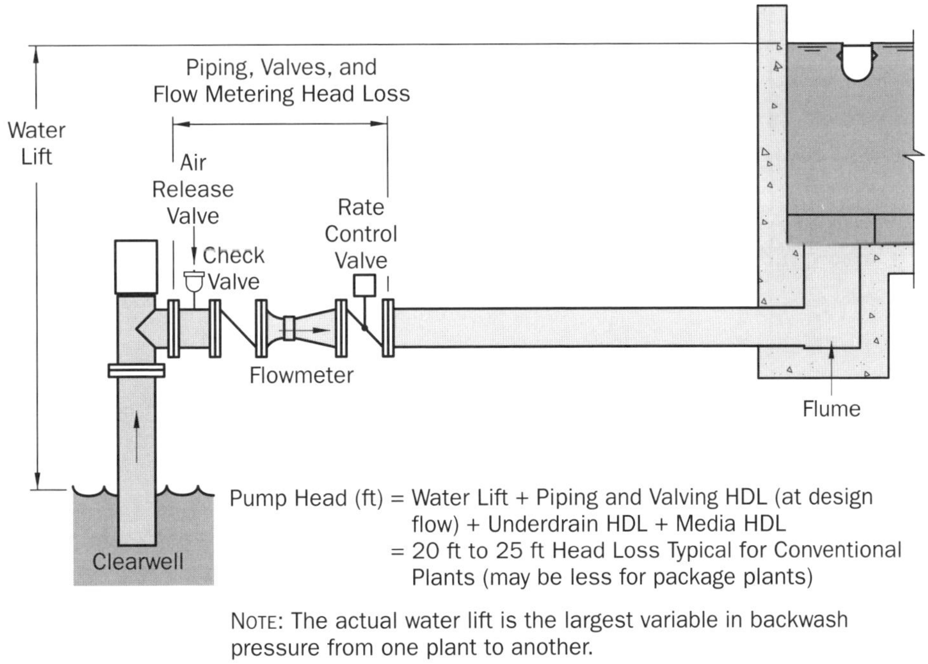

FIGURE 7-15. Pumped system head loss

for larger values because the total water lift can vary considerably. As a result, the backwash pressure should be calculated for each specific application.

High-Rate Condition

As filters become larger, the backwash flow will increase accordingly, while the pressure required may remain relatively constant. The result is a high-flow/low-head condition. For smaller filters, selecting a pump to meet those criteria may not be difficult. However, for larger filters, it can be more challenging. When the flow is thousands of gallons per minute, it may be difficult to obtain a pump that will produce that flow at only 20 ft to 25 ft of pressure. For example, Figure 7-16 depicts a typical pump curve wherein the actual pressure produced by the pump is greater than needed by the backwash system for high rate. In this case, a rate control valve is needed to reduce the pressure to the desired amount.

Reducing the pressure with a valve involves wasting a considerable amount of energy. The alternative would be to provide a slower-speed motor (1,150 rpm, for example) or a variable-speed drive. If a 1,150-rpm motor can provide the necessary pressure, it may provide a cost-effective solution. However, slower-speed motors can often be relatively expensive. In some cases, a cost analysis may be desirable to determine whether it is less expensive than a standard motor with a variable-speed drive.

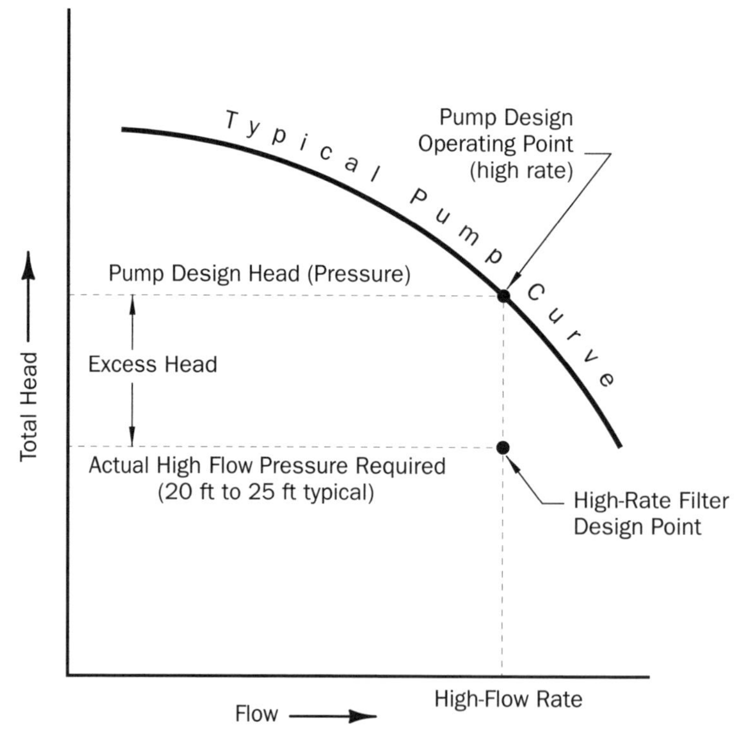

FIGURE 7-16. Backwash pump curve, high-rate design

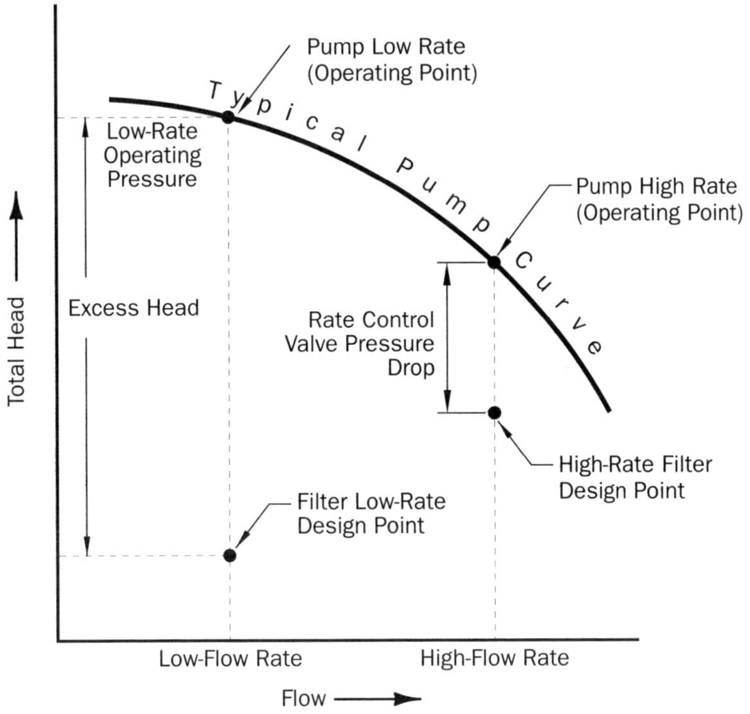

FIGURE 7-17. Backwash pump curve, low-rate flow

Low-Rate Condition

When a low-rate wash is planned in addition to the normal high rate, two different operating conditions must be satisfied, as shown in Figure 7 -17.

The lower flow rate may be only one-third to one-fourth of the high rate. As a result, the pressure required for a low-rate wash may be considerably less, especially because the low-rate wash may be started at a lower water level.

When a fixed-speed pump is used, the excess pressure at the low-rate condition will have to be reduced by a rate control valve. However, the requirement of satisfying two different flow and pressure conditions may require the use of two pumps or make a variable-speed pump more attractive. In either case, care must be taken to ensure that the operating limits of the pump are not exceeded. Refer to the manufacturer for additional information on this issue.

Backwash Pump Sizing Summary

It is desirable that the backwash pump deliver flow and pressure close to the design point. However, it is not always possible or practical. The design of the rate control/pressure-reducing valve then becomes very important. In addition, the proper sequencing of the backwash components is critical. For example, if the valves were opened first, a significant hydraulic shock condition could occur when the pump is turned on.

It is strongly recommended that new control systems should be started up using a bare filter, before the underdrain, support gravel (if used), or filter media are installed. Optimization of the controls can then be done without damage to the system.

Hydraulic shock from the backwash pump is one of the most common causes of filter failures. Even underdrain systems have been ruptured by hydraulic shock. Therefore, great care must be taken in the selection of the pump, rate control/pressure-reducing valve, and also the control system.

Selecting the Rate Control Valve

Butterfly Valves

Package plants and other smaller filters may use hydraulic diaphragm valves for on/off service as well as for rate control. However, medium to large filter plants commonly use butterfly valves for rate control. The following discussion assumes the use of butterfly valves for rate control.

In Chapter 3, the need for a large flume entry pipe was discussed. However, the bulk of the backwash delivery piping is usually smaller for economy and for ease of installation. For sizing the pipe, velocities of 5 ft/sec or less are recommended at the maximum design backwash rate. However, the backwash rate control valve may have to be smaller than the pipe in order to function properly.

It is recommended that butterfly valves used for rate control be sized to operate in the 30° to 60° open range at the design flow. Flow control is much less accurate outside that range. The manufacturer should be consulted for valve flow data.

Low- /High-Rate Control

When a low-rate and a high-rate backwash flow is used, sizing the rate control valve is more complicated. Some designers have provided two different sizes of valves for the two conditions. However, it is common for only one to be provided.

If the valve is sized for the high-rate flow, accurate throttling in the low range may be difficult. As shown in Figure 7-17, there may be significant excess pressure at the low-rate condition. Care must be taken to ensure that the rate control valve does not cavitate under those conditions. A pressure differential of less than 20 psi is recommended for a butterfly-type flow control valve. The manufacturer should be consulted for additional information on this issue. The discussion of backwash pump sizing in the previous section addresses this also. In some cases, a compromise may be desirable in which the rate control valve may be slightly large for the low-rate flow and slightly small for the high-rate flow.

Actuators

Smoothly operating valve actuators are another necessity. The author prefers electric gear-driven actuators. They operate smoothly and are consistent. In addition, they are easily adapted to computer control and can be provided with diagnostic capability.

However the valve is sized, a slow, smoothly operating actuator is mandatory. A quarter-turn actuator with a speed in the range of 60 to 120 sec is recommended. Pneumatic and hydraulically operated valves may not operate slowly and smoothly. If those actuators are provided, whatever accessories are available that can be used for speed control should be included.

Backwash From the Distribution System

When backwashing from the distribution system, a rate control/pressure-reducing valve capable of operating with a high differential pressure will be required. A differential of 80 psi or more is not uncommon in this application. The backwash system pressure requirements should be subtracted from the system pressure to determine the actual valve differential, as shown in Figure 7-18. System pressure may also fluctuate because of operating demand on the water storage reservoir(s).

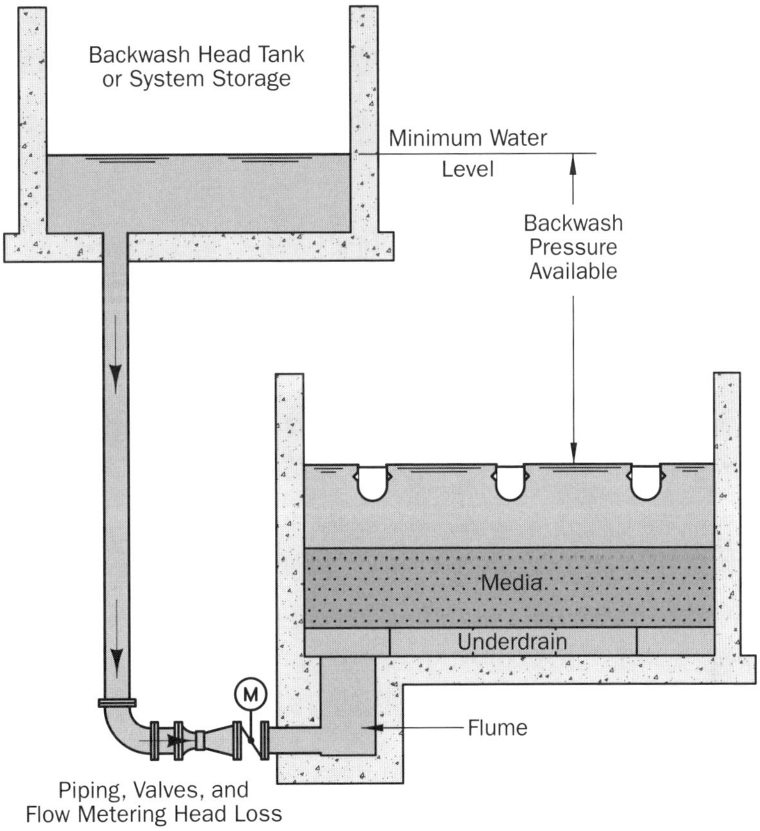

Backwash Head (ft) = Piping HDL + Underdrain HDL + Media HDL
(at highest flow rate design)

Valve Differential = Available Pressure – Backwash Head

FIGURE 7-18. Gravity system head loss

FIGURE 7-19. Blown clay tile (missing crossbar)

Several manufacturers make specialty valves for operating with a high pressure differential. A positioner is also recommended to allow the valve to be modulated to control the flow.

Hydraulic Shock

The previous sections of this chapter have stressed the importance of using the proper sequencing during the backwash cycle. Equipment failures, large rate-control valves, fast valve actuators, or control sequencing problems can cause hydraulic shock. As an example, opening the valves and then starting the backwash pump can create a shock wave that can be very disruptive to a filter. Figure 7-19 shows a result of such an event.

The sudden transition from still water in the filter to a high velocity is sufficient to upset support gravel and destroy the strongest underdrain in one event. Plastic underdrains have been torn apart and porous plates have even been torn off by hydraulic shock (Figure 7-20).

Although equipment failures can cause hydraulic shock, most such problems can be resolved by sizing rate control valves properly, providing slow valve actuators, and using the proper sequencing of events during backwash. A backwash troubleshooting guide is provided in Figure 7-21.

FIGURE 7-20. Blown plastic underdrain (with porous plate)

Problem	Probable Cause	Proposed Solution
Hydraulic shock	• Controls sequencing	• Revise controls
	• Improper valve size	• Change valve
	• Improper valve operation	• Revise controls
	• Improper valve speed	• Change actuator
	• Valve or controls failure	• Repair as required
Uncontrolled air	• Controls sequencing	• Revise controls
	• Lack of air release valves	• Add valves and revise controls
	• Improper air release valves	• Add check valves
	• Leaking air release valves	• Repair valves
	• Leaks in other components	• Repair as required
	• Valve or controls failure	• Replace equipment and make repairs as required. Add safeguards to controls if possible.

FIGURE 7-21. Backwash troubleshooting guide

Summary

In addition to cleaning the media, it is extremely important that each component of the backwash system be optimized for efficient operation. Efficient operation of the backwash system includes air release valves, rate control valves, proper sizing of the backwash pump, and the proper sequence of operation.

The most common cause of filter damage is from hydraulic shock or uncontrolled air generated by the backwash system. These problems are usually the result of improper sequencing of controls, poor equipment selection, or both. It is highly recommended that the equipment be selected and the controls designed by someone familiar with filter operations, as discussed in this handbook.

CHAPTER 8

Filter Controls

Introduction

A general discussion of controls is included in this chapter with the intent of providing the investigator with guidelines for understanding how filters operate and how to identify and correct typical problems.

Typical Controls Problems

When modulating a valve, it is rarely possible for the valve to stop at the exact point that satisfies the control requirements. Because of this, it is common procedure to establish an acceptable error called a *deadband.* Using a deadband, an acceptable valve position might be several percentage points on either side of the desired set point.

Control problems can include programming errors, such as having no deadband or one that is too small. In either case, small movements of the valve may continually overshoot or undershoot the set point or deadband. The result is excessive wear and possible failure of the valve. Other control problems include the lack of proper sequencing to eliminate air in the backwash piping and valve actuators that move too fast.

Process Decisions

There are many control modes for operating filters, including constant effluent rate, declining rate, constant level, and equal loading. The choice of operating mode is usually a process decision made when the plant is originally designed. The method of operation chosen depends on a number of issues, such as raw water quality, hydraulic loading, peaking factors, and effluent quality requirements. For example, declining rate operation might be used in a wastewater treatment plant where the flow varies, while constant effluent rate operation might be used in a potable water treatment plant that operates at a fixed speed.

Flexibility

When possible, it is recommended that more than one operating mode be designed into the plant controls. For example, when using a constant effluent rate method of operation, it may be possible to program the controls to allow declining rate operation as an option. Flexibility of operation is always desirable, and it may be beneficial to the plant staff to be able to choose more than one method of operating the plant.

Plant Control Modes

Regardless of which control mode is used, the plant controls must be properly designed and operated to ensure that the plant will function as desired. Several common control modes are discussed in the following sections.

Constant Rate

In the constant rate method of control, water is provided to multiple filters using a common influent channel, with a flooded inlet to each filter, as shown in Figure 8-1. An effluent rate control valve and an effluent flowmeter are provided for each filter, as shown in Figure 8-2. A set-point flow is established for each filter by dividing the plant influent flow by the number of filters on-line. Each filter's effluent rate control valve modulates to that set-point flow. As the filter accumulates solids and the head loss increases, the effluent rate control valve opens to maintain the flow set point.

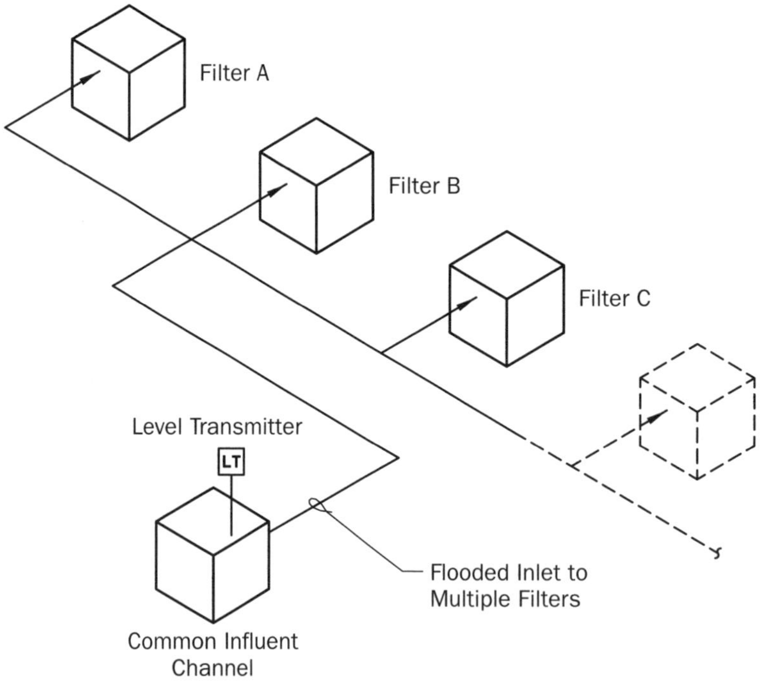

FIGURE 8-1. Constant-rate and declining-rate flow diagram

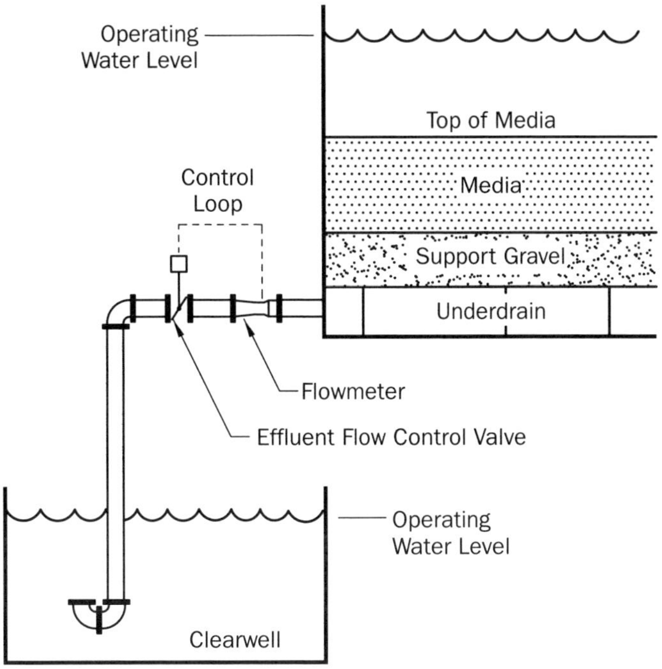

NOTES:
1. As head loss increases, the flow control valve modulates to maintain a constant flow rate.
2. When the effluent flow control valve reaches 90° open, the filter must be backwashed.

FIGURE 8-2. Constant-rate filter

Especially with a small number of filters, a level transmitter in the influent channel may be provided to override the individual filter control, as shown in Figure 8-1. If the level in the common influent channel rises or falls outside of a previously established deadband, the flow set point for each individual filter can be indexed up or down to maintain the influent channel level. However, for very large filter plants, long influent channels can have significant differences in level from one end to the other, which may complicate or prevent the level override from being used.

A major design consideration for using this method of control is how to handle the flow during backwash. The options are discussed in the following sections.

Reduce the Plant Flow

The plant flow could be reduced by the operating rate of the filter to be backwashed. When the filter is ready to go back on-line, the plant rate can be increased to normal. The chemical feed and raw water supply equipment would have to be sized and selected properly to be able to operate in variable flow conditions for this option to function properly.

Maintain a Spare Filter

If the plant flow allows, a spare filter could be kept in standby mode until another filter goes into backwash. The standby filter could then come on-line to maintain the plant flow.

Increase the Filter Rate

Using this option, the operating rate of each of the remaining filters could be increased proportionally, while a filter is taken off-line for backwash. If the filter rate is normally low, this option may be viable. The choice of which option to use during backwash for constant effluent rate control depends on many things, including the plant flow rate, the number and size of filters, the type of equipment available, and the ability to control the plant flow.

Although the initial capital costs may be slightly higher for this method, it provides the operators with a greater degree of control over the operation of the filters. A disadvantage is that additional space may be required in the pipe gallery for the flowmeter and the effluent rate control valve, especially if the flowmeter requires straight runs of piping. Although the primary emphasis of this handbook is for gravity filters, this method of control will also work for pressure filters.

Declining Rate

The declining rate mode of operation uses the same basic flow diagram as the constant-rate mode, as shown in Figure 8-1. However, there is no effluent flowmeter and no effluent rate control valve, as shown in Figure 8-3. An open/close automatic valve may be included to isolate the filter for backwash purposes.

Using this mode of operation, the effluent valve is set in a fixed position and does not modulate. As the various filters are backwashed, put back on-line, and accumulate solids over a period of time, the flow rate for each filter changes. The cleanest filter takes a higher flow rate than the average for the other filters. The flow rate for that filter declines over time as it accumulates solids. The influent channel level is maintained by adjusting the frequency of backwashing of the various filters. Declining rate operation normally works best with a larger number of filters, typically six to eight or more. When one filter goes off-line to backwash, the flow rate through the remaining filters is increased to make up the difference. The backwash frequency is determined on the basis of maintaining nearly constant level in the influent channel. If there are only a few filters, the incremental increase can be significant and could result in lower effluent quality.

If a filter plant is designed with an effluent rate control valve and a flowmeter to run in constant rate mode, it can be converted to declining rate by bypassing the modulating function of the rate control valve. As a result, it may be convenient to provide both the effluent flowmeter and a rate control valve and provide the software logic for the plant to be able to run in either mode. The operators will then be able to choose the mode of operation that best fits the current needs of their plant.

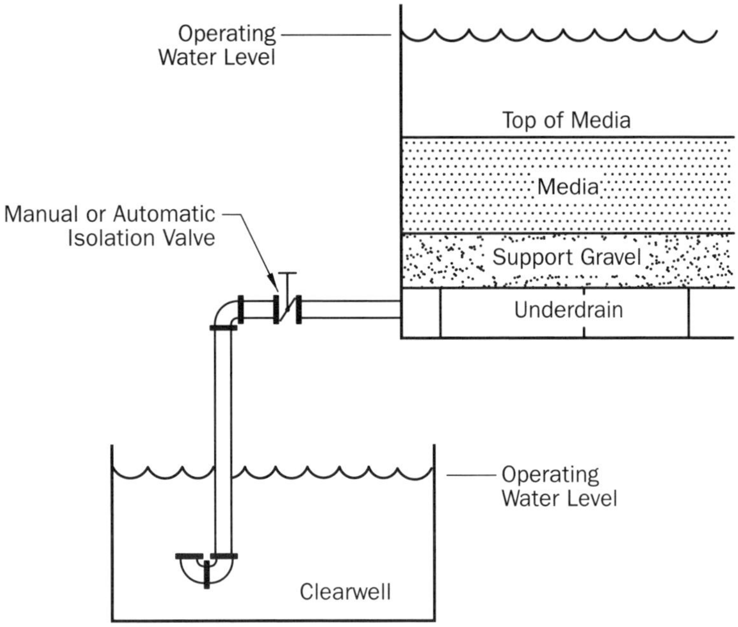

NOTES:
1. The effluent flow rate varies with the cleanest filter operating at the highest rate.
2. The effluent valve stays in position and does not modulate.
3. The operating water level is the same for all filters.
4. Filters are backwashed with a time interval with the intent of maintaining a nearly constant level.

FIGURE 8-3. Declining rate filter

Equal Loading/Constant Level

Instead of a common flooded inlet, the equal loading method splits the water equally to all filters, with individual weirs as a part of this mode of control, as shown in Figure 8-4. The same type of operation can be achieved by using a flowmeter and rate control valve on the influent to each filter. Package plants occasionally use this method of control. The level is then maintained in each filter by modulating its effluent rate control valve to produce a constant level. With the exception of the influent weir or flowmeter, the equipment is little different from the constant-rate or declining-rate mode of operation. Instead of modulating with a constant effluent rate through a flowmeter, a level probe is used to match the effluent flow with the influent.

When backwashing a filter, the options are similar to that of the constant-effluent-rate mode of control. Either the plant rate is reduced, a spare filter can be brought on-line, or the rate is allowed to increase in the remaining filters.

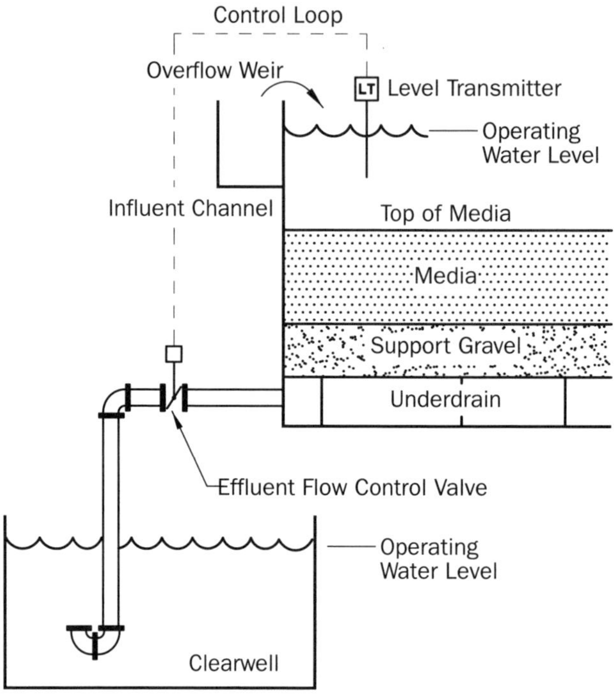

NOTES:
1. Weirs are used to split the flow equally to all filters.
2. The level probe is used to modulate the effluent rate valve to maintain a constant level in each filter separately.
3. The filter is backwashed when the level cannot be maintained.

FIGURE 8-4. Equal loading/constant-rate filter

Equal Loading/Variable Level

Another variation of an equal loading method of filter operation also uses an influent weir to split the flow equally to each filter. However, instead of maintaining constant level, the level is allowed to rise to maintain flow as the head loss increases. The result is much deeper filters, while using less equipment and having simpler controls (Figure 8-5). Equal loading filters often operate with a series of siphons that are actuated by solenoid valves.

A plant of this type is normally built with at least four filters, although six or eight are better. When the level rises to a maximum-allowed-level set point, the filter is backwashed by simply closing the influent valve and opening the backwash waste valve. The normal flow from the other filters in the plant are used to supply the water for backwash. As a result, the plant must be operating in order to backwash any given filter. The backwash driving head, and therefore the flow rate, can only be adjusted by raising or lowering the control weir in a gravity plant with this mode of operation. If only four filters are provided, the available backwash rate may be lower than desired.

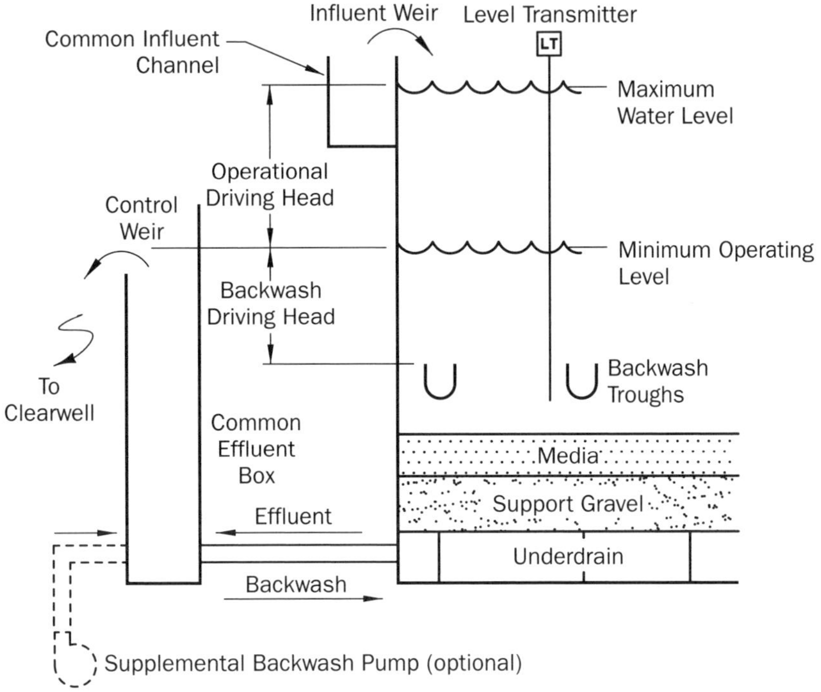

NOTES:
1. Weirs are used to split the flow equally to all filters.
2. As head loss builds up, the water level rises to maintain a constant effluent flow.
3. A minimum of four (4) filters is normally required for this control mode.
4. Deep filters are typically required.
5. Filters are backwashed when the maximum water level is reached.

FIGURE 8-5. Equal loading/variable-level filters

In some cases, a supplemental backwash pump is used to provide additional flow to the common effluent box for backwash purposes.

Besides the expense of constructing very deep filters, this method of operation has several process issues that must be considered. For example, there is significant free fall from the common influent channel down to the operating level of the filter. Free fall can break up any residual floc from the clarifier and make the filter less efficient. In addition, the free fall may result in entrained air, which can cause short filter runs, as discussed in Chapter 2.

The advantage to this type of operation is the simple controls and the fact that it is usually not possible to have uncontrolled air in the backwash water or hydraulic shock.

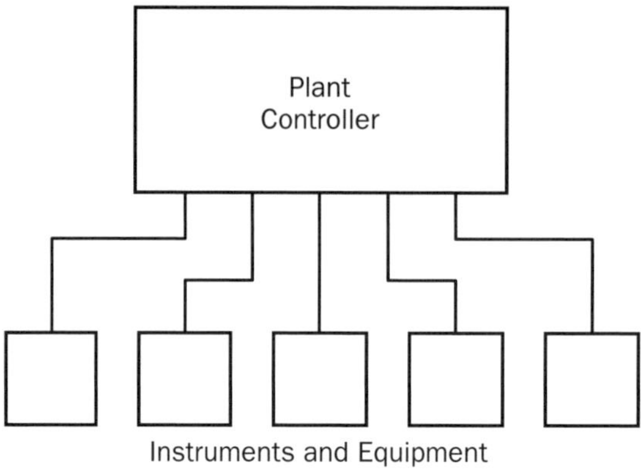

NOTE: All instruments and plant equipment generate discrete or analog signals that are hard wired directly and individually to the plant controller.

FIGURE 8-6. Hard-wired system

Communication Types

When a central control system is used, there are a number of communication and wiring methods that are used.

Hard-Wired System

In the past, the traditional method of communication between field instruments and equipment and the plant controller was to hard wire each component individually to the plant controller, as shown in Figure 8-6. Communications for individual components are relatively fast with this method, but it involves the use of a considerable amount of wiring to connect the components to the central plant controller. The controller then polls each component separately for data collection and control, which can add to the response time for any function.

Distributed Input/Output and Control

As a method of reducing the amount of wiring, remote terminal units (RTUs) or remote network panels (RNPs) can be installed at strategic locations for data collection. The local RTU/RNP may also have control capability that would speed response time for some activities of the control system. Using this method, each RTU/RNP polls the equipment connected to it, and the plant controller polls each RTU.

An alternate arrangement for connecting the RTUs in a "daisy chain" is also shown in Figure 8-7. The daisy chain method allows communication to travel in either direction, in case of a break in the wire.

Field Bus

The field bus is an improved method of communication that can have a number of different protocols, including DeviceNet, Modbus, RTU, Ethernet IP, or Foundation Field Bus, as shown in Figure 8-8. Using this method of communication, all of the

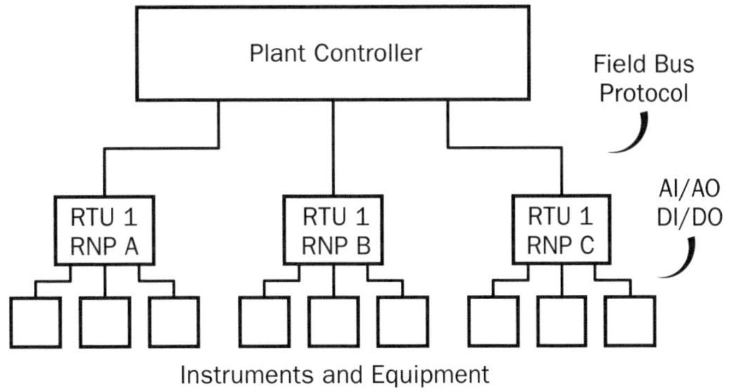

NOTES:

1. All instruments and equipment generate discrete or analog signals that are converted to a digital field bus communication protocol at remote panels.
2. Remote panels may be RTUs for distributed control systems or RNPs for systems with central control.
3. Results in less overall wiring.

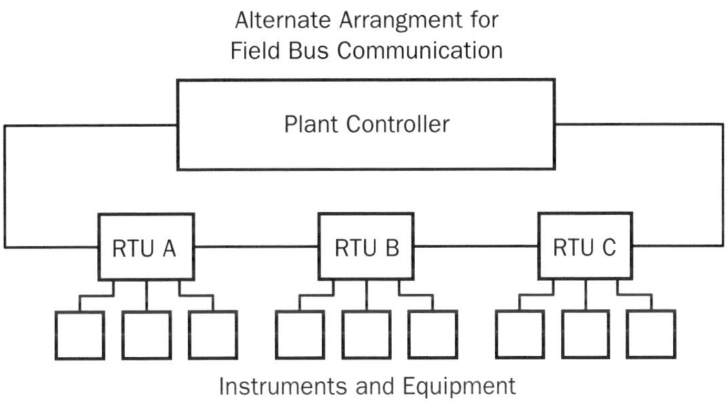

FIGURE 8-7. Distributed input/output and control

instruments and components are on-line and can communicate with the plant controller at any time. As opposed to polling each component of the system separately in sequential order, signals are only sent when the state of that component changes. The *report-by-exception method* increases the speed of communication, because every component does not have to be polled each time.

Many equipment and instrumentation devices are being provided with at least one field bus protocol for communication purposes. Because of the compatibility of equipment and its speed, this method of communication is becoming more common.

Methods of Equipment Control

Depending on the instrumentation and other equipment available, there are a number of different types of control that can be implemented, including manual operation with the plant controller, automatic operation, local control, and manual override.

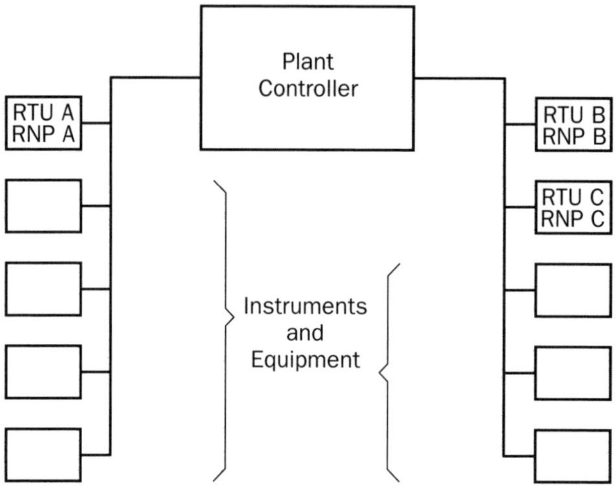

NOTES:

1. Only instruments and equipment that cannot communicate directly using the chosen field bus protocol are connected to remote panels.
2. Communications are report-by-exception with a background "heartbeat" to ensure that a connection is maintained.
3. Results in faster overall communication.

FIGURE 8-8. Field bus

Manual Operation

There are plants where the operators choose to run the plant manually, while other plants are fully automatic. The choice is usually one of preference or the availability of equipment.

The following are several variations of manual control:

- Each valve must be operated locally by hand wheel or lever, and each pump or other piece of equipment is started independently.
- Each filter has a manually operated console where each valve, pump, or other piece of equipment is activated independently.
- A fully automated control panel could be provided, which is also capable of operating each piece of equipment independently in the manual mode.

With any control system, it is occasionally necessary or desirable to operate the filters manually. In those cases where manual operation is practiced, there should be a posted procedure outlining the exact sequence of events and the time intervals for each. Every operator should backwash the filters in exactly the same way. A consistent procedure is necessary to ensure that the system operates as desired. For example, opening valves out of sequence can cause damage to the filters.

Automatic Operation

The plant controller initiates and conducts the backwash cycle when in automatic operation. A backwash cycle is typically initiated automatically by filter head loss, high turbidity in the effluent, or elapsed time. Once begun, the controller activates each component in the cycle for the established time interval according to the operating program. In this manner, the filters will be backwashed in the same manner every time, unless there is an equipment failure.

It is highly recommended that the operator observe the backwash process to ensure that the cycle is operating as desired. Even with an automatic system containing many alarm functions, it is possible for a communication failure or other equipment malfunction to occur that could result in damage to the filter. In addition, the operator will be able to gauge whether the filters are cleaned properly by observing the backwash.

When a new control system is installed, it is highly desirable to conduct trials on a filter that is empty. It always takes time to "debug" a control system, and by using an empty filter, it will be possible to optimize the operation of the control system before the rest of the plant comes on-line.

Local Control

Many pieces of equipment may be capable of local control. If the plant has the capability of fully automatic control, a piece of equipment would require a selector to choose either automatic control or local control. Once in local control, another selector could be used to activate the equipment.

The ability to have local control capability is very useful if the automatic controller is not operating. Electric valves are commonly provided with these features, and it can be provided for almost all other equipment.

Manual Override

In the event of a power or controls failure, it is very important to be able to operate valves manually to isolate the various plant processes. Many types of equipment, such as pumps, cannot be operated manually. However, valves can operate manually if they are properly equipped. It is recommended that all automatic valves be equipped with manual override to allow emergency operation.

Diagnostics

Many pieces of equipment, including automatic valves, are available with a number of diagnostic tools. These diagnostics can be made accessible to the main plant controller, especially if DeviceNet or other equivalent type of communication is being used. The availability of diagnostics is highly desirable for the operator's use and is one of several reasons why electric valves are preferred by the author. Pneumatic or hydraulic valves do not normally have these features.

Data Collection and Reporting

When electromechanical controls are used, paper chart recorders are typically provided. The recorders may be either circular or strip-chart type and are used to record flow, effluent turbidity, and sometimes reservoir level.

With the advent of computerized SCADA systems, the data from all instrumentation can be stored electronically without the need for paper chart recorders. The data recorded may include influent, effluent, and backwash flows; raw turbidity; individual filter turbidity; final effluent turbidity; influent and effluent pH and temperature; head loss from each filter; clearwell and reservoir levels; particle counters on raw and finished water; and streaming current monitors. In addition, computer systems allow various types of data to be overlaid, such as head loss and turbidity, or particle counting. Comparing these data types may help the operator significantly in the operation of the plant.

If the data is stored electronically, it is critical that a backup system is provided to record data so that it will not be lost during a power outage or equipment failure. In addition, the stored data must be in a form that is readily accessible and usable by the operator. If these issues are satisfied, electronic data recording and reporting becomes a highly desirable feature. The SCADA system can also be programmed for generating the necessary daily, weekly, and monthly reports normally required for water or wastewater filter systems.

Integration

Integration involves assembling the control equipment and developing the necessary software to control all the plant equipment and instrumentation components. The methods used for programming the equipment can have a significant impact on the response time of the control system. For example, if all of the components are polled individually and all plant operations and calculations are done sequentially, it can result in a lengthy response time for control functions. Lengthy response times can result in erratic valve operation, flooding, process problems, and excessive loss of water during backwash.

Response time is defined as the time required to activate or determine the status of equipment. The time required for equipment to perform its function, such as a valve opening or closing, is an operational issue and is not related to response time. Reporting the valve position would be a response-time issue, however.

It is highly recommended that the control system be programmed for multitasking operation with the goal of obtaining a response time not longer than one of two seconds. The problems mentioned previously could occur if there is a lengthy time delay in the control system. Therefore, it is very important that the control system have a fast response time.

In terms of response time, water plant operations do not need to be in the millisecond range like some manufacturing processes. However, a system with a response time in the range of 2 sec or less will waste less water and be more cost-effective to operate.

Miscellaneous Controls Issues

There are a number of filter problems that can occur as a result of controls issues or related equipment.

Valve Speed

If the valves open or close too quickly, it can result in a sudden change in flow with the resulting hydraulic shock or could make it difficult to modulate properly to level or flow set points. A minimum actuator speed of 60 sec for 90° of travel is recommended. Longer actuator speeds in the range of 120 sec are also common. In addition, smooth operation is desirable. Some hydraulic and pneumatic actuators are subject to opening in a series of surges, which can cause hydraulic shock, even if the overall cycle time is acceptable.

Rate Control Valves Too Large

Large rate-control valves may make accurate modulation difficult. A butterfly valve should be in the 30 to 60 percent open range in order to properly control flow. If the valve is over or under the desired range, small changes in valve position can make large changes in flow, which can in turn result in considerable hunting-and-seeking and excessive actuator wear. In addition, it may also be difficult to reach the proper set point.

Uncontrolled Air

Uncontrolled air is one of the most common problems in filter plants, as discussed in Chapter 7. It should therefore be assumed that air will enter the backwash piping system, and the control sequence should be adjusted to eliminate that air before water is sent to the filter. For example, it is common to start vertical turbine pumps and run them against a closed valve for a few seconds in order to drive all of the air out of the system at the high points using air release valves. The actual time necessary should be determined by observation. Once the system is purged of air, the backwash cycle can proceed normally. Chapter 7 also has a discussion of backwash pump operation, and Chapter 9 contains a discussion of purging air from backwash piping.

Deadbands

It is common practice in treatment plants to have valves modulate to achieve a set-point flow or level. There are several methods used to control the modulation of a valve, including hunting and seeking, and proportional control. When a valve is modulated to achieve a set point using either method, it is very difficult to hit the set point exactly. However, process considerations hardly ever require exact numbers. As a result, it is common to allow the valve to stop if it reaches a value close to the set point. For example, a deadband could be established whereby the valve could stop if a value is reached that is ±5 percent of the set point. If used, the other plant processes must be capable of operating within the range of the established deadbands.

If the modulation deadbands are set too narrow, the rate control valves may never be able to adjust themselves properly. Continual modulation of the effluent rate control valve is a typical problem that can be caused by small deadbands. Excessive modulation by the valves can result in heavier than normal wear on the actuators and possible instability of the system.

Hunting-and-Seeking

A simple method of modulation involves hunting-and-seeking or trial and error, which is used primarily with electric valves. For example, a valve may be programmed to rotate a set amount. When it stops, the flow or level achieved is compared to the dead band. If the dead band is not achieved, the valve will rotate again for the set amount of travel. If the dead band is achieved, the valve stops. If the valve overshoots the set point, then the amount of travel can be cut in half, and the valve can then back up that amount. It should then hit the deadband in one try. If the valve still has not stopped within the deadband in five or six tries, either the valve travel is too great or the deadband is too small. When the valve is reacting to a process change, the amount of valve travel required may not be great, and the modulation should be as previously described. If the valve must travel for a longer distance, a different procedure may be required.

Electric valves typically have a limit of how many times per hour they can be activated. Exceeding that amount may result in excessive wear on the actuator. Using the hunting-and-seeking method, it is typical for the valve to move between 1° and 3° per cycle. If the valve is closed (0°) and has to open to approximately 43° to begin service, it is not practical to have it advance that far in 1°- to 3°-increments. In that case, it is much more efficient to have the valve open to 40° in one cycle, for example, and then hunt-and-seek to the dead band. Programming the controls to operate in this manner is relatively easy if the valve always operates the same way.

Proportional Control

Proportional–Integral–Derivative (PID) control is a method of process control commonly used to control the operation of a valve through the feedback of a flowmeter or a level probe. Proportional control has been practiced for many years with hydraulic and pneumatic valves. Those systems attempt to control the valve to a percentage of full open and may use a deadband. With the advent of computer controls, a more accurate method of control is possible. However, adding PID control capability will typically add to the expense of a valve.

There are several adjustable factors to be considered in PID control:

- Proportional gain allows the PID control to be tuned, which changes the magnitude of the output signal.
- An integral factor allows the output to the valve to be adjusted depending on the magnitude of the difference between the actual value and the set point.
- A derivative factor adjusts the output to the valve based on the rate of change of the actual process value compared to the set point.

All of these factors are important in fine-tuning the modulation of a valve. If the PID controls are not optimized, the valve could experience excessive hunting-and-seeking, with the associated additional wear on the actuator. Therefore, these factors should be easily adjustable by the operator in case of process changes. However, a comprehensive description of each of the factors and their impact on the process should be explained in depth in the plant O&M manual, which will be discussed in more detail later in this handbook.

Low Backwash Rates

Low backwash rates can be a particular problem, especially with the equal loading type of control mode where water from the on-line filters is used to backwash another filter.

Typical filtration rates for conventional gravity plants are in the range of 4 to 5 gpm/ft^2. If the effluent flow from three filters is used to backwash a fourth, then the available backwash flow would be three times the average rate, or 12 to 15 gpm/ft^2. Depending on the media selection, backwash rates normally vary from 12 to 20 gpm/ft^2. If the required backwash rate is higher than 15 gpm/ft^2, the actual backwash rate will not be high enough to properly clean the media. In this case, at least five filters would be required so that four filters could be used to backwash the fifth.

If a filter is not kept clean, mudballs can occur, which can greatly reduce the effectiveness of the media and shorten its life. Pressure filters can also have the same problem if there are four filters or less.

Time Intervals

Sufficient time must be allowed for each stage of the backwash cycle to ensure that each filter is adequately cleaned. The operator should observe the backwash cycle to ensure that the filter is clean and that no water is wasted. All time sequences should be adjustable by the operator to allow the most efficient operation of the plant.

During the design process, additional time delays should be provided in the backwash cycle, even if unused. The ability of the operators to add delays or change the time intervals adds greatly to the flexibility and efficiency of the control system.

Summary

The type of controls and equipment provided can have a dramatic impact on the operation of filters. It is very important for these issues to be addressed by the designer and communicated to the operators so that they understand the mechanics of how the filters are supposed to function.

Even when all the issues have been addressed, problems can occur during startup of new control systems or when there has been a major modification. For that reason, it is strongly recommended that trials be conducted on controls using an empty filter. Upsets can be avoided if the underdrain, media, and support gravel (if used) are installed after the controls are validated.

Problem	Probable Cause	Proposed Solution
Excessive valve hunting-and-seeking	• Narrow deadband • Proportional control out of adjustment • Rate control valves too large	• Change deadband • Modify proportional controls • Provide a smaller valve
Uncontrolled air	• Failed valve(s) • Improper controls sequence	• Replace/rehabilitate valves • Modify controls to purge air
Long response time	• Programming issue	• Evaluate programming • Evaluate system components
Hydraulic shock	• Failed valve(s) • Improper controls sequence • Actuators too fast	• Replace/rehabilitate valves • Check controls sequence • Replace actuators

FIGURE 8-9. Controls troubleshooting guide

A troubleshooting guide is included in Figure 8-9. Most filter problems can eventually be traced back to problems with the controls, or equipment selection, or both.

CHAPTER 9

Gravity Filter Troubleshooting Procedures

Introduction

The recommended procedure for troubleshooting a gravity filter for mechanical failures or upsets is outlined in this chapter. If problems are identified, it is recommended that the investigator refer to the troubleshooting guide at the end of the individual chapters for potential causes and solutions. A discussion of troubleshooting pressure filters is contained in Chapter 10.

Whether the investigations are conducted by plant operators or a consultant, a good understanding of the information contained in this handbook is necessary. In addition, the investigator should have a considerable amount of familiarity with plant processes and equipment. However, the troubleshooting process is a very good learning tool for entry-level personnel, in the company of a more experienced investigator.

Review Categories

There are several categories of investigations that may be performed on a filter plant, including peer review, a periodic review, and troubleshooting.

Peer Review

Peer review involves a plant evaluation by an independent investigator who is not normally involved with the operation of the plant. The purpose of a peer review is to verify that the proper operating procedures are followed in the plant and to make recommendations for possible improvement.

Periodic Review

A periodic review is recommended for regular intervals. The intent of this review is to provide an early warning of potential problems, so they may be resolved before irreversible damage has been done to the filter.

Troubleshooting

If there are known problems in the plant, a full-scale troubleshooting investigation should be made to identify and resolve the specific problems, as discussed in this chapter.

Plant Data Sheet

Before troubleshooting a filter, it is often beneficial to determine the operating parameters of the plant. It is recommended that the investigator fill out the filter plant data sheet (Figure 9-1) to the extent possible before beginning troubleshooting.

A. Design Flow:	Summer	__________		
	Winter	__________		
	Average	__________		
	Peak	__________		
B. Type of Plant:	Domestic Wastewater	__________		
	Industrial Wastewater	__________		
	Potable	__________		
		Slow Sand		__________
		Package Plant		__________
		Conventional		__________
C. Raw Water Source:	Lake	__________		
	Impoundment	__________		
	Stream/River	__________		
	Well	__________		
	Domestic Wastewater	__________		
	Industrial Wastewater	__________		
D. Chemical Feed:	Coagulants:	Alum		__________
		Polyaluminum Chloride (or similar)		__________
		Ferric Chloride		__________
		Other		__________
	Polymers:	Anionic		__________
		Nonionic		__________
		Cationic		__________
	pH Buffers:	Lime		__________
		Soda Ash		__________
		Sodium Hydroxide		__________
	Recarbonation:	CO_2		__________

(Figure continued next page)

FIGURE 9-1. Filter plant data sheet

	Disinfection:	Chlorine—Gas		______
		Chlorine—Liquid		______
		Chlorine—Tablet		______
		Chlorine—Powder		______
		UV		______
		Chloramine		______
		Other		______
	Other:	Powdered Carbon		______
		Potassium Permanganate		______
		Ozone		______
E. Flocculation:	None	______		
	Hydraulic	______		
	Mechanical:	Turbine		______
		Paddle		______
	Air	______		
	Other	______		
	Detention Time at Design Flow	______	min	
F. Sedimentation/Clarification				
Conventional:	None		______	
	Circular		______	ft^2
	Rectangular		______	ft^2
	Square		______	ft^2
	Settling Tubes		______	ft^2
	Inclined Plates		______	ft^2
	Trough Spacing		______	ft
	Basin Coverage of Troughs		______	%
	Ballasted/High Rate		______	gpm/ft
	Other		______	
G. Package Plant Clarification:	Settling Tubes		______	
	Granular Media (roughing filter)		______	
	Ballasted High Rate		______	
	Other		______	

(Figure continued next page)

FIGURE 9-1. Filter plant data sheet (continued)

H. Filters:				
1. Filter Sizing:	Number of Filters			
	Dual Bay	______	ft ×	______ft
	Individual	______	ft ×	______ft
	Size (each bay)	______	ft^2	
2. Underdrain Type:	Dual Parallel Lateral:	Clay		______
		Plastic		______
		Porous Plate		______
	Folded Plate	______		
	Wedge Wire	______		
	Pipe Lateral	______		
	False Deck Type:	Strainers		______
		Wheeler		______
		Cone		______
		Porous Plate		______
	Flume Type:	False Deck Plenum		______
		Center-Fed End		______
		Center-Fed Middle		______
		End-Fed Middle		______
		End-Fed End		______
3. Valve Data:	Rate Control Valve Type:	Butterfly		______
		Diaphragm		______
		V-Port		______
		Other		______
	Actuator Types:	Electric		______
		Pneumatic		______
		Hydraulic		______
	Actuator Speed	______	sec	
	Valve Operation:	Smooth		______
		Jerky		______
		Discontinuous		______
		Other		______
	Type of Air Valves Used:	Vertical Turbine		______
		Air Release		
		Combination Air/ Vacuum Release		
	Year Valves Installed			______

(Figure continued next page)

FIGURE 9-1. Filter plant data sheet (continued)

			ES	UC	Depth, in.
I.	Media Type:	Anthracite	______	______	______
		Sand	______	______	______
		Garnet Sand	______	______	______
		Coarse Sand Monomedia	______	______	______
		Activated Carbon	______	______	______
		Greensand	______	______	______
		Other	______	______	______
J.	Support Gravel			Layers	Depth, in.
				______	______
				______	______
				______	______
				______	______
				______	______
				______	______
				______	______

K.	Height From Top of Media to Bottom of Wash Trough:		______	ft
L.	Control Type:	Constant Rate	______	
		Constant Level	______	
		Declining Rate	______	
		Equal Loading	______	
		Other	______	
M.	Filter Rate:	Normal	______	gpm/ft^2
		Maximum Demand	______	gpm/ft^2
		During Backwash	______	gpm/ft^2
N.	Backwash Rate:	High	______	gpm/ft^2
		Low	______	gpm/ft^2
O.	Backwash Supply:	Vertical Turbine	______	
		Centrifugal	______	
		System Pressure	______	
		Low Pressure Gravity	______	
	Pump Control (if used):	Full-Voltage Start	______	
		Variable Speed	______	
		Slow Start	______	
P.	Air Scour (if used):	Rate	______	cfm/ft^2
		Duration	______	min
	Type:	Over-the-Top	______	
		Air Grid	______	
		Through Underdrain	______	
		Plenum Feed	______	

(Figure continued next page)

FIGURE 9-1. Filter plant data sheet (continued)

	Sequential	______________	
	Concurrent	______________	
Q. Surface Wash (if used):	Straight Line	______________	
	S-Shaped	______________	
	Subsurface	______________	
	Rate	______________	gpm/ft^2
	Duration	______________	min
	Duration During High-Rate Wash	______________	min
	Number of Sweeps per Bay	______________	
	Diameter	______________	ft
R.	Date Plant Built	______________	
	Date Last Rehabilited	______________	

FIGURE 9-1. Filter plant data sheet (continued)

Optimize Pretreatment

Before investigating the filter for mechanical failures, the operator should attempt to optimize the pretreatment processes first to eliminate those issues from consideration. If the pretreatment process is operating efficiently, the investigator should proceed with troubleshooting the filters. However, if the investigation is a peer review, all processes including pretreatment should be investigated thoroughly.

Typical Potable Water Treatment Plant Processes

A flow diagram of a typical conventional potable water treatment plant is shown in Figure 9-2. The basic pretreatment processes include chemical feed, flash mixing, flocculation or coagulation, and sedimentation/clarification. Following filtration, the rest of the plant processes include disinfection, pH adjustment (if necessary), and the necessary instrumentation and controls to operate the plant.

Feeding chemicals into the raw water stream conditions the water and improves the efficiency of the other plant processes for the removal of organics, turbidity, harmful pathogens, and to improve taste and odor. The chemicals may include one or more polymers, a coagulant, a pH adjustment chemical, and powdered carbon. Disinfectants are also sometimes added to the raw water, although that is typically dependent on the need for disinfection contact time and the likelihood of producing unwanted site products, such as trihalomethanes.

Once the chemicals are added, a flash- or rapid-mix step is added to adequately disperse the chemicals in the water for maximum effectiveness. For the flash-mixing step to be efficient, it is recommended that at least 3 ft of head loss or equivalent energy input be provided at any flow rate. Static mixing, hydraulic jumps, free fall, and in-line mechanical mixers are used for this purpose.

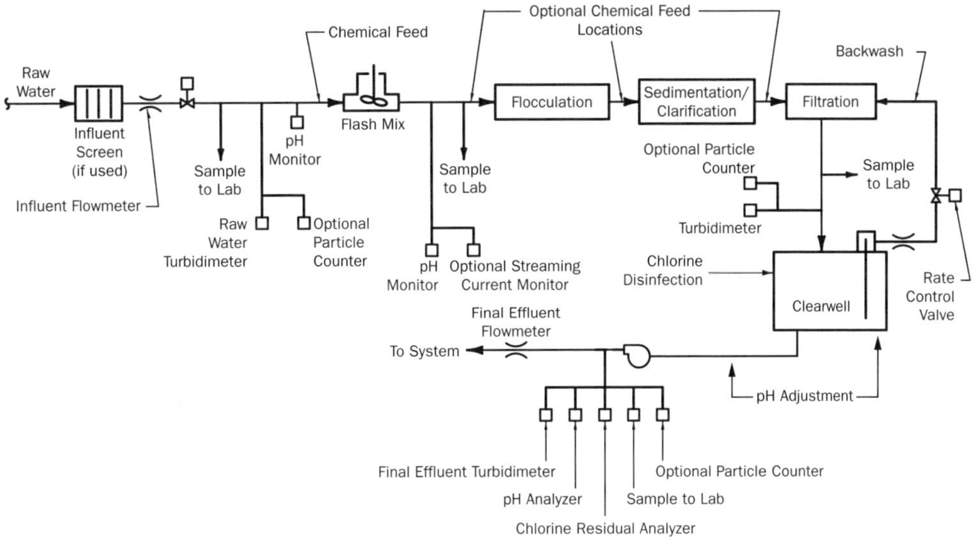

FIGURE 9-2. Typical potable water treatment plant processes

The next step is to provide slow mixing and adequate detention time for flocculation and coagulation to occur. The intent of this step is to make discrete solid particles agglomerate to a size large enough to either settle out or be filtered.

The conditioned water flows to a sedimentation basin/clarifier where the settleable solids are removed. Any residual solids are removed by the filter. It is very important that these pretreatment processes be optimized so that most of the solids are removed before filtration. As the pretreatment efficiency increases, the filter run time between backwashes also increases, reducing the cost of operation.

Wastewater Treatment Plant Processes

The plant processes for wastewater treatment will be considerably different than for a potable water plant. However, there may still be a chemical feed, a clarifier, and a filter. For the purposes of this handbook, pretreatment in wastewater treatment plants is not included separately.

Instrumentation

An individual plant may not have all of the components shown on Figure 9-2. However, it is highly recommended that the instrumentation shown should be provided. In addition, it is often useful to have the capability of feeding the various chemicals at different locations. For example, there are treatment plants that require seasonal changes in polymer for proper operation.

The recommended instrumentation includes turbidimeters on the raw water, the discharge of each filter, and the final combined effluent. pH meters are recommended before and after chemical feed and on the final effluent water. Streaming current monitors and particle counters may also be provided before and after chemical feed and on the combined plant effluent.

Package plants and some conventional plants may operate at a fixed flow rate most of the time. However, with many plants the flow rate may change from day to day or even more often. As a result, it may be desirable to provide variable-speed chemical feed pumps that can be flow-paced by the influent flowmeter. However, when the flow changes, it is highly recommended that the plant operators verify that the plant is operating as desired.

Procedure

Throughout the remainder of this chapter, there are checklists containing questions to be answered by the investigator during the evaluation of the various processes and pieces of equipment. If the answer to any of them is "yes," further investigation may be required to identify and resolve potential problems. In addition to answering the questions, the investigator should keep a complete set of notes regarding all observations. The questions and notes can be used for writing a report or memorandum of the investigation, and for comparison to past and future evaluations.

A complete set of the checklists is contained in Appendix B.

Interview Operators and Plant Management

As a group, plant operations personnel are typically very conscientious and try to do the best job possible with the equipment they have. Plant operators do not always know the design theory behind the equipment and materials used in their facility. However, they are an extremely valuable resource because they know what works and what does not at their facility. Because of the operators' knowledge, designers should spend as much time as possible in the field learning from them and observing equipment and processes in operation.

The plant operators may also be able to provide a clear definition of any problems. Depending on what problems there may be, the investigator may not have to follow all of the troubleshooting outline. The choice of a complete or partial evaluation will depend on the investigator's experience and knowledge. However, if a partial evaluation is conducted, it may be necessary to go back and perform a complete evaluation at a later date once a clear definition of the problem is available.

Define the Problem

Defining the problem consists of interviewing the plant operators and management, evaluating the pretreatment processes, and inspecting the filters. Once the problem(s) has been identified and documented, the next step is to determine the causes and make recommendations for repairs.

When beginning a troubleshooting evaluation of a plant, it is always helpful to identify the known problems first. Answers to the questions in Table 9-1 should be provided by the plant operators.

TABLE 9-1: Define the problem

	Yes	No
Are the filter runs short because of high effluent turbidity?	☐	☐
Are the filter runs short because of rapid buildup of head loss?	☐	☐
Are the filter runs short because of rapid increase of particle counts?	☐	☐
Is the effluent turbidity not as good as it should be?	☐	☐
Are there visible problems in the filter?	☐	☐
Is the air distribution not as good as it should be (if air scour is used)?	☐	☐
Are there irregularities in the media surface?	☐	☐
Is the filter media not being cleaned properly?	☐	☐

When problems are identified, the investigator should focus the evaluation toward any components or processes that could cause the problem.

Pretreatment

Water entering a slow-sand filter plant may receive little or no pretreatment. On the other hand, extensive pretreatment may be required for other types of filters. Whether the water to be filtered is for potable uses or tertiary treatment for wastewater, the pretreatment provided is site specific for that water. Pretreatment should be reviewed frequently, whether or not any specific problems have been identified.

Typical Failures

Failures caused by problems in the pretreatment processes may include improper use of the primary coagulant and/or polymers, or inadequate coagulation, flocculation, or settling. In addition, process problems can occur with fluctuating raw water conditions, such as pH, turbidity, and the presence of algae. These various modes of failure can result in excessive use of chemicals, upset media, poor treatment efficiency, or taste-and-odor problems. Resolving pretreatment problems may involve frequent jar testing and optimization of chemical feed, changing chemicals, repairing or replacing equipment, adding or replacing clarifier effluent troughs, adding settling tubes in the clarifier, or adding entirely new processes, such as flash mixing or coagulation.

Flash Mixing

Adequate flash mixing is important for efficient operation of the chemicals. Enough energy must be imparted to the water to thoroughly disperse the chemicals. Typically, it is recommended that the energy input be approximately 3 ft of water head, as a minimum.

Items to be reviewed are shown in Table 9-2

Adequate flash mixing equipment should be provided if not currently available.

TABLE 9-2: Flash mixing

	Yes	No
Has the flash mixing equipment broken down?	☐	☐
Is flash mixing equipment needed?	☐	☐
Is the energy input less than 3 ft of water head?	☐	☐

TABLE 9-3: Coagulation/flocculation

	Yes	No
Is the contact time for chemicals short?	☐	☐
Are more chemicals required than indicated by jar tests?	☐	☐
Do jar tests indicate that flocculation is needed?	☐	☐
Is the floc (if visible) entering the clarifier visually differently from that obtained in jar tests?	☐	☐

Coagulation/Flocculation

High-rate package plants may not have a flocculation stage of treatment. However, in a conventional plant, adequate detention time and mixing may be necessary for the chemicals to operate efficiently.

Items to be reviewed are shown in Table 9-3.

Some plants rely on adequate contact time for the chemicals to react. However, if the plant processes are not working as expected, mechanical flocculation may have to be added.

Chemical Feed

Package plants and conventional treatment plants both rely on the chemical feed systems being optimized in order to produce a quality effluent. In some cases, wastewater treatment plants may also use chemicals to achieve the desired effluent quality. Optimizing the chemical feed systems is commonly done by either jar testing or using a streaming current monitor (SCM), or both. An SCM measures the electrical charge of the water. The intent is to use adequate coagulant to neutralize the charge.

pH Adjustment. In some cases, it is necessary to buffer the raw water for the coagulant to function properly. Soda ash and lime are commonly used for that purpose. Once coagulation and flocculation have been achieved for most normal processes, the pH of the water should not be changed until after filtration. There are some exceptions, such as increasing pH for improving manganese removal and special cases of taste-and-odor removal. Besides enhancing sedimentation and clarification, coagulation and flocculation also condition the water for filtration. Adding a buffering chemical before the filter could result in the destruction of floc and a reduced efficiency of filtration. In addition, it may cause calcification of the media or the underdrain.

TABLE 9-4: Calibrating chemical feed pumps

	Yes	No
Are calibration cylinders required?	☐	☐
Do the pump settings change with time?	☐	☐

Final pH adjustment for delivery to the distribution system should be done in the clearwell. If there are mixing or sampling problems in the clearwell, those issues should be resolved there.

Calibrating Chemical Feed Pumps. Calibrating the pumps frequently is an important part of ensuring that the proper amount of chemicals are being fed. The heads of positive displacement diaphragm pumps, for example, can become fouled with highly viscous polymers or unmixed chemicals. They can also be fouled with contaminants. In either case, it is recommended that the chemical pumps be calibrated at least weekly and a new feed chart be prepared each time to use in setting the proper feed rates. The suction piping of each chemical feed pump should include an appropriately sized calibration cylinder for that purpose.

Items to be reviewed are shown in Table 9-4.

Streaming Current Monitor. During rainstorms or other variable raw water conditions, it is usually necessary to conduct jar tests more frequently. Under those conditions, once a day would be the minimum, while once per hour may be required to keep up with changing raw water quality. An SCM is a valuable tool for verifying that the proper amounts of coagulants are fed into the water. However, the author's preference is that an SCM only be used for a monitor and not for a controller. There are raw water conditions that can cause inaccurate results when using an SCM. In addition, the set point for an SCM may need to be changed to meet seasonal raw water quality conditions. Therefore, it is necessary to conduct frequent jar tests to verify the set point for the SCM. It is suggested that weekly recalibration be conducted. In addition, the SCM may require frequent cleaning if using recycled wastewater.

Jar Testing. Jar testing is highly recommended for optimizing chemical feed systems. During steady-state raw water conditions, no changes in the chemical feed systems may be required. However, it is recommended that jar testing be conducted at least once a week, while once per day (or more) would be preferred. Frequent jar testing ensures that the chemical feed systems are optimized to the extent possible. In addition, it will maintain the operator's skill and consistency in performing them. Appendix A discusses jar testing procedures.

Jar testing relies on visual results, which can be difficult to assess when the raw water quality is very high. During periods of high turbidity, a heavy, fluffy floc might be expected. However, when the raw water quality is very high, the optimum chemical feed may not produce visible results in a jar test. In this case, a vacuum filter test may be helpful.

TABLE 9-5: Jar testing

	Yes	No
Are jar tests being done less than weekly?	☐	☐
Do the jar tests results differ from how the plant works?	☐	☐
Is the raw water pH subject to change?	☐	☐
Is the raw water turbidity subject to rapid changes?	☐	☐
Does the jar test yield different results from the streaming current monitors (if used)?	☐	☐
Is the primary coagulant or polymer less efficient during different seasons of the year?	☐	☐

Jar Testing With a Vacuum Filter. Conduct a jar test using a range of chemical dosages that bracket the expected range. If no results are visible, process enough of the contents of each jar through a vacuum filter to produce a sample to test for turbidity. The vacuum filter provides a better approximation of the full-scale treatment process than jar testing alone. If the turbidity of the filtered sample is acceptable, the chemical dosages used in that jar can be translated to the full-scale plant. If not, additional jar tests will be required.

Items to be reviewed are shown in Table 9-5.

Historical Records. Maintaining historical records is another important function in operating a plant. For each significant change in raw water quality, it is recommended that turbidity, alkalinity, raw water pH, water temperature, and all chemical feed rates and/or chemical changes be recorded in terms of parts per million on the date that the changes were made. It is suggested that this data be filed according to raw water pH. When there is a change in the raw water quality, historical records are helpful in quickly zeroing in on the proper operating parameters.

Appendix C has a sample data sheet for historical records.

Clarification

Clarification is an important part of the pretreatment process to remove solids and reduce the load on the filters. Optimization of the clarification process is typically dependent on the original equipment design and the optimization of the chemical feed systems. The goal is for the clarified effluent to be free of solids that might contaminate the filter and to have an overflow turbidity of 1.0 or less.

Plug Flow. Most conventional clarifiers operate with the flow entering one end and exiting the other. A circular clarifier operates in essentially the same manner, wherein the water flows in through the center and radiates out to the sidewalls. If the only collection trough is on or near the endwall, the clarifier operates in a plug-flow mode.

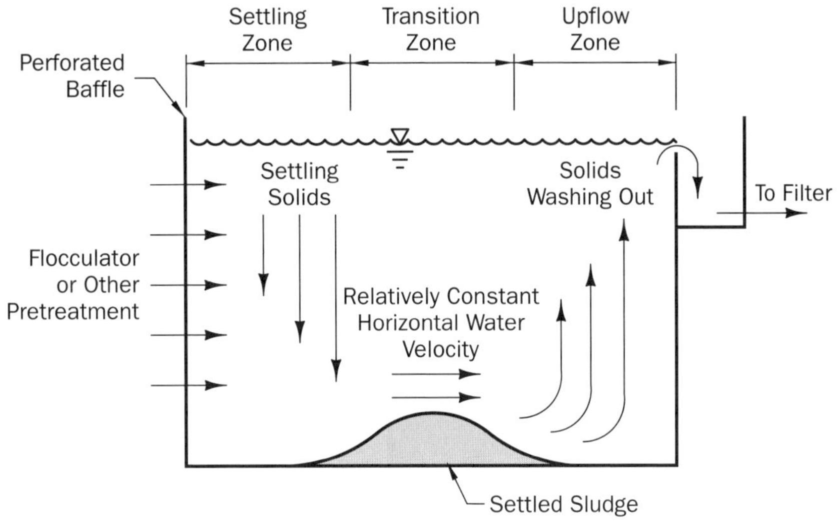

FIGURE 9-3. Plug-flow clarifier

A plug-flow clarifier, as shown in Figure 9-3, will typically have an initial settling zone where the solids are beginning to sink. Solids will usually be visible on the surface in this zone. The transition zone that follows may be visually free from heavy solids on the surface. However, in the final upflow zone, heavy solids may again be visible as they are washed up along the endwall and over into the collection trough, where they enter the filter.

The high uplift velocity of water is caused by the water having to exit the clarifier through the collection trough at the end. The solution to this problem is to install longitudinal collection troughs.

Longitudinal Collection Troughs. A significant improvement in settling can be achieved by installing longitudinal collection troughs (in the direction of flow) or a series of cross troughs, as shown in Figure 9-4. As part of this design, a downturn baffle is provided to route the incoming water to the bottom of the clarifier. Once the water is in the clarifier, the flow differs dramatically from the plug-flow design. Weirs on the troughs regulate the flow such that a small amount of water is picked up uniformly along the length of the troughs. In that way, there is very little horizontal velocity remaining at the end of the basin. The entire length of the clarifier can be used for settling, which greatly improves efficiency. A discussion on trough spacing is included in the following sections.

Settling Tubes. One of the simplest methods of improving the solids removal efficiency and capacity of a clarifier is to add settling tubes. However, in order to be effective, there must be adequate surface area, and the collection troughs should be on 6-ft centers or less and should cover the entire settling area.

Settling tubes consist of rows of tubes approximately 2 in. square, inclined at an angle of 60° from the horizontal. Every other row of tube is slanted in the opposite

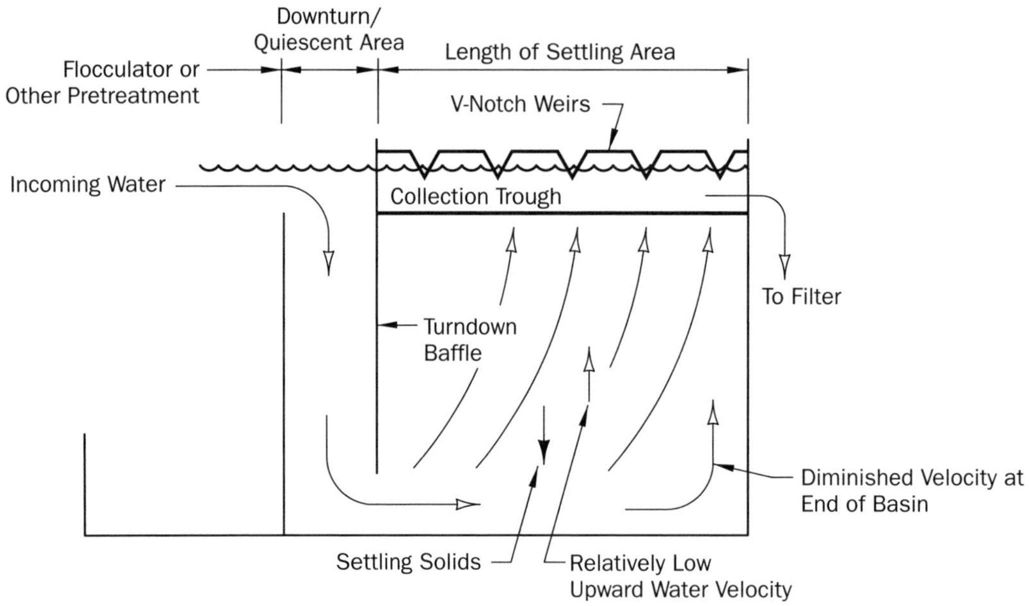

NOTE: Clarified water is gradually collected along the length of the settling area so that little, if any, horizontal velocity is left at the end of the basin to wash solids out.

FIGURE 9-4. Clarifier hydraulics

direction to mitigate short-circuiting. Tube bundles approximately 2 ft to 3 ft wide, 10 ft to 12 ft long, and 20 in. to 36 in. deep may be constructed of PVC or ABS plastic by solvent- or spot-welding channels between vertical sheets. They may also be constructed by joining corrugated sheets slanted in alternate directions. Please note that tube bundles are highly flammable and should be protected accordingly.

In a standard clarifier, a solid particle must sink to the bottom of the clarifier in opposition to the flow of water which is generally upward. A solid particle is not fully removed from the flow stream until it reaches the bottom of the clarifier. On the other hand, once water enters a bank of settling tubes, the maximum settling distance is only a few inches, as shown in Figure 9-5. When the solids settle out to the bottom of the tube, they accumulate and slide out the bottom, while the water flows continuously upward along the top of the tube, as shown in Figure 9-5. The result is a highly efficient settling process that reduces the solids loading on the filters and allows higher flow rates through the clarifier.

Trough Spacing. The settling tubes have no head loss and have little or no effect on the clarifier hydraulics other than having rows of tubes in alternating directions. The collection troughs, in large part, dictate the flow pattern through the clarifier. In addition to collecting water uniformly along the length of the clarifier, the spacing between troughs is also important. If the distance between the troughs becomes too great, then it will promote short-circuiting through the settling tubes, as shown in Figure 9-6. The maximum spacing between collection troughs in the clarifier is recommended to be 6 ft

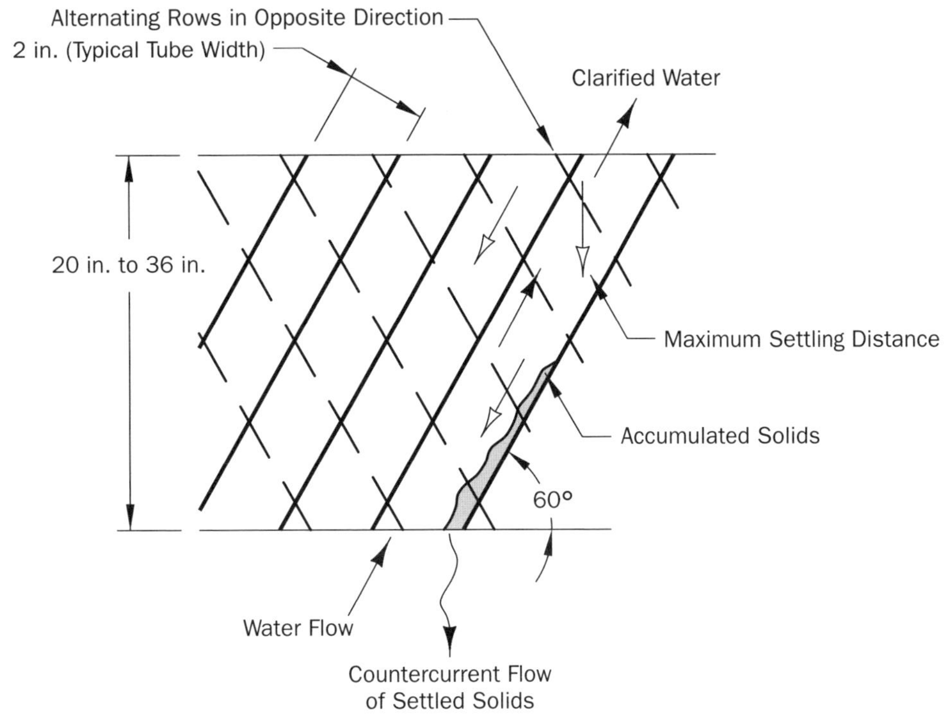

FIGURE 9-5. Settling-tube theory

or less. The maximum horizontal distance for the water to reach a trough is then 3 ft or less, also shown in Figure 9-6.

Settling-Tube Overflow Rate. The allowable overflow rate for settling tubes depends on the raw water quality and the pretreatment efficiency. Heavy, dense solids typically settle quickly during the winter. However, a light, fluffy floc typical of summer conditions may settle slowly and require a higher surface area. Overflow rates for settling tubes of up to 3 gpm/ft^2 have been used, and 2 gpm/ft^2 is common. However, for a very conservative design, an overflow rate of 1 gpm/ft^2 for the area covered by settling tubes is recommended for very light, fluffy floc. Higher rates could be allowed if tests show they will work or if there are similar successful plants on the same water.

Higher rates will often result in less efficient settling and higher solids-loading on the filters. However, providing settling tubes will usually result in better plant performance.

Settling Tube Anchorage. Because the settling tubes are made of plastic, they have nearly neutral buoyancy. As a result, the upward flow of water in the clarifier can cause the tube bundles to float. In addition, pluggage or partial pluggage of the tubes with solids could result in an upward force on the tubes.

To prevent the tubes from floating, it is customary to anchor the tubes to their supports, as shown in Figure 9-7.

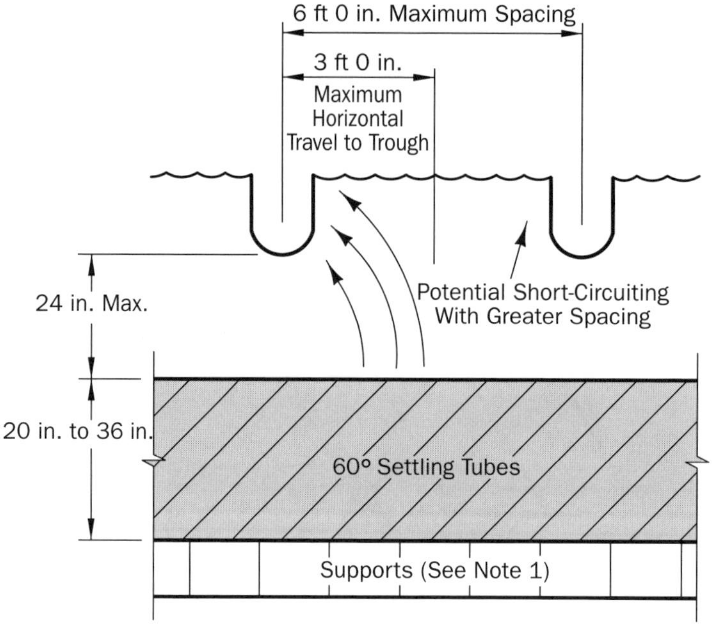

NOTES:

1. Supports to be less than 2 in. wide and designed for 10–15 lb/ft^2 surface loading or per manufacturer's recommendations.
2. Settling tubes to be not more than 2 in. wide, solvent-welded into bundles not more than 24 in. wide and 12 ft long. Adjacent rows of tubes to slope in opposite direction.

FIGURE 9-6. Settling-tube design details

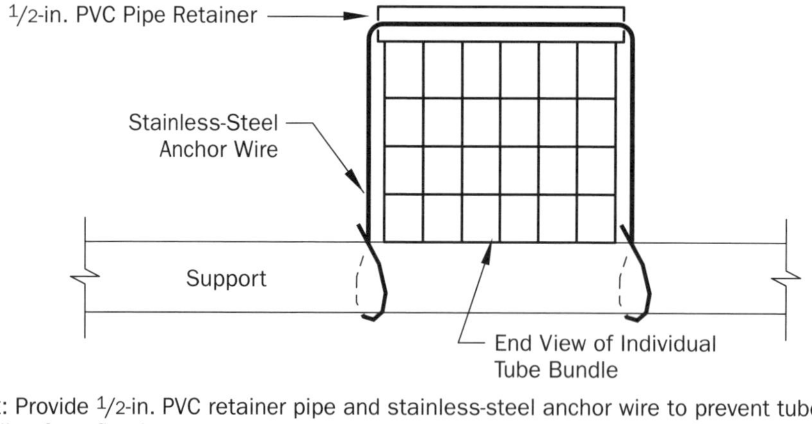

NOTE: Provide 1/2-in. PVC retainer pipe and stainless-steel anchor wire to prevent tube bundles from floating.

FIGURE 9-7. Settling-tube anchor details

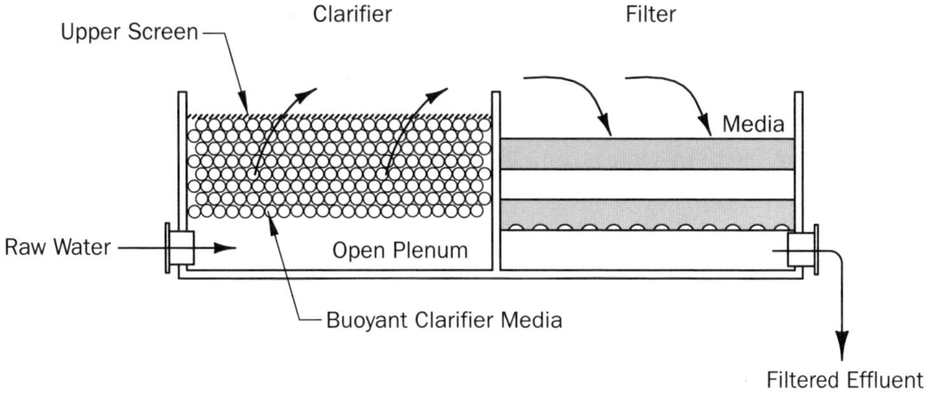

FIGURE 9-8. High-rate package plant using upflow buoyant media clarifier

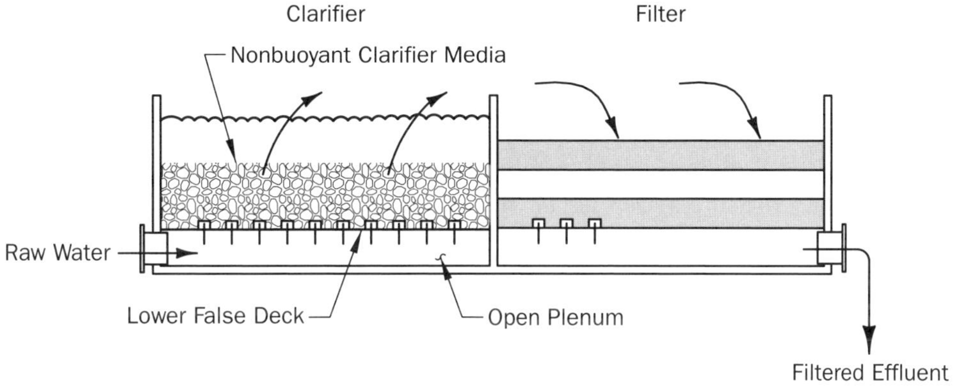

FIGURE 9-9. High-rate package plant using nonbuoyant clarifier media

Package Plants. The concept of a package plant involves the manufacturer selecting and packaging most of the equipment needed for a complete installation. The packaged equipment may include tankage (usually steel or aluminum), filter media, filter internals (underdrain, air scour, or surface wash), clarifier internals, automatic valves, chemical feed equipment, controls, and blowers and compressors. Preengineered package plants can be an economical solution for small water treatment plants.

Some high-rate package plants use a granular media clarifier, which is essentially a roughing filter. These roughing filters are advertised to operate at rates of 8 to10 gpm/ft^2 or higher. One manufacturer uses floating plastic beads for this purpose (Figure 9-8), while others use a nonbuoyant silica or garnet gravel (Figure 9-9). For the filters to function properly, an influent pressure of 4 to 5 psi may be required. The manufacturer can provide recommendations.

In the case of floating beads, a screen is provided on the upper surface to prevent them from washing over into the filter. In addition, provision should be made to prevent the loss of floating beads backward in the influent pipe when the plant is turned off. The floating beads accumulate head loss against the retaining screen, and the clarifier is flushed when a terminal value is achieved.

The heavier-than-water granular media also backwashes on head loss. However, because the heavier-than-water media is not restrained on the downstream side (upper), it may be prone to turbidity breakthrough. Whenever the raw water quality changes or the water chemistry changes, it may be necessary to reset the head loss at which the clarifier is flushed to prevent breakthrough.

Package plant manufacturers normally rate their units for low- to moderate-turbidity raw water conditions. For high turbidity applications, the manufacturer should be contacted for recommendations. In some cases, a pilot study may be warranted to determine the suitability of the equipment.

Very High Rate Clarifiers. Very high clarifier rates in the range of 40 to 60 gpm/ft^2 may also be achieved by passing the chemically treated water upward through a fluidized bed of sand. In that case, care must be taken so that excessive chemicals do not overflow to the filter. The manufacturer should be contacted for recommendations on solids loading capacity and other process issues.

Clarifier Summary. Regardless of the type of clarifier used, the intent is to remove as much suspended solids as possible before the water reaches the filters. The lower the solids loading on the filters, the longer they will run and the less backwash wastewater generated. Efficient pretreatment and clarification will reduce the overall cost and improve efficiency of operating the filters.

Items to be observed when reviewing the clarification process are shown in Table 9-6.

Observe Normal Clarifier Operation

In this section, the investigator should observe the operation of the clarifier to determine the effectiveness of the pretreatment. Problems with the pretreatment will

TABLE 9-6: Clarifier design summary

	Yes	No
Are the collection troughs more than 6 ft apart?	☐	☐
Is the overflow rate higher than accepted standards?	☐	☐
Are significant amounts of solids overflowing the clarifier?	☐	☐
Are the filter runs shorter than they should be (significantly less than 24 hr)?	☐	☐
Is the clarifier effluent turbidity greater than 1 ntu?	☐	☐
Does the clarifier operate in plug-flow mode?	☐	☐
Are there collection troughs at only one end?	☐	☐
Are there any clarifier equipment problems or failures?	☐	☐

TABLE 9-7: Observe normal clarifier operation

	Yes	No
Is there free fall between the sedimentation basin/clarifier and the filter?	☐	☐
Is there visible floc overflowing from the sedimentation basin/clarifier?	☐	☐
Is the floc over the filter the same size and volume as that in the sedimentation basin/clarifier?	☐	☐
Is the filter effluent quality poor?	☐	☐
Is there visible contamination in the sedimentation basin/clarifier discharge or on the surface of the filter (wood chips, pine needles, leaves, plastic, etc.)?	☐	☐
Is more water coming over the trough weirs in some areas than others?	☐	☐

be readily apparent by either heavy solids carrying over the weirs or poor filter effluent quality.

Items to be reviewed are shown in Table 9-7.

The investigator may make recommendations for improving the pretreatment processes, even if they are not directly related to any filter problems that may exist.

Observe a Filter Backwash

Once the pretreatment issues have been evaluated, the main task of investigating the filter can begin.

If a filter has problems, irregularities will often be apparent during backwash. For example, a good filter will clean up uniformly and have a smooth transition between events with no visual problems such as hydraulic shock or uncontrolled air.

For comparison purposes, one of the better plant filters and one of the poorer ones should be investigated. Differences between the two (if any) can be useful in helping to determine the source of problems that may exist. Fill out the checklist in Table 9-8 for each filter.

Once the evaluation is complete, the investigator may choose to extend the backwash and probe the gravel, take core samples, or proceed directly to inspecting the media surface. As an alternative, the investigator may decide to inspect the media surface first to determine if probing the gravel or core sampling is necessary.

Probe Gravel

In general, the life of the filter media is dependent on the stability of the support gravel, if used. If a direct media underdrain retention system is used, the investigator may proceed to the next section. If there are significant mounds, depressions, or heavy mudballing, probing the gravel is not necessary. It can be assumed from those observations that the gravel has been compromised. If the investigator is unsure of the condition of the filter or if monitoring changes over a period of time is desired, probing the gravel may be beneficial. When probing the gravel, a level variation of more than ±1 in.

TABLE 9-8: Observe filter backwash

	Yes	No
1. General		
Are clean areas and dirty areas visible on the water surface during backwash?	☐	☐
Are bursts of air or boils visible on the water surface?	☐	☐
Are fine bubbles widespread across the filter during high-rate wash?	☐	☐
Is the apparent backwash rate too high?	☐	☐
Too low?	☐	☐
Is there any visible loss of media?	☐	☐
Are there any apparent problems with the sequence of events or controls?	☐	☐
Was there floating scum or other contaminants that did not wash out of the filter?	☐	☐
Is the amount of media expansion too high?	☐	☐
Too low?	☐	☐
Is any uncontrolled air apparent during the high-rate backwash?	☐	☐
Did the high-rate backwash flow taper off gradually at the end of the cycle?	☐	☐
Was the backwash wastewater still dirty when the cycle was terminated?	☐	☐
Is the distance between the troughs too great?	☐	☐
Is air released at the beginning of low-rate wash (if used)?	☐	☐
Is additional air released at the beginning of high-rate wash?	☐	☐
2. Surface Wash		
Was there any problem with the rotation of the surface wash sweep?	☐	☐
Was any media loss apparent during surface wash?	☐	☐
Were there any problems with the surface wash sequence? (Refer to Chapter 7)	☐	☐
3. Air Scour (if used)		
Did the air come on "hard?"	☐	☐
Did the air come on in one corner or end of the filter and move to the other end?	☐	☐
Was there any jump of water level just before the air came on?	☐	☐
Was there any irregularity of the air pattern?	☐	☐
Did the air turn off before the low-rate water wash started?	☐	☐
Were substantial amounts of air still coming up through the media when the water began weiring over the wash trough?	☐	☐
Did the air turn off with the water level too high?	☐	☐
Did the air turn off with the water level too low?	☐	☐
Were there still high amounts of air in the water when the backwash flow ramped up to the high rate?	☐	☐
Was there any boiling or irregularity of air distribution during the concurrent wash step (if used)?	☐	☐

is typically an indication of a serious problem. Chapter 4 has a description of probing the gravel. As an alternative, a limited excavation could be conducted, as discussed in this chapter and in Chapter 5. In addition, historical records should be kept to compare the results over time.

Core Sampling

There are several different methods that have been used to extract a core sample of the media from a filter. The author's preferred method is described in Chapter 5. Core sampling using that method should be conducted near the end of a backwash cycle, when the filter media is clean. A low-rate backwash of approximately 5 gpm/ft^2 should be used during core sampling to loosen the media slightly.

The purpose of this type of core sampling is to determine the general condition of the media. The method of core sampling described in Chapter 5 does not necessarily capture all of the media. Therefore, the actual depths of each layer may be different. A limited excavation may be required to determine the actual depths of the various media layers. At the same time, the investigator should be careful not to punch a hole into the gravel. Core sampling is only recommended for dual media or multimedia, with a maximum depth between 30 in. and 36 in.

Analyzing the Sample

A visual inspection should be made of the media while it is still in the core sampler. Different characteristics can be evaluated by dumping the media out of the core sampler onto a flat surface.

In the Sampler. The core sampler should be constructed of clear 2-in. PVC piping. While the media sample is still inside the core sampler, the relative depths of the various layers should be measured. For anthracite and sand, the approximate middle of the intermixing zone between them should be measured. For garnet sand (if used), the depth of the remaining material should be measured. It should be noted that one half of the original depth of garnet will typically be intermixed with the silica sand. The media sample should also be inspected to determine if there are any visible mudballs or other contaminants below the media surface.

On a Flat Surface. After analyzing the media in the sampler, the contents should be emptied onto a clean flat surface for a physical inspection. If the media grains appear sticky or are not clean, it may be helpful to shake or stir the sample in a jar of water. By doing so, the investigator will be able to determine the relative cleanliness of the media. If there is a substantial amount of solids in the media, it is not getting clean. In this case, an appropriate investigation should be made.

After analyzing the media in and out of the core sampler, the checklist in Table 9-9 should be completed.

Physical Media Characteristics

The various layers of media are normally purchased using characteristics such as effective size and uniformity coefficient, as discussed in Chapter 5. When taking core

TABLE 9-9: Analyze media on a flat surface

	Yes	No
Are contaminants present below the surface?	☐	☐
Is the media sticky?	☐	☐
Is the media dirty?	☐	☐
Are there significant amounts of sand fines up in the anthracite?	☐	☐
Is there a significant accumulation of coarse grains of anthracite?	☐	☐
Is there significant intermixing between the anthracite and the sand?	☐	☐
Have significant amounts of anthracite, sand, or garnet sand (if used) been lost?	☐	☐
Are the grains of anthracite worn smooth?	☐	☐

samples, it is usually not necessary to retest for those items when analyzing the filter for upset conditions. The exception would be when the media bed and support gravel (if used) are intact and in good condition but there is an excessive amount of fines on the surface or there are significant process problems with removal efficiency. However, the investigator should first evaluate the efficiency of the chemical feed and pretreatment processes.

Over a long period of time, the grains of anthracite can be worn smooth. Sand is typically more durable. In either case, worn grains of media do not normally contribute to upsets in the filter. They are more likely to result in a degradation of filter efficiency.

There are cases where severe problems with dissolved air in the raw water have resulted in significant loss of the upper portion of the anthracite during backwash. If so, this will be readily apparent in the core sample. There will be a much larger coarse fraction than would normally be expected.

Inspect the Media Surface

After backwash, the water should be drained below the media surface to allow visual inspection. If walking on the surface is desired, all the water must be drained out of the filter. Depending on the results of this inspection and the client's needs, other filters may or may not be inspected.

Most of the characteristics listed in Table 9-10 represent serious problems with the filter. If a filter has any of these characteristics (Figure 9-10), an investigation should be made to resolve these problems as soon as possible.

Excavation

Limited Excavation

If probing shows that the gravel is relatively level, it may not be necessary to investigate that filter any further. On the other hand, if there are high or low spots in the gravel, a limited excavation could be made to verify the condition of the gravel surface.

TABLE 9-10: Inspect media surface

	Yes	No
Is the surface of the media irregular?	☐	☐
If so, does the surface of the media have large high and low areas?	☐	☐
Are ridges visible?	☐	☐
Are mounds visible?	☐	☐
Are sloped areas visible?	☐	☐
Are craters or "rat holes" visible?	☐	☐
Are cracks visible?	☐	☐
Are there depressions along the walls?	☐	☐
Are there residual solids or contaminants on the media surface?	☐	☐
Does the media need to be skimmed?	☐	☐
Are there clean areas and dirty areas?	☐	☐
Are mudballs present?	☐	☐
If so, are they in concentrated areas?	☐	☐
Are the mudballs large? Record all observations as shown in Chapter 5.	☐	☐
In a filter with surface wash, are the unswept areas still dirty?	☐	☐

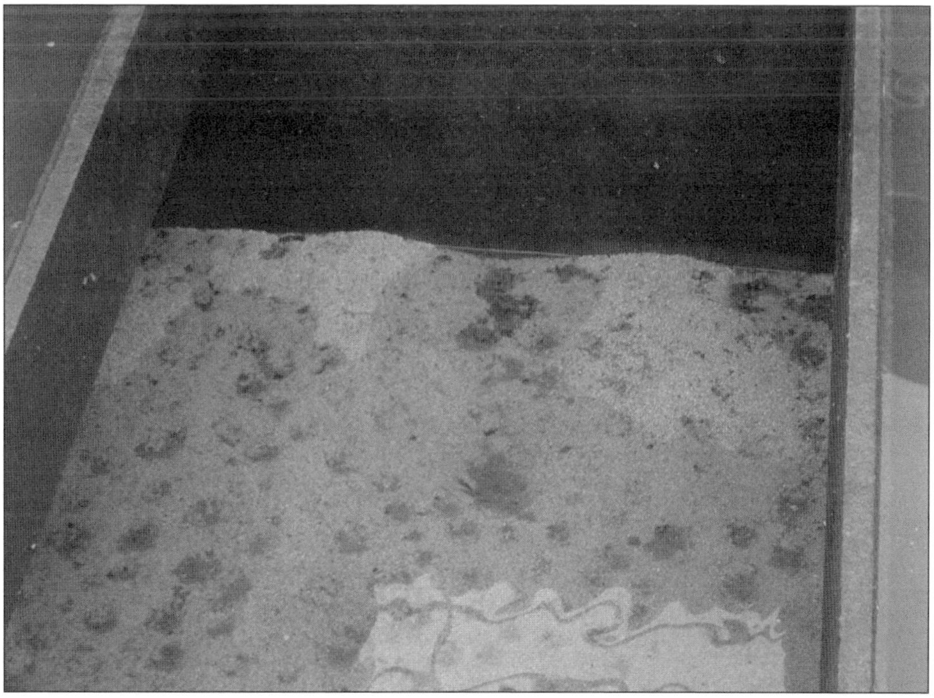

FIGURE 9-10. Upset monomedia sand

FIGURE 9-11. Excavation

The goal for a limited excavation is to dig down through the media to expose an area about 2 ft^2 on top of the gravel (Figure 9-11). More is usually not practical because of the volume of media that must be excavated. The media will dry out rapidly, and it is usually necessary to maintain a minimum 1:1 slope to prevent sloughing. As a result, a large quantity of material must be excavated to expose the surface of the gravel. Chapter 5 has more extensive descriptions of excavations.

During the limited excavation, a number of issues can be evaluated, as shown in Table 9-11.

When making a limited excavation of the media, care should be taken not to disturb the gravel when it is exposed. Hand excavation may be necessary when close to the gravel to prevent disturbances. In addition, walking boards should be used on the surface of the gravel once it is exposed. Sketches should be made and pictures taken of all irregularities that are uncovered. Samples should be taken of the media, mudballs, or any contaminants that are present. An example of plastic contamination on top of the remaining gravel is shown in Figure 9-12.

TABLE 9-11: Limited excavation

	Yes	No
Is there any irregularity in the depth of the various media layers?	☐	☐
Is the interface between the media layers irregular?	☐	☐
Are there mudballs in the media?	☐	☐
Are there mudballs on the surface of the gravel?	☐	☐
Does the surface of the gravel have grains from lower levels?	☐	☐
Is the surface of the gravel irregular?	☐	☐

FIGURE 9-12. Plastic debris and upset underdrain

Irregularities in the surface of the media, such as mounding, depressions, or grains from lower levels on top, represent severe problems.

Complete Excavation

A complete excavation can provide additional information for troubleshooting a filter. However, that can only be done when the media is to be replaced. Issues that can be identified during a complete excavation include gravel intermixing, gravel sizing problems, location and extent of any disruptions, location and extent of underdrain fouling, contamination, and general condition of the underdrain. Chapter 5 has a description of complete excavation. The investigator should ensure that all problems related to the filters are resolved before replacing the media.

Flume/Clearwell

The presence of craters or depressions on the surface of the filter generally indicates that media is being lost through the underdrain. Occasionally, high velocity backwash water coming up through a crater will blow some sand or garnet sand up into the backwash trough. However, the smaller size and higher density of sand and garnet sand generally result in their being lost into the underdrain and into the clearwell. Anthracite is usually washed out to waste during backwash because of localized high velocities in upset areas of the media. The flume and clearwell should be inspected to determine how much media is lost, and the condition of the underdrain, if possible.

Media Trap

Depending on the availability of access into the flume, it may be desirable to install a media trap on the backwash/effluent piping, as shown in Figure 9-13. Any media entering the flume in substantial quantities will eventually migrate out through the backwash/effluent piping. Because of its weight, the media will travel along the bottom of the pipe, and some will fall down into the media trap, which consists of a 1-in.- to 2-in.-pipe connection and valve with a container underneath. Once a week, the operator should open the trap valve and drain a small amount of water into the bucket or container. If media is lost, it can be detected in the container.

Flumes

If there is media in the trap or if there are craters or other irregularities in the media, the flume or plenum space under the filter should be inspected. Larger plants often have an inspection port in the flume. If one is not available, the backwash/effluent piping may have to be taken apart for access, if possible. The presence of media in more than trace amounts will probably require that the flume be cleaned when the media is replaced.

Clearwell

Inspecting the clearwell may require the plant to be shut down. Some clearwells may be divided into compartments or separate tanks. Simply looking through a sight

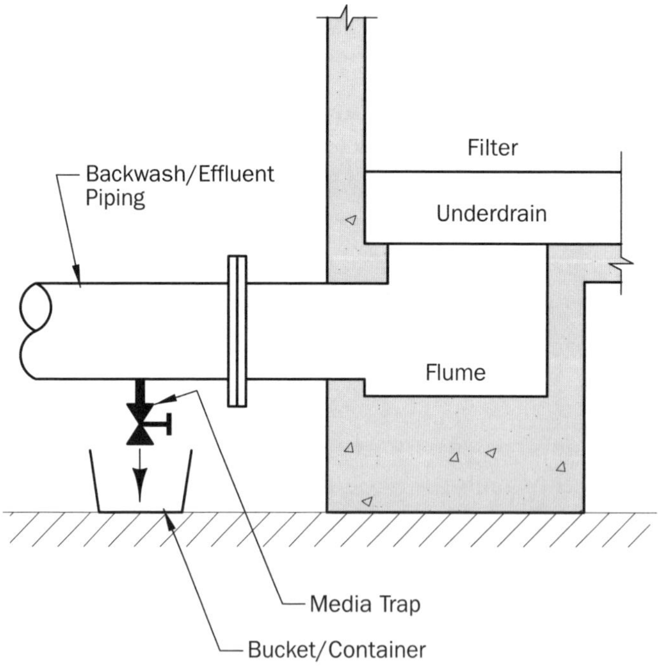

FIGURE 9-13. Media trap

glass in a fixed location may not be adequate. Any place where water is discharged to the clearwell should be inspected for lost media.

Complete Excavation

Another method of inspecting the flume is available when the media is completely excavated and removed. In the case of a dual parallel lateral underdrain, a hole can be bored in the top of the primary chamber of the underdrain over the area of the flume. A long rod can be used to probe down through the underdrain through the openings in the bottom of the flume block and down into the flume. Using this method, media can often be detected if there is any in the flume. When finished, the hole should be threaded and sealed with a pipe plug. Probing of this type may be necessary in several locations to ensure a complete evaluation of the flume.

The questions in Table 9-12 should be answered for both the flume and clearwell.

The results should be recorded and pictures taken wherever possible. The presence of large amounts of media or any support gravel (if used) in the flume or clearwell usually represents a major failure that needs immediate attention.

Structural

During the inspection of the filter, any apparent structural problems should be noted. These problems may either be above the media surface or below. Typical problems include the following:

- Sidewall erosion caused by surface wash, if used (Figure 9-14)
- Construction joint separation (Figure 9-15)
- Cracks or leaks in the walls
- Cracks in the filter floor or underdrain

Inspection Results

After performing the inspections, it should be possible to reach a conclusion as to whether or not there are filter problems and what the potential causes might be. The checklist shown in Table 9-13 should be completed.

If there are no significant problems in the filters, the data should be recorded and stored for comparison with future investigations.

TABLE 9-12: Flume and clearwell

	Yes	No
Is there media in the trap (if provided)?	☐	☐
Is media present in the manifold or flume?	☐	☐
Is there more than a trace?	☐	☐
Is there a large amount (a cubic foot or more)?	☐	☐
Is support gravel present (if used)?	☐	☐

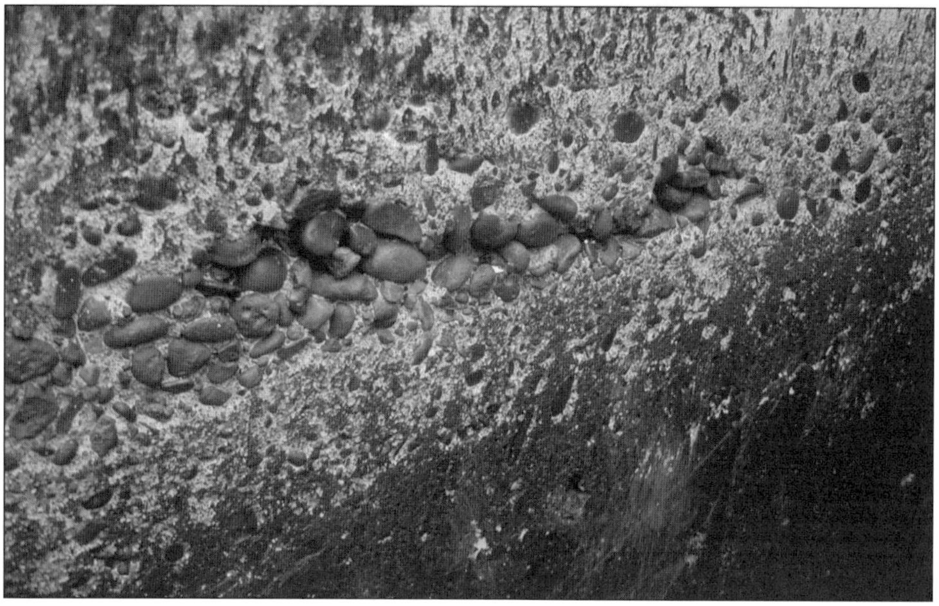

FIGURE 9-14. Sidewall erosion caused by surface wash

FIGURE 9-15. Construction joint separation

TABLE 9-13: Inspection results

	Yes	No
Are there pretreatment problems?	☐	☐
Is there significant contamination in the filter?	☐	☐
Are there any structural failures?	☐	☐
Is the gravel upset?	☐	☐
Is the filter media fouled or upset?	☐	☐
Are the failures slight?	☐	☐
Are the failures significant?	☐	☐
Does the failure appear to be caused by:		
• Hydraulic shock?	☐	☐
• Air scour?	☐	☐
• Uncontrolled air?	☐	☐
• Something else?	☐	☐

If significant problems are identified, an investigation will be required as soon as possible into the probable causes and methods of repair. Once the conditions causing the problem(s) are identified and resolved, the media and support gravel (if used) should be removed and replaced. Delaying replacement for a substantial length of time creates the risk of media leaking into the underdrain, flume, and clearwell.

Causes of Failures

The causes of filter failure originating within the filter box may include incompatible media and support gravel (if used), support gravel incompatible with air scour (if used), failure of direct media retention underdrains (if used), excessive loss of media caused by inadequate freeboard, and underdrain failure.

Although the above issues can and do cause failures, most problems are caused by equipment and controls outside the filter box, such as hydraulic shock, uncontrolled air, equipment failures, pretreatment failures, air scour failures, or control malfunctions.

The following guidelines are provided to guide the investigator in resolving filter problems.

Chemical Feed

The chemical feed systems in a conventional plant are essential to the proper operation and should be calibrated and optimized frequently. During seasons of the year when the raw water quality remains relatively constant, optimization should occur at least weekly. However, when the raw water quality is changing, some tests should take place daily or hourly.

Raw water conditions usually change seasonally. In addition, heavy rains can cause landslides in the watershed, exposing different soil types, and logging or recreational activities can also change raw water quality. Any of these events could result in a requirement for a different polymer or other chemical changes. For example, some treatment plants use one polymer in the summer and another in the winter.

Each type of chemical has different uses and limitations. For example, alum is pH sensitive and may require the addition of a buffering chemical, such as soda ash or lime, to work properly. The use of a polymer is highly recommended to help ensure that very small particles, such as *Giardia* and *Cryptosporidium*, are removed effectively. In some plants, both a coagulating and a filter-aid polymer may be required for maximum removal efficiency.

The operational staff needs a high degree of flexibility to run the plant. For example, multiple feed points are recommended for polymers, as well as the main coagulant. A spare chemical feed system is also recommended in case a different polymer or coagulant is needed in the future. It should also be plumbed to multiple feed points.

Several other conditions can occur that will cause the treatment plant to operate differently from the jar tests, including scale-up problems and water temperature.

Scale-Up Problems

Scale-up problems are difficulties in translating the jar test results to operation of the plant. Differences can occur from chemical pumps being fouled and out of calibration, inaccurate flowmeters, or math errors. In addition, plant processes naturally work differently than jar test simulations. The recommended goal is to have the plant operating within 10 percent of the jar test results. If there are scale-up problems, the operator should recalibrate the chemical feed pumps and the flowmeters to ensure they are operating accurately.

Temperature Effects

The water temperature can exert a significant influence on the time required for coagulation and flocculation. For example, chemical reactions take longer in colder water. As a result, seasonal changes in temperature usually require treatment changes. Whether the treatment plant processes are inside or outside, there can also be changes in temperature during treatment, depending on the source of the raw water, weather conditions, the type of basin construction (concrete, steel, or aluminum in some cases), and the surface area of the plant. If these changes are significant, they may have to be taken into account in the design of the plant processes.

Temperature can cause a major impact on jar testing. If a water sample is allowed to sit for any length of time in a heated space, it can increase in temperature, which results in the production of faster reactions than the plant may experience. Jar testing should take place immediately after taking a raw water sample, and a new sample should be provided for each separate jar test. For this reason, it is recommended that raw water be piped into the lab area and allowed to run continuously in order to ensure a fresh sample. Some operators have taken the jar testing equipment outside to ensure that the sample temperature remains the same as that of the water from a plant that is installed outside. The lighted base typically used with jar testing equipment can also raise the water temperature in the jars. The lights should only be turned on at the end of the test to view the results.

Operating by pH

When alum is used as the coagulant, there is a relatively easy way to determine if there is a scale-up problem between the jar tests and the plant. For any given raw water, alum typically coagulates in a relatively narrow range of pH. Therefore, when conducting jar tests, it may be beneficial to measure the pH in the jar that produces the best results. When operating the plant, measure the pH in the flocculator or clarifier after the pH has stabilized from changes in the alum feed. If the pH is not the same in the plant as in the jar tests, an adjustment to the alum feed may be required.

Plant Flow Rate

In addition to temperature, plant processes are affected by large changes in the plant flow rate. For example, higher flow rates mean shorter detention times for chemical reactions, which may not be acceptable in the winter, because chemical reactions are slower in cold water. In general, plant processes tend to work better at lower rather than higher rates. Therefore, in the winter the plant should be run for a longer period of time at a lower rate to achieve optimum results. The same is true of a filter operating with a light, delicate floc. Removal efficiencies will be better when the plant is run at lower rates. If a plant must be run faster to meet demand, a tougher floc may be necessary, which might require a higher polymer feed or even a different polymer for that purpose.

When it is necessary to change the plant flow rate, great care must be taken to ensure that the chemical pumps are adjusted appropriately. Some plants adjust chemicals automatically by pacing them off the raw water flow. However, a pump calibration curve may not be exactly linear. Therefore, it is highly recommended that the operator closely monitor plant performance when the flow is changed and make any chemical adjustments necessary.

Contamination

Contamination in the filter is typically the result of either solids passing through the clarifier or being windblown directly into the filter. If the contamination is passing through the clarifier, an attempt should be made to improve the raw water supply or the clarifier efficiency. Scum baffles or scum removal may be required in the clarifier. In the case of windblown contamination from pine needles or leaves, trees in the immediate vicinity of the plant may need to be cut down. As an option, some filters have been covered or had screens installed over them to prevent airborne contamination.

As long as contamination is present, the filters should be skimmed frequently to remove solids before they have a chance to become waterlogged and sink to the bottom. The surface of the media should be inspected for contamination monthly as a minimum and should be skimmed as required.

Hydraulic Shock

If there is a hydraulic shock event, it can be identified using the pressure gauge close to the filter, as shown in Figure 9-16. If the pressure gauge close to the backwash pump remains relatively constant while the local pressure gauge fluctuates, the problem may be in the rate control valve. If both pressure gauges fluctuate, the valves may be opening too soon or too quickly.

The important issues to be evaluated while troubleshooting hydraulic shock are shown in Table 9-14.

Hydraulic shock is a common source of problems in filters. If hydraulic shock is present or suspected, the following steps should be taken.

Backwash Sequence

Develop an actual sequence of backwash. Several coworkers may be required to assist the investigator in observing all of the equipment associated with the backwash process, including pumps, valves, blowers (if used), air release valves, and pressure gauges, as shown in Figure 9-16. The exact sequence of events and the time interval for each should be determined. The intent is to determine if the backwash sequence is according to that recommended in Chapter 7. Depending on the number of coworkers, it may take several backwash cycles to accurately determine the sequence and time interval for all of the components.

Automatic Valves

Incorrect operation of the automatic valves can easily cause hydraulic shock. It is common for backwash pumps to have considerable excess pressure because of the

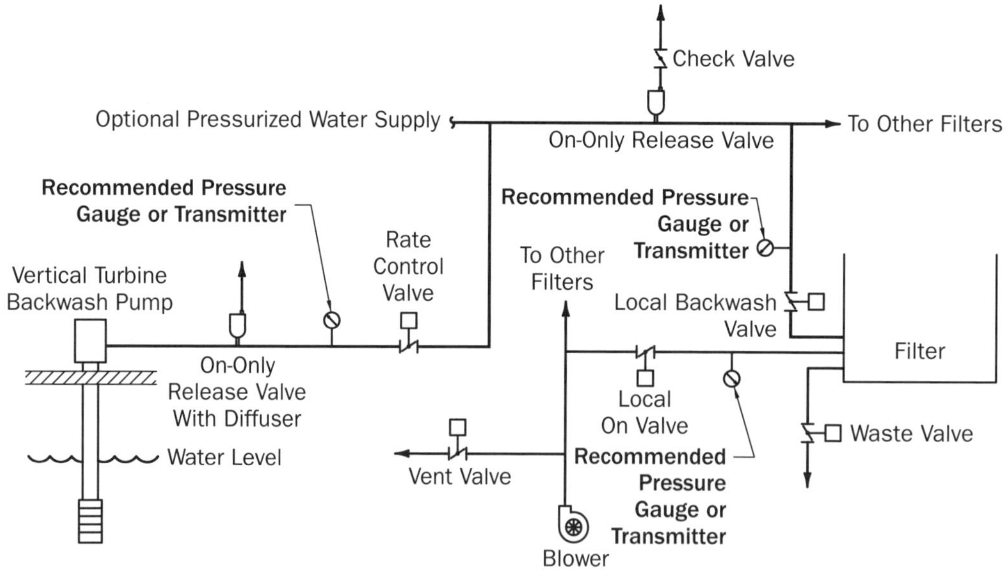

FIGURE 9-16. Backwash system

TABLE 9-14: Hydraulic shock

	Yes	No
Do pressure gauges on the backwash piping fluctuate rapidly?	☐	☐
Are any of the valves or controls malfunctioning?	☐	☐
Is the backwash pump allowed to come on with the valves open?	☐	☐
Is air still discharging from the release valves when backwash water is flowing into the filter?	☐	☐
Does the water level get low enough in the clearwell to allow a vortex to be generated next to the backwash pump?	☐	☐
Are there surges in the pressure gauges on the backwash system?	☐	☐
Do the valves travel from full open to full closed in less than 60 sec?	☐	☐
Do the rate control valves hunt-and-seek excessively?	☐	☐
Is the actuator linkage on the valves worn, broken, or loose?	☐	☐

problems that occur in matching high flow with low head conditions. If this is the case, smooth operation of the valves is even more important. Repair or replace the valves as necessary.

Automatic Controls

Automatic controls (if used) should operate the same way every time once the plant is commissioned. The exception is when instruments, valves, etc., fail. Alarms should be provided to alert the operators of any of these problems. Modify the controls sequence and time intervals as required to function as described in Chapter 8.

If the filters are backwashed manually, a recommended sequence, including time intervals, should be posted for use by the operators for consistency.

Valve Failure

Valve failures can be caused by fast-acting actuators, valve/actuator linkage failures, limit switch failure, or seat failures.

The cycle time for a valve to travel 90° can be easily timed. If the actuator opens too fast or is not smooth, it should be replaced. For example, some pneumatic actuators are prone to opening in jerky stages. Electric gear-driven actuators with a minimum cycle speed of 60 sec are recommended.

Linkage failures can also be easily detected, because the valve will not respond to the controls properly or at all. In this case, repairs will obviously be required.

Limit switch failure is normally characterized by the valve not being able to stop when it reaches either the full-open or full-closed position. It is recommended that the plant operators be trained in adjusting and setting the limit switches in case a failure occurs.

Seat failures will prevent a valve from closing completely, may result in an actuator failure because of excessive torque, or may allow the valve to leak. In any case, the seat or the entire valve will have to be replaced.

Eliminating Hydraulic Shock

Hydraulic shock during backwash can usually be eliminated by ensuring that the pumps and valves are operating in the proper sequence and that the valves operate slowly and smoothly. For example, the backwash pump should start up against closed valves to eliminate air. After that, the local isolation valve and rate control valve can open slowly to deliver water to the filter. Chapter 7 has a more in-depth discussion of sequencing the operation of the various equipment components during backwash.

Commissioning

When commissioning a plant, it is highly recommended that the controls sequence be modified as necessary to operate as previously described and in Chapter 7. While commissioning controls, water should be run through a filter that has not had filter media and support gravel (if used) installed. Once the controls are functioning properly, the media can be installed and the plant put into operation. Following this procedure will allow any necessary modifications to be made for the elimination of hydraulic shock, without damaging the support gravel (if used) and media.

Uncontrolled Air

Uncontrolled air occurs during the backwash cycle and is easily identified as large concentrated eruptions of air and water. Uncontrolled air can be very disruptive to support gravel and should be addressed during the design of the plant. However, resolving uncontrolled air problems in operating filters is relatively straightforward once the problem has been identified. Uncontrolled air can also be caused by improper sequencing during manual operation of the plant. If manual operation is contemplated, a written procedure should be posted for the operators to use.

Common sources of uncontrolled air are from vertical turbine pumps, leaking air-release valves, improper air-release valves, piping or fitting leaks, vortex in gravity tanks (if used), and controls failures. In the design of a filter plant, it is prudent to assume that air will eventually enter the backwash system and cause problems if not eliminated. For example, vertical turbine pumps are commonly used in backwash systems and are one of the worst sources of uncontrolled air. However, if the plant is designed correctly, uncontrolled air from any source can usually be eliminated by proper sequencing of the controls and use of air release valves.

Air Release Valves

Over a period of time, air release valves can begin to leak water. If they do, they may also leak air inward. In that case, repairs will be required.

During backwash, an improperly operating air-release valve can be easily detected by holding your hand over the discharge connection. If a suction or inflow of air is detected, the valve is not working properly or is not designed properly. If the seats are worn, an air release valve may also leak water during normal operation.

A rubber-seated check valve should be installed on the discharge of each air release valve to prevent air from reentering the piping during low pressure conditions. In addition, seats and floats should be replaced as required or replaced completely with the proper valve. Chapter 7 has a more complete discussion on air release valves.

Pump Operation

When a plant uses the equipment shown in Figure 9-16, the following pump startup sequence is recommended to eliminate uncontrolled air:

1. The rate control valve limit switches should be set so that the valve is slightly open (approximately 5 percent) when in the closed position.
2. The backwash pump should be turned on and run against the local backwash valve in the closed position. Having the rate control valve partly open will allow the pump to pressurize all the piping.
3. The pump should run against the closed valve until all of the air release valves quit discharging air. The exact time required should be determined in the field with an additional contingency and programmed into the controls.
4. Once the air is eliminated, the local backwash valve should be opened.
5. The rate control valve can now safely modulate to the design set point. The valve should have a slow, smooth-moving actuator, as discussed previously.

Chapter 7 has a more in-depth discussion of backwash sequencing.

Vortex

If the backwash supply comes from a gravity tank and it is generating a vortex, eliminating uncontrolled air is more complex. The options are to install a vortex baffle, as shown in Chapter 7, raise the operating water level of the tank (if possible), or provide additional storage capacity.

Pressurization

To mitigate the uncontrolled-air problem, some treatment plants connect an optional pressurized water supply to the backwash piping, as shown on Figure 9-16. The intent is to keep the backwash piping pressurized at all times to eliminate air from entering the system. Pressurizing the backwash line is an option, but the method of control previously described will resolve the problem in most plants. If uncontrolled air is still an issue, there may be other problems, such as open channel flow in the backwash piping or leaking air-release valves.

The items in Table 9-15 should be reviewed if uncontrolled air is present.

Surface Wash (If Used)

The surface of the media should be relatively clean after backwash. The exception is unswept areas, as shown in Figure 9-17. The backwash sequence using surface wash

TABLE 9-15: Uncontrolled air

	Yes	No
Is uncontrolled air visible during backwash?	☐	☐
Are the air release valves leaking water?	☐	☐
Do the air release valves allow air in?	☐	☐
Do any of the pipe fittings or appurtenances leak?	☐	☐
Are additional air release valves needed?	☐	☐
Do the controls need to be modified to eliminate air?	☐	☐
If a gravity backwash is used, is a vortex baffle needed?	☐	☐

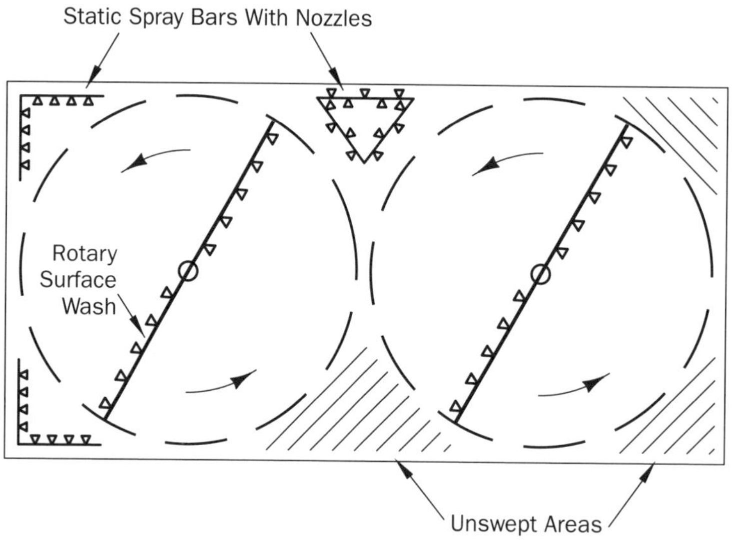

FIGURE 9-17. Surface wash coverage

should be in accordance with Chapter 7. For example, if the surface wash runs too long during the high-rate backwash, media can be washed out and lost in the backwash wastewater. The surface wash should operate for 1 to 2 min in advance of the backwash and not more than 1 min into the high-rate backwash cycle. No intermediate low-rate backwash is necessary.

For maximum effectiveness, the surface wash arm should be flat and level, with not more than 2 in. from the center line of the sweep arm to the media surface. As a practical rule of thumb, the surface wash nozzles should drag the media surface when the arm rotates.

TABLE 9-16: Surface wash

	Yes	No
Are some of the nozzles not working?	☐	☐
Have any of the caps come off the nozzles?	☐	☐
Are there large unswept areas where the media is not being cleaned?	☐	☐
Has it been longer than a year since the bearings were checked?	☐	☐
Are the sweep arms out of level?	☐	☐
Do the sweep arms have trouble rotating?	☐	☐
Is the surface of the media more than 2 in. below the center line of the sweep arms?	☐	☐

If the surface wash sweeps do not appear to be operating properly, the main bearings and individual nozzles may need to be cleaned or replaced. Chapter 10 has a discussion of maintenance of surface wash arms.

Either air scour or surface wash sweeps must be used to clean the media. Otherwise, the life of the media will be dramatically shortened. Surface wash sweeps work reasonably well over the area they cover. However, rotary sweeps do not normally reach into the corners of the filters. On larger filters, there is another dead spot on the sidewall in between the sweeps, as shown in Figure 9-17. Static spray bars have been successfully added on some large filters to help improve the surface cleaning efficiency of rotary washers, as also shown in Figure 9-17. Static spray bars should be piped to the same water supply as the rotary surface washers. If the flow of water available for surface wash is limited, the nozzle orifice size may have to be modified to allow a greater number of nozzles on the rotating wash arm as well as the static spray bars.

The items in Table 9-16 should be reviewed if surface wash is used in the filter.

Air Scour

A number of different types of failures can occur during air scour backwash. The following discussion assumes the use of dual parallel lateral underdrain with the air header and J-pipes underneath in the flume. A discussion of problems common to other underdrain types is contained later in this chapter.

J-Pipe Test

One J-pipe provides the air for an entire lateral. They must be designed to operate under pressure and flow conditions specific to each plant, as discussed in Chapter 6. During construction, several tests should be made with the J-pipes and the air scour system as noted in the following section.

Initial J-Pipe Test. The J-pipes (if used) should be tested during construction before the underdrain is installed. The filter should be filled with water to a depth between 6 in. and 12 in. above the discharge of the J-pipes, and the blower started. The high volume of air coming through each J-pipe will throw a plume of water into the air. The plume from each of the J-pipes should be roughly equivalent and should remain relatively constant. If there are any significant differences or fluctuations, the test should be terminated and the manufacturer contacted to make the necessary modifications. If there are differences or fluctuations and the blower is determined to be discharging the proper amount of air during this test, the air header and J-pipes may have to be replaced. Incidently, crimping the ends or welding an orifice over the end of the J-pipes probably will not solve any distribution problems.

When the underdrain is installed, the J-pipe must be well up in the primary chamber, as shown in Figure 6-6. A 1-in. bed of base grout is normally recommended to bed the underdrain. The base grout is also used to level the floor. Depending on how far out of level the floor is, more than 1 in. of base grout may be required.

After the floor is leveled and the J-pipes are installed, the height of the J-pipes above the floor should be measured. If the height is within the tolerance recommended by the manufacturer, installation of the underdrain may proceed. If the height is too low, the J-pipes may have to be modified or replaced. Otherwise, the air distribution in the lateral may be affected.

Final J-Pipe Test. Once the underdrain is installed, a similar test should take place with 6 in. to 12 in. of water. If this test demonstrates unequal air distribution between the various laterals, it is an indication that the J-pipes may be too large in diameter. If this is the case, the best alternative may be to increase the speed of the blower and deliver more air to the various J-pipes. The additional flow of air may be adequate to generate enough head loss to produce equal distribution. Increasing the blower speed may involve changing drive pulleys and perhaps the motor. If the blower has enough capacity, it may be less expensive to make these modifications than to tear out the underdrain and replace the air header and J-pipes. However, any changes should be coordinated with the manufacturer.

During installation, some manufacturers and contractors cover the ends of the J-pipes with tape to prevent contamination. The tape should be removed before installing the underdrain. If it is not, that lateral will have no air at all. In this case, a hole may have to be cut in the top of the underdrain to remove the tape. The hole in the underdrain can usually be repaired by hot-air welding.

The checklist in Table 9-17 should be prepared during startup and reevaluated periodically.

Purging the Air Piping

A complete test of the air scour system should be made once the underdrain is in place but before the media is installed. The air should come on smooth and even, all at once across the filter. If the air comes on hard in one end and spreads out slowly across the filter, the controls used to modulate the flow may need to be adjusted. Air scour coming on hard can be detrimental to support gravel (if used).

The intent is for the air to come on in two stages. The first is to modulate the rate control valve to deliver a small amount of air into the piping to push the water out.

TABLE 9-17: J-pipe tests

	Yes	No
Is the initial J-pipe test uneven?	☐	☐
Is the final J-pipe test uneven?	☐	☐
Does the air come on hard?	☐	☐
Does the air pattern change over time?	☐	☐
Does the pressure relief valve activate and reset during air scour?	☐	☐

The rate control function for a positive displacement blower is provided by a vent valve, as shown on Figure 6-6. A centrifugal blower uses an inlet rate control valve, as shown on Figure 6-15. After a small amount of air appears in the filter, the rate control valve can be modulated to deliver the design rate of air to the filter. If the rate control valve cannot be adjusted accurately enough to achieve that goal, it may be too large, or its actuator may be too fast. As a general rule of thumb, the rate control valve should be one or two pipe sizes smaller than the air distribution piping. If the rate control valve is too large, it may have to be replaced. The manufacturer should be contacted for recommendations. Chapter 6 has a more detailed discussion of air rate control valves.

Blower Type

Positive displacement blowers are recommended for air scour in filters, because the discharge pressure is more nearly the same at any of the operating conditions. A centrifugal blower would be acceptable, if it can function properly in the three phases of air scour, including purging, full air scour at the low water level, and the highest pressure which occurs at the final high water level. Chapter 6 has a more detailed discussion of blower types. Problems with blowers are usually associated with the size of the rate control valve or the speed of its actuator.

Blower Pressure

The blower must be designed to produce the required air rate at the maximum back pressure that will normally occur at the bottom of the wash trough. When using coarse monomedia, the design point may be at the lip of the wash trough. Lack of proper pressure would result in poor air distribution at the end of the concurrent cycle.

Pressure Relief

The pressure relief valve is designed to protect the air piping system in case of any malfunction that would allow the blower to operate against closed valves. It is recommended that the pressure relief valve be set higher than the normal operating air pressure as recommended by the manufacturer. If the relief valve repeatedly goes on and off during normal operation, it is set too low and should be recalibrated or replaced. If this occurs, the air system should be evaluated to determine if there are any irregularities causing additional head loss and to make sure the system components are properly designed and operated.

Height of Piping

The air distribution piping in the pipe gallery should be as high as possible to prevent the intrusion of water. The individual air valve for each filter can be at any point on the drop leg from the distribution piping to the filter inlet.

If the air scour starts too hard or fast, one of the reasons may be that the water purge stage (if provided) may not be long enough for the amount of piping.

The elevation of the distribution piping and the amount of water that enters the piping will dictate the length of time required for the purge cycle. Controls should have the capability to extend that time duration as required.

Underdrain Failures

Most underdrain failures are caused by controls or equipment failures outside of the filter. Those failures caused by the underdrain itself can be identified by type, as noted in the following sections.

False Deck Underdrains

Various false deck underdrain types are discussed in Chapter 6. Not all of them use nozzles or strainers, but the following comments are relevant for most other false deck underdrains.

Nozzles/Strainers (if used). Nozzles/strainers are sometimes subject to fouling by the media or excess chemicals. In addition, nozzles sometimes break or are cracked during installation.

False Deck Structure. False decks are composed of metal in package plants, and either cast-in-place concrete or pre-cast panels in conventional plants. Metal false decks can experience structural failure from flexure caused by repeated backwash cycles and lack of proper support. Other false deck types can fail because of problems with the grout or excess head loss due to pluggage of the nozzle/strainer.

Grout. One of the more common failures of concrete false decks involves the grout between the sections or individual underdrain components. For example, carborundum porous plate underdrains are grouted together to form a continuous false deck, and therefore have a very high linear footage of grout.

Excess Head Loss. Any of the false deck underdrain types can be ruptured because of excessive backwash head loss, plugging, an accumulation of media in the plenum space, grout failures, or a buildup of other debris. As a result, it is common for false deck underdrains to have a pressure relief pipe connecting to the plenum space. Water passing through the pressure relief pipe is an indication of a potential failure and warns the operators to take corrective action before a catastrophic failure occurs.

Dual Parallel Lateral Underdrains

Because they are made of HDPE, the dual parallel lateral underdrains are relatively resilient. However, they are still subject to problems with grout, seal angles, installation temperature, calcification, and hydraulic shock or uncontrolled air. These underdrains are described in more detail in Chapter 6.

Grout Failures. One of the more common problems with the dual parallel lateral underdrain is failure of the grout. Both the plastic and the clay tile versions of the dual parallel lateral underdrain are laid in rows approximately 1¼ in. to 1½ in. apart. Grout is poured into these rows to take up the space and lock the underdrain in position. A retainer strip is laid over the flume for the edge of the blocks to sit on, with the intent of preventing grout from falling down into the flume, as shown in Figure 9-18. If the retainer strip fails or is not installed, grout could fall into the flume and affect the filter hydraulics. A worse condition would occur if a hole or gap is left in the grout that connects to the flume, as also shown in Figure 9-18. In this case, backwash water coming through that hole or gap could upset the gravel (if used) and could result in loss of the media downward. If a substantial amount of the media is lost into the flume, it could be backwashed up into the other laterals, causing widespread pluggage and eventual filter failure. Such a gap could also be created if the grout were not mixed properly and a weak spot occurred over the flume.

If a filter has been performing poorly, an inspection of the flume is recommended, if possible. The intent is to determine if there is a grout problem visible or whether media has been lost.

Seal Angle. With either the clay tile or plastic version of the dual parallel lateral underdrain, a seal angle is usually installed on top of the underdrain and against the wall, as shown in Figure 9-19. The wall behind the angle is caulked to provide a watertight seal. If the concrete wall is not straight, there could be a considerable gap behind sections of the seal angle, as shown in Figure 9-20. In that case, the caulk might not hold, and a hole could be blown through it during backwash, which would disrupt the support gravel (if used) and result in the loss of media and failure of the filter. Here again, a considerable quality control effort is required during construction of the wall to make it as straight as possible and also to ensure that the seal angle is properly installed. If there is a gap behind the angle that is too large for caulk, an epoxy grout may be required.

Installation Temperature. The temperature during installation of the underdrain can have a negative impact on the operation of the plant. When installing the clay tile block, the biggest issue is the consistency of the grout and the speed at which it cures at a given temperature.

When installing the plastic dual parallel lateral underdrain, it should be noted that an individual 3-ft-long block may expand or contract ¼ in. or more. If the underdrain is installed on a very hot day, it will be expanded to its maximum length. If it is allowed to cool or freeze, the underdrain lateral could contract and rupture the joint between the blocks. Similar problems could occur due to installation on a cold day. If

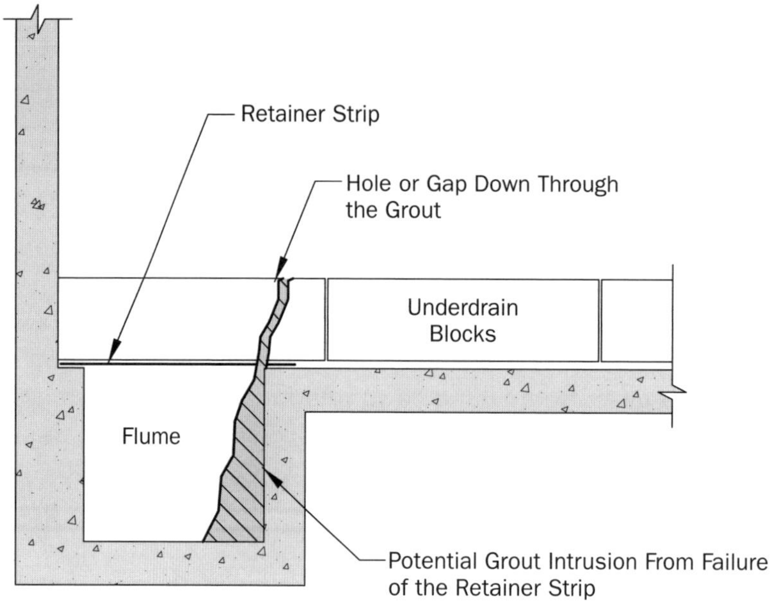

FIGURE 9-18. Grout intrusion to flume

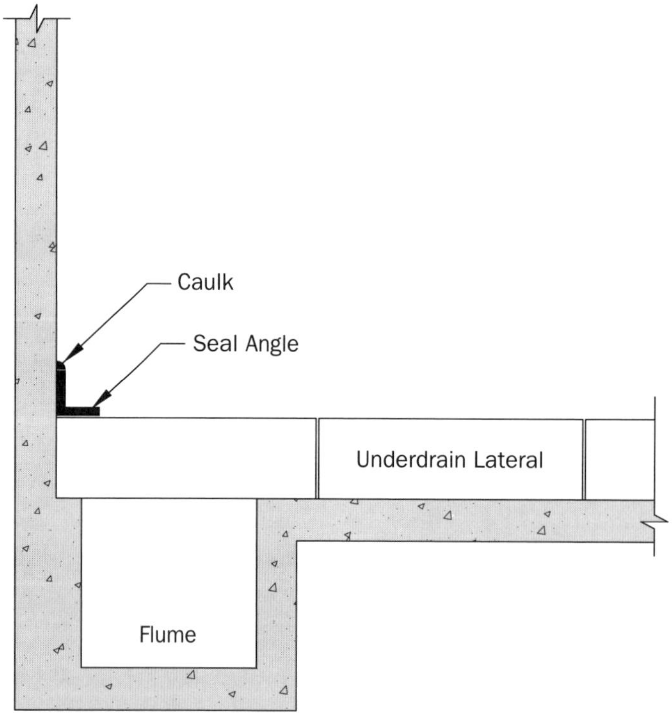

FIGURE 9-19. Seal angle

FIGURE 9-20. Seal angle failure (construction)

the underdrain is to be installed on a very hot or a very cold day, the manufacturer should be consulted for the proper installation techniques.

Calcification

Both the clay tile underdrain and the porous plate cap are prone to calcification. Calcification of the plastic underdrain block is usually less of an issue. The top deck orifice holes may be cleaned by drilling them to the proper size, as shown in Figure 9-21. However, once the clay tile block develops extensive interior calcification, it may be necessary to completely replace the underdrain, whereas a porous plate cap can be removed and a new one installed if necessary.

Direct Media Retention Underdrain

Pluggage of direct media retention underdrains can occur from fines in the media, inadequately treated raw water, calcification, contamination in the clearwell or flume backwashing into the underdrain, or chemical feed problems. To mitigate these issues, the following steps are recommended:

- Ensure that the direct retention underdrain is specifically designed for the media to be used. For example, some direct retention underdrains may not be compatible with the garnet sand used in multimedia. In addition, be sure that the media always stays wet. Support gravel and direct media retention systems

FIGURE 9-21. Drilling top deck orifices

are normally designed using the concept that wet sand bridges. Dry media might sift down through support level gravel and might also foul a direct media retention underdrain.

- The pretreatment systems should be continuously optimized to ensure that the filter works properly. There should be minimal problems with raw water or chemical fouling if the pretreatment systems are optimized. However, adding a polymer to the backwash water could cause fouling of a direct media retention underdrain. The manufacturer should be contacted for recommendations.
- Direct media retention systems may be subject to calcification depending on the raw water quality and the plant processes. It is recommended that the plant processes be optimized to prevent calcification in the filter, if possible.
- In the case of a porous plate cap over a plastic underdrain, some plant operators have determined that calcification of the cap is less of a problem than for support gravel. The cap can be removed and replaced relatively easily, while support gravel may have to be jackhammered apart. As a result, it is important to clean the clearwell and each filter flume very carefully to eliminate contamination before commissioning the plant. In addition, it is equally important to ensure that the plant processes are optimized to the extent possible at all times.

- When using the porous plate cap on top of the plastic dual parallel lateral underdrain, application of the caulk sealing the plate to the block is very important. If the caulk is applied at the job site and left exposed to sunlight, it can bubble up, creating weak spots that the media can penetrate. As with most forms of failure, if the media is allowed to penetrate the underdrain and the flume, a major failure of the filter could occur. It may be better to have the porous plates installed on the underdrain blocks at the factory where the environment is more easily controlled.
- Different manufacturers attach porous plate caps to their underdrain in different ways. However, it is recommended that the underdrain have a recessed area for the porous plate with a vertical lip as shown in Figure 6-23.

Common Failures. Many, if not all, of the common failures discussed in this handbook, such as hydraulic shock, can cause catastrophic failures within any underdrain system. Because replacement of the underdrain and filter media represents a substantial cost, it is extremely important that all of the various plant components and controls be optimized at all times to prevent problems.

Underdrain Replacement

If a filter failure occurs with an older underdrain system, it may be desirable at that point to consider complete replacement with an air scour underdrain of the dual parallel lateral type. Even though replacing the underdrain and filter media represents a substantial cost, the plastic underdrain is more resilient and is likely to last longer. In addition, air scour will provide a more efficient cleaning system for the filter media than can be achieved by hydraulic backwash underdrains.

The issues in Table 9-18 are indicative of possible underdrain failures.

TABLE 9-18: Underdrain failures

	Yes	No
Are craters apparent along the sidewalls of false deck underdrains?	☐	☐
Has the clean bed head loss increased over time?	☐	☐
Is the air pattern uneven?	☐	☐
Is there media in the flume or media trap?	☐	☐
Are there grout failures between rows of underdrain blocks?	☐	☐
Is calcification a potential problem?	☐	☐
Are there gaps behind the seal angles?	☐	☐
Are the joints ruptured between underdrain blocks in the same row?	☐	☐
Was the underdrain installed on a very hot or cold day?	☐	☐

Troubleshooting Report

Once the evaluation is complete, a report should be written to discuss the issues. The basic report should consist of the following items as a minimum:

- Filter plant data sheet (Figure 9-1)
- Statement of the problem
- All the checklists contained in this handbook. (It should be noted that a complete set of all the checklists is included in Appendix B for the investigator's use.)
- Observations and conclusions
- Recommendations for repair or rehabilitation
- Summary statement

Depending on the needs of the owner, cost estimates may or may not be included in the report. However, cost estimates provided at this stage would normally be only budgetary in nature. To generate more accurate cost estimates, a detailed preliminary engineering design report should be written. Again, depending on the needs of the owner, identification of the responsibility for the various issues may also be included. Most problems result from equipment or controls failure and design issues. If there is any issue with the plant operators, it is normally that they have not been properly trained. In the author's experience, the operators are usually willing and eager to learn but may lack the necessary information and training.

CHAPTER 10

Pressure Filters

Introduction

Pressure filters are a specialty application of granular media filtration. They consist of an enclosed steel vessel in either a vertical or horizontal configuration, as shown in Figures 10-1 and 10-2. The benefit of pressure filters is that they can be installed "in line" with the raw water pumping system, sized to push the water through the filters, and deliver the treated water to the desired process or distribution system with no other pumping required. Because pressure filters are normally "in line," their operating pressure is obviously higher than atmospheric.

Treatment of well water for the removal of iron and manganese is a typical potable water application for pressure filters.

In addition to potable applications, there are numerous uses for pressure filters in commercial and industrial applications. Whether it is for potable water or industrial/commercial water applications, the design and selection of media and support gravel (if used) is very similar, or should be, to that for conventional gravity filters.

Basic Design

Whether a pressure filter is of a vertical or horizontal configuration, the basic descriptive design issues are the straight-side shell length, the diameter, the head type/depth, and the design pressure. Other issues specific to either vertical or horizontal pressure filters will be discussed in the following sections.

Straight-Side Shell Length

The length of the steel shell of the tank between the heads is called the straight-side shell length, as shown in Figures 10-1 and 10-2. The side shell is formed of steel rolled to the appropriate diameter and welded to the heads.

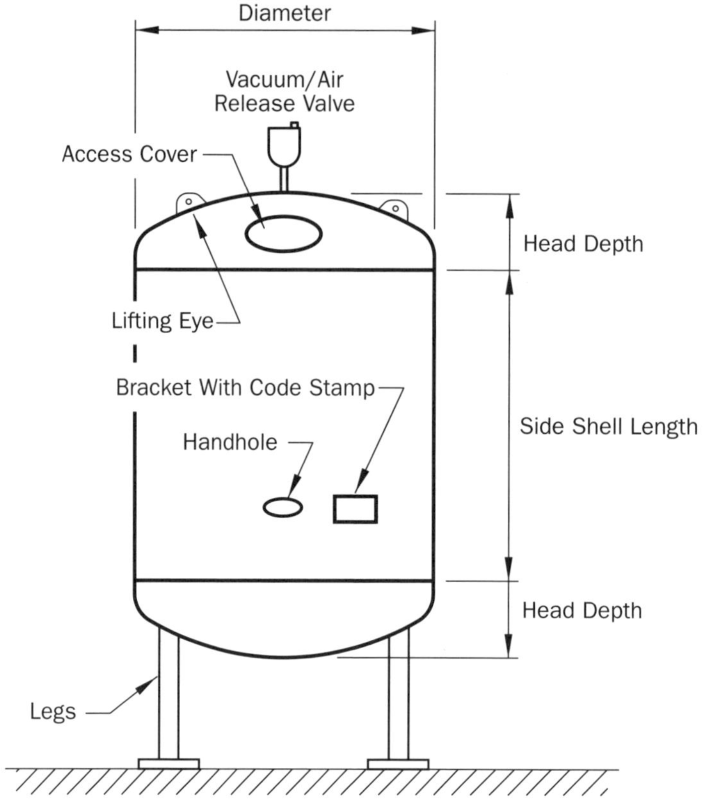

$$\text{Filter Area} = \frac{\pi \text{ Diameter}^2}{4}$$

FIGURE 10-1. Vertical pressure filter

For vertical pressure filters, the straight-side shell length is determined by the media load to be used, the depth of support gravel (if used), and the design and location of the underdrain.

For horizontal pressure filters, the diameter and the straight-side shell length are used to determine the filter area, as shown in Figure 10-2.

Diameter

The size of a pressure filter is a function of the plant flow rate, the media type, the number of filters used, and the method of backwash. These factors are tailored to fit the specific application. Each of these issues is addressed separately, depending on whether vertical or horizontal pressure filters are used.

The diameter of pressure filters varies widely, depending on the application. For example, many *point-of-use (POU) pressure filters* are less than 12 in. in diameter. For commercial, industrial, or larger potable applications, pressure filters may be 10 ft to 12 ft in diameter or larger. However, because of shipping considerations, most

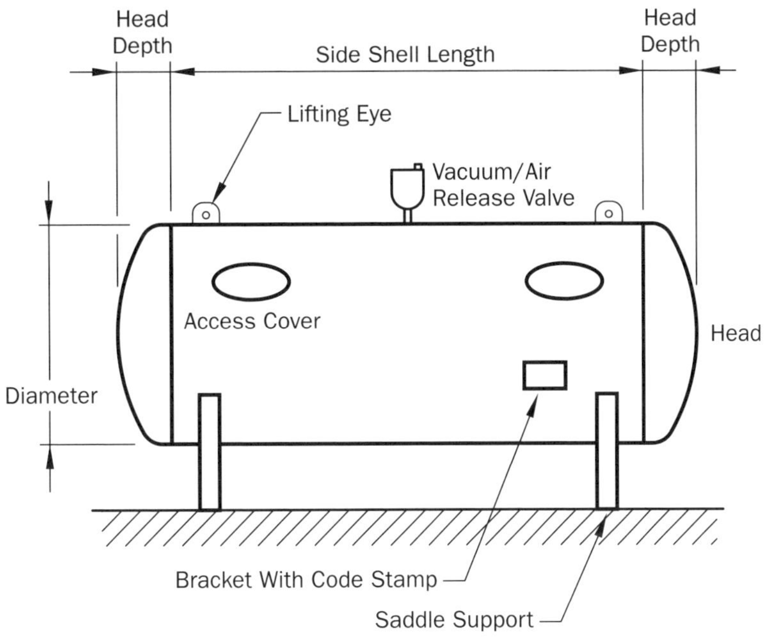

FIGURE 10-2. Horizontal pressure filter

pressure filters are 10 ft in diameter or less. Low pressure *can-type filters* (Figure 10-3) have been built much larger, although they are typically field erected.

Head Types/Depth

Cylindrical or spherical shapes resist pressure best. However, a hemispherical head on either end of a pressure filter would greatly increase its height or length. Shallower heads are typically used for this reason.

The depths of several types of heads are shown in Figure 10-4 for a typical 8-ft-diameter pressure filter. It should be noted that in order to resist the pressure, shallower heads are usually made of thicker metal for the same pressure rating. In addition, the shallower heads are usually more expensive. Therefore, the most common head in use for larger pressure filters is the semi-elliptical design. Pressure filters using shallower heads are generally designed to fit in space-restricted locations.

Design Specifications

The inlet pressure for a pressure filter is the sum of the clean bed and operational head loss plus the pressure needed by the system. Once that has been determined, it is common to increase that amount by 10 or 20 percent to an even value for design purposes. For example, if the total pressure requirement was 80 psi, the design pressure might be 100 psi for conservatism.

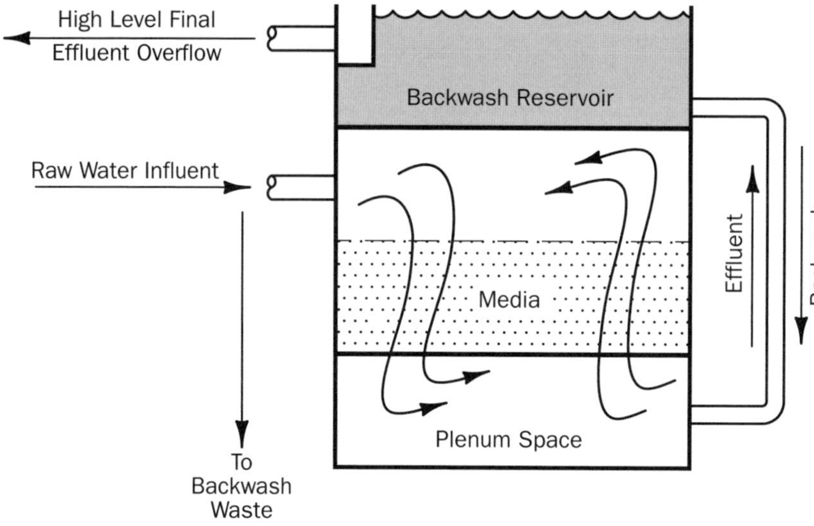

FIGURE 10-3. Low pressure, over-and-under, can-type pressure filter

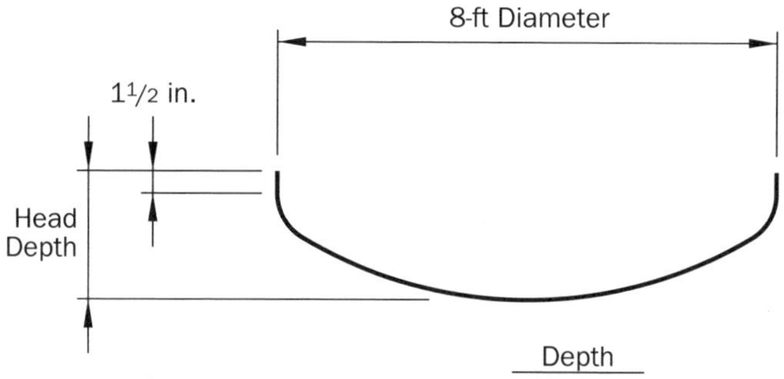

	Depth
Standard Dished and	15.4 in.
Flanged	20.0 in.
100 percent Crown Radius	13.8 in.
85 percent Crown Radius	25.5 in.

FIGURE 10-4. Pressure tank head types

Once the design pressure is established, it is recommended that the tank shell and heads be designed and constructed according to Section IX of the ASME Boiler Code. The code requires a testing and inspection process that helps ensure the integrity and serviceability of the filter. Tanks designed, tested, and inspected according to this process will have a permanent metal tag or *code stamp* affixed to a bracket on the side of the tank as shown in Figure 10-1 and 10-2. The tag will contain the manufacturer's name and the design and test pressures that were used.

There are many manufacturers who produce pressure filters, especially for POU and other smaller applications, who do not use the referenced code. These tanks are typically built of higher tensile strength steel and are lighter weight than those built according to the referenced code. However, for larger applications, particularly those for municipal and industrial use, code tanks are recommended for safety purposes. When troubleshooting a plant, locate the code stamp and compare the design pressure with the plant pressure. If there is no code stamp, the owner's manual should be consulted or the manufacturer contacted for the appropriate information.

Operational Design

In addition to the basic physical structure, there are many operational design issues that must be considered in the design and troubleshooting of a pressure filter.

Unit Flow Rate and Head Loss

The unit flow rate for any filter is normally dependent on the media design, the raw water quality, the pretreatment efficiency, and the intended service. Typical rates for dual media and multimedia are 4 to 6 gpm/ft^2. Other media types, such as greensand or activated carbon, will have different unit flow rates, usually lower depending on the raw water quality. The normal operational head loss for a filter is usually in the range of 8 ft to 10 ft of water pressure.

Because of their nature, pressure filters often have a relatively high inlet pressure. As a result, it is not uncommon for pressure filters to operate at much higher flow rates and differential pressure (head loss) than the normal limits previously mentioned. The fact that additional driving head is available does not change the physical aspects of solids removal. If a high quality treated water is required, such as for potable use, it is recommended that the unit flow rate be within the limits discussed previously for gravity filters. Depending on the application, higher rates for industrial or commercial use may produce acceptable water quality.

In regard to the differential pressure or head loss, it is recommended that the actual values should not greatly exceed the 8 ft to 10 ft range. The continued use of excessive head losses can contribute to compaction and mudballing of the filter media.

Valves and Controls

Small- to medium-size POU industrial and commercial pressure filters commonly use hydraulically actuated diaphragm valves and electromechanical controls, which are perfectly acceptable for these applications. Some types, even very large pressure filters, may use manual valves, which are also acceptable.

When automatic valves and controls are used, especially for larger potable pressure filters, it is recommended that electric valves and controls be used similar to those for conventional gravity filters. It is the author's opinion that electric valves and controls will provide better service and help produce a higher quality water. They will also help the media last longer.

Air/Vacuum Release Valves

In previous chapters, the use of air release valves has been discussed, particularly on backwash supply lines. Air release is also required for pressure filters, although there is less probability of air entering the system. However, vacuum release is also required for pressure filters.

Pressure filters can be designed to resist high internal pressure but are not able to withstand a very high vacuum. As a result, a combination air/vacuum release valve is a necessity for each pressure filter or each cell of a multicell horizontal pressure filter.

The worst-case design for vacuum release would occur with a catastrophic failure of the effluent or backwash piping, which would allow the tank to drain. Calculations should be made to determine the initial worst-case flow out of the filter. The vacuum release valve should be capable of allowing a similar volume of air in, with the same pressure drop.

Smaller POU filters and others of similar size may not need an air/vacuum release valve. However, they are recommended for all other types of pressure filters.

Corrosion Allowance

For pressure filters in applications where long life is desirable, it is recommended that an extra thickness of metal be provided for the tank shell and heads. An extra $^1/_{16}$-in. thickness is suggested for that purpose. For example, if the design thickness of the shell is $^3/_8$ in., an extra $^1/_{16}$-in. corrosion allowance will result in a $^7/_{16}$-in.-thick shell.

A Manhole/Handhole

Access to the inside of a pressure filter is required for the installation of filter media and routine maintenance. POU filters and other smaller units may have only a very small pipe connection for access. Slightly larger filters may have a handhole on the upper head and another on the lower side shell.

A manhole or a handhole are similar in construction. The only difference is size. They consist of an elliptical hole in the tank, with a reinforcing ring around the outside. A removable cover fits against the inside of the tank, with clamps and bolts on the outside to keep it in place. A handhole is typically 4 in. high and 6 in. wide, while the minimum size for a manhole is 11 in. × 15 in. However, it is recommended that manholes be a minimum of 12 in. × 16 in. for better access.

For larger pressure filters, at least one manhole and one handhole should be provided for access and ventilation when a worker is inside. For multicell horizontal pressure filters, a manhole and handhole should be installed in each cell. For larger filters, two manholes might be provided.

Surface Wash

Smaller pressure filters do not use surface wash. However, for larger filters, it should be considered mandatory unless air scour is provided. As with any filter, surface wash will greatly extend the life of the media.

Pipe Connection Sizes

It is recommended that the inlet/backwash waste and effluent/backwash piping connections be sized for velocities not greater than 5 ft/sec. Higher velocities will result in higher head loss and possible erosion of the connections.

Coatings

The interior coating of a pressure filter should be designed for the application. For example, any filter used for potable purposes will obviously have to have a suitably approved coating or lining. Commercial- or industrial-grade pressure filters may be glass lined, rubber lined, or have plastic coatings for the service intended. Some smaller filters are made of fiberglass. Exterior coatings on steel tanks may consist of a primer and an acrylic polyurethane or equivalent suitable for the service.

Seismic Bracing/Anchors

Seismic bracing will be a requirement for most pressure filters, except the smaller POU types. The applicable codes should be reviewed to determine the requirements.

Bracing for a pressure filter will likely be extensive, because of their weight and height. Bracing for a vertical pressure filter will be especially heavy, because of their high height-to-width ratio.

The bracing will most likely have to be through the floor, with heavy anchor bolts and an extensive foundation. Because of their weight and height, bracing a pressure filter to a building structure would be difficult.

Backwash

Some smaller POU filters may be backwashed with raw water. Backwashing with raw water is not recommended for larger filters. Some larger filters may backwash from the distribution system, although other installations have multiple filters designed so that the remaining on-line filters can backwash the others.

It should be noted that during filtration or backwash, most pressure filters remain pressurized.

Backwash From Distribution. Figure 10-5 illustrates a single filter designed to produce water directly to the distribution system. A filter operating in this manner only needs two automatic valves, plus a pressure-reducing valve on the backwash waste pipe. When in filter mode, the influent valve is open and the backwash valve is closed. Water is pumped through the pressure filter and directly to the distribution system. To backwash the filter, the influent valve is closed and the backwash waste valve is open. Water comes backward from the distribution system through the filter, flushing solids to waste. The pressure-reducing valve is used to reduce the pressure to atmospheric and also acts as a rate control valve.

The requirement of this method of backwashing is that the distribution system must have enough storage to backwash the filter.

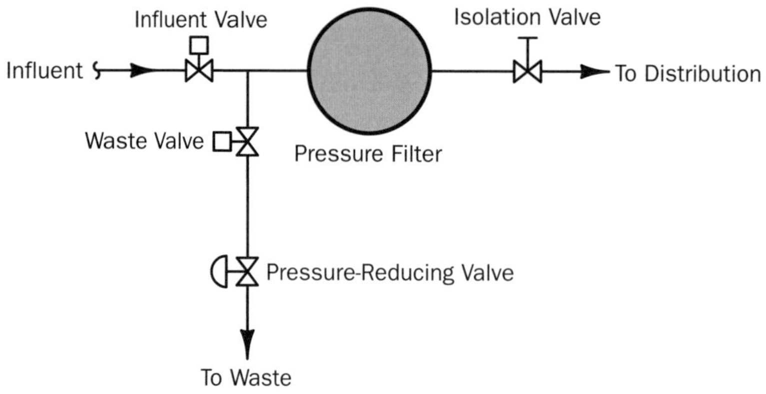

Backwash Procedure

1. Close influent valve.
2. Open waste valve.
3. Backwash for a set time.
4. Close waste valve.
5. Open influent valve and return to service.

FIGURE 10-5. Backwash using high service line

Multiple Filters. For larger installations or where the storage in the distribution system is limited, multiple filters can be provided so that the effluent from the remaining on-line filters can be used to backwash the other unit. Figure 10-6 illustrates a system with four filters. Any given filter is backwashed from the effluent of the other three. For example, to backwash Filter A, the following procedure could be used:

1. Close the common effluent valve.
2. Close Filter A influent valve.
3. Open Filter A waste valve.
4. Backwash for a set time period.
5. Close Filter A waste valve.
6. Open Filter A influent valve.
7. Open the common effluent valve and return to service.

For this system to operate properly, it is important that the backwash flow rate be tailored to the specific media in use and the water temperature. For example, if the four filters in Figure 10-6 were operating at 4 gpm/ft^2 with greensand media, three filters would be able to provide a total of 12 gpm/ft^2 to backwash the fourth filter. In the case of greensand, 12 gpm/ft^2 is adequate. However, if the filters had multimedia and were operating at 5 gpm/ft^2, a total of 15 gpm/ft^2 would be available to backwash the fourth filter. For applications using cool well water, that rate may be adequate. However, it would not be adequate for warmer water. An additional filter might be required for warmer water whereby four filters could be used to backwash a fifth.

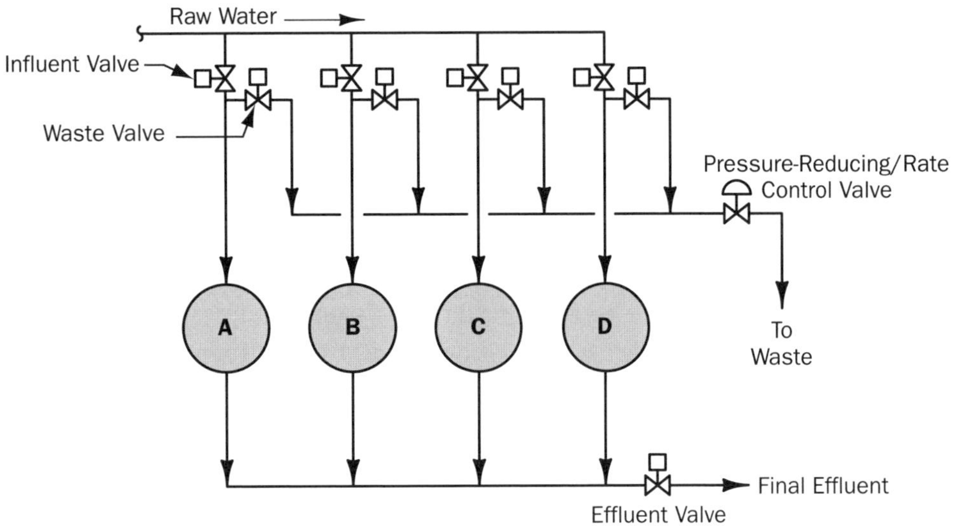

FIGURE 10-6. Backwash using multiple filters

Backwash Waste. The ability to handle the backwash wastewater is often a limiting factor in the design of pressure filters. For example, if the backwash waste is discharged to a sewer system, then the sewer must be capable of handling the full backwash flow rate of a filter. If the limits of the sewer system are exceeded, smaller filters will have to be provided.

On the other hand, if a backwash waste pond is being used, it must be of adequate size to settle the water from all the filters. The size of the pond may dictate whether or not the filters are all washed sequentially at the same time, or whether there is an interval between washes. Any interval must be adequate to settle the solids and decant the supernatant before the next filter backwashes.

Filter Mode

Pressure filters are often viewed as an inexpensive method of providing treatment. As a result, most installations have very simple valves and controls. When multiple filters are used, they are most often operated in declining rate mode during filtration, where the cleanest filter takes the most flow. Operating in declining rate mode is acceptable when treating well water for iron and manganese, for example. However, if surface water is treated, it is recommended that additional valves and controls be added to allow the filters to operate in constant-rate mode. It is the author's opinion that a higher quality of water will be produced in this case.

It should be noted that when backwashing one filter of a multiple filter installation, the unit flow rate on the filters still in service will rise unless controls are provided to throttle the incoming raw water. If a constant raw water flow is provided, the designer needs to make sure that the remaining filters will operate successfully at the higher rate when one is off-line in backwash.

Vertical Pressure Filters

Vertical pressure filters usually have either top or bottom connections, or side connections. As an alternative, a filter could have a top inlet and a side outlet. The choice of location for the connections is usually a function of the filter's size and the type and elevation of the underdrain.

Top/Bottom Entry

A cross-section of a vertical pressure filter with top and bottom connections is shown in Figure 10-7. Smaller filters typically have connections as shown in this figure, although the smallest have a combination connection on the top. For lack of any formal definition, it is assumed that the term "larger" refers to filters larger than 3 ft in diameter.

For larger filters, it is the author's opinion that bottom connections are used primarily so that the underdrain can be installed as low as possible in the head. The intent is to minimize the side shell height. In these cases, the underdrain typically does not cover the entire cross-sectional area of the head, as also shown in Figure 10-7.

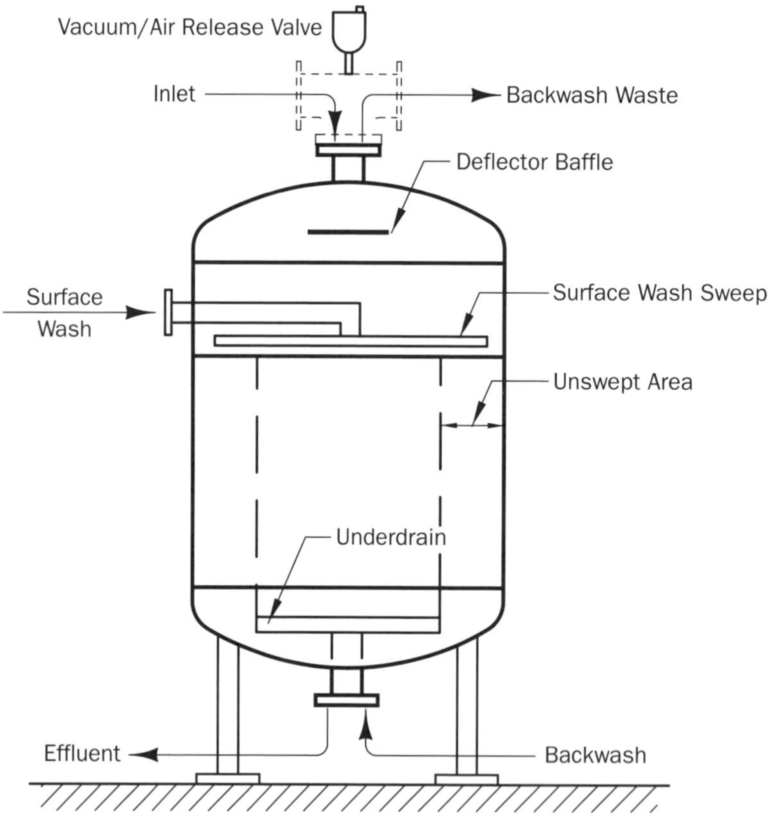

FIGURE 10-7. Vertical pressure filter with top/bottom connections

Using a small underdrain is not recommended practice. However, it can be mitigated to some extent by using additional support gravel. Generally, however, when a small underdrain is installed, a direct media retention system is used without support gravel. When this is done, a large edge condition can occur, as discussed in previous chapters. For that reason, it is recommended that a bottom entry not be used on larger filters.

A bottom entry should not be used unless the underdrain covers the entire cross-sectional area of the filter. In addition, the piping with a bottom entry adds to the height of a pressure filter, which also adds height to the building (if used).

Top Inlet

The inlet connection also functions as a backwash waste discharge. For a larger filter, having the backwash waste discharge from one point at the top means that some of the wastewater must travel laterally to the discharge point. There can be some inefficiency in the discharge of solids if the wastewater has to travel laterally (horizontally). For top discharges, especially if the top side shell is short, a deflector baffle may be provided to help prevent media from being washed out. As the wastewater moves up toward the top of the tank, the reduced cross-sectional area caused by the curvature of the head results in a dramatic increase in velocity of the wastewater. As a result, any expanded media that rises above the side shell has a high probability of being washed out of the filter. In that case, the deflector baffle will be of little use.

Side Connections

Side piping connections for larger filters are shown in Figure 10-8. It should be noted that both the influent/backwash waste and the effluent/backwash connections should be at least 1 in. from the weld seam between the straight-side shell and the head for structural purposes. The inlet/backwash waste pipe may connect to one central collector/distributor in the middle of the filter, or may branch out to multiple collector/distributors as shown.

A pipe header/lateral underdrain is shown at the lower connection. It should be noted that a direct media retention underdrain could be used if it covered the entire cross-sectional area of the filter.

When the lower connection is in this position, the volume occupied by the lower head is *unswept*. If left empty, the lower head would have to be filled with the coarser gravel used with the pipe lateral underdrain. Even so, that volume would be prone to accumulating slimes and could potentially cause taste-and-odor problems over a period of time. Therefore, it is common practice to fill the lower head with concrete on-site, before the header lateral, support gravel, and media are installed. In the case of an industrial installation with highly corrosive water, it may not be feasible to use concrete unless it is coated with a corrosion-resistant material.

Side Shell Height

A calculation of the desired straight-side shell height for a vertical pressure filter is shown in Figure 10-9. For calculation purposes, an 8-ft diameter filter is assumed, with 8-in. connections. Concrete fill is also assumed in the lower head.

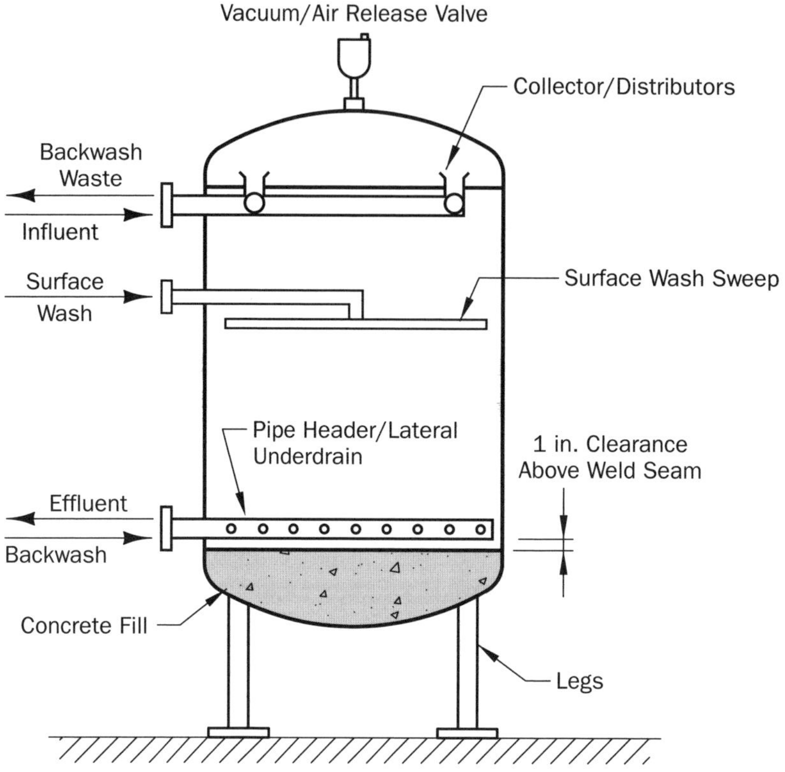

FIGURE 10-8. Vertical pressure filter with side connection

Starting at the bottom, 6 in. of 1 1/2 in. × 3/4 in. gravel is assumed to be required to provide 1 in. of cover over the top of the laterals. Three inches of three other sizes of gravel are used, including the barrier layer. It should be noted that this gravel design may change, depending on the media used.

Thirty inches of media is assumed, which could be dual media, multimedia, greensand, or activated carbon. In some cases, greater depths of media may be needed, which would require a higher side shell height.

An allowance of 9 in. is provided for media expansion, with 9 in. of freeboard above that. The intent of the freeboard allowance is to prevent the loss of media during backwash. As the backwash wastewater rises, the velocity will increase because of the restriction caused by the influent/backwash waste piping and the reduction in area because of the curvature of the head. Therefore, the allowance for freeboard is below the influent/backwash waste piping.

Following the previous recommendations, the minimum straight-side shell length is recommended to be 72 in. for any large filter. Adjustments to that will be required if greater depths of media are used, or if different gravel gradations are used.

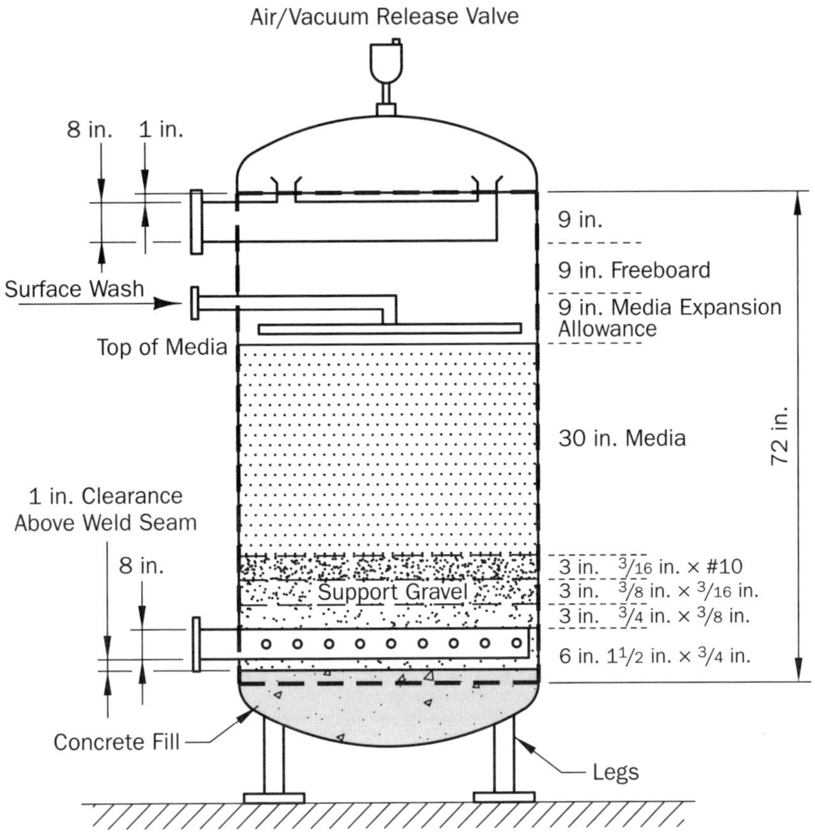

NOTE: Assume 8-ft diameter pressure filter with 8-in. piping.

FIGURE 10-9. Vertical pressure filter, straight-side shell height

Horizontal Pressure Filters

Plan and elevation views of a typical horizontal pressure filter are shown in Figure 10-10. When the tank is not subdivided internally, the influent/backwash waste, surface wash, and effluent/backwash connections typically come out one end of the tank. Also, it should be noted that there are two manholes for access and ventilation in this illustration and no handholes.

Horizontal pressure filters are typically 8 ft or 10 ft in diameter, and the side shell length may be between 16 ft and 40 ft long. Smaller applications usually use vertical pressure filters, while longer tanks are not practical because of difficulties in shipping. If more surface area is required, multiple units are normally provided.

Surface Area Calculations

The filter surface area for a horizontal pressure filter is normally calculated as the diameter times the straight-side shell length. Even though the heads increase the area slightly, they are normally not included because of the difficulty of backwashing those

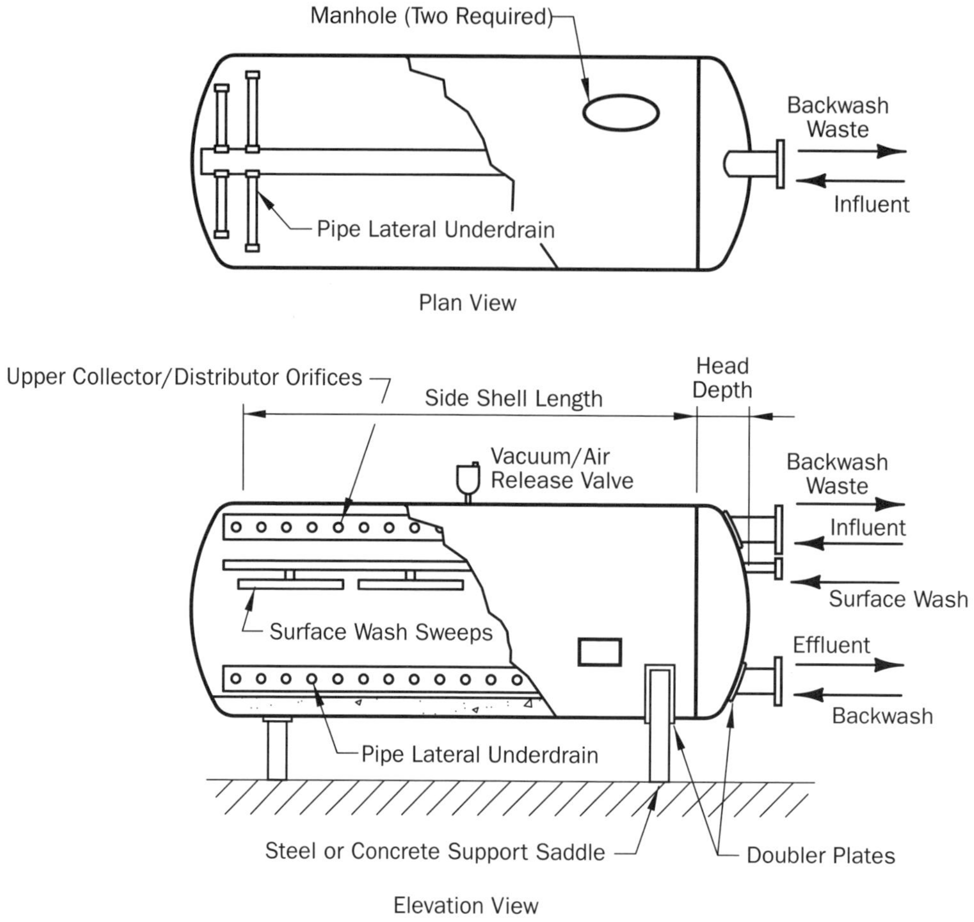

FIGURE 10-10. Horizontal pressure filters

areas. However, the extra volume added by the heads must be included in calculations for support gravel (if used) and filter media volume.

Doubler Plates

Because of the size and weight of horizontal pressure filters, a doubler plate is recommended where the support saddles are located. They are also recommended on the influent/backwash waste and effluent/backwash piping connections. Doubler plates on the surface wash piping may or may not be provided, at the designer's discretion.

Supports

Supports for horizontal pressure filters generally consist of either steel or concrete saddles. The number of saddles depends on the tank length and diameter. The use of concrete saddles will require an asphaltic mastic or felt to be laid on the saddle for the tank to set on.

Steel saddles may either be welded to the tank or separate. If separate, an asphaltic mastic or felt is also recommended between them and the tank.

Whichever type of saddle is used, seismic bracing must be taken into account, as previously discussed.

Diameter

The minimum recommended diameter for a horizontal pressure filter is 8 ft. Smaller diameters do not have adequate height for the underdrain, support gravel (if used), and filter media. In addition, the acceleration of the water upward during backwash is such that media would tend to wash out with smaller diameters.

A cross-section of a typical 8-ft-diameter horizontal pressure filter is shown in Figure 10-11. The filter media is assumed to be 30 in. deep and is centered on the tank center line.

The use of direct media retention underdrains would create a considerable unswept area on both sides of the tank, which would create the edge condition problem discussed in Chapter 6. Therefore, it is recommended that support gravel always be used in horizontal pressure filters. The use of gravel minimizes the unswept area, and essentially eliminates an edge condition problem.

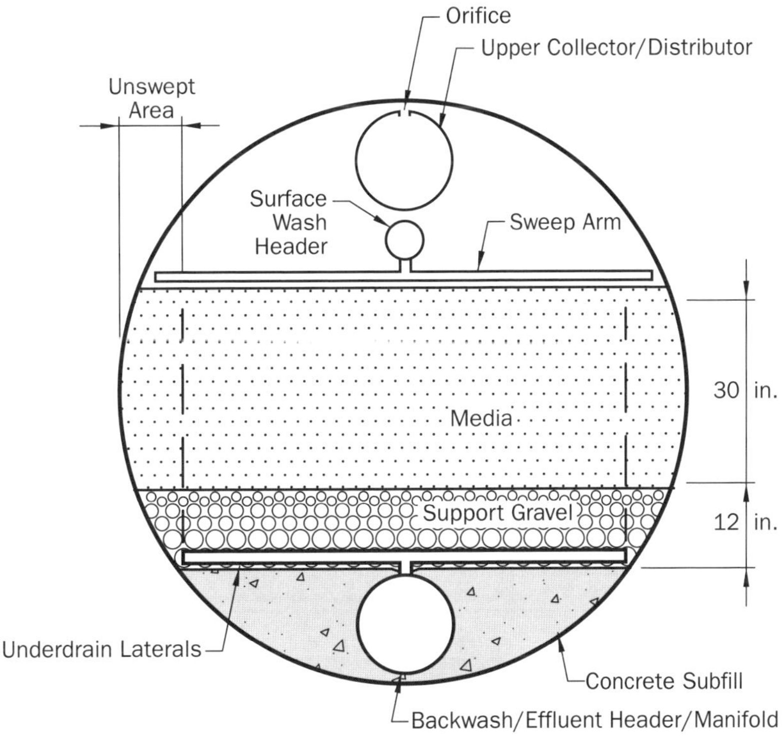

FIGURE 10-11. 8-ft-diameter horizontal pressure filter

The underdrain shown in Figure 10-11 is a modified pipe lateral design where a tee on each lateral connects to the header pipe. The tee configuration allows the lateral to be at a higher elevation, and the header pipe does not block the distribution of backwash water.

Using this design, a concrete subfill is recommended below the laterals for the same reason as in vertical pressure filters.

A cross-section of a 10-ft-diameter horizontal pressure filter is shown in Figure 10-12. The features are essentially the same as for the 8-ft-diameter tank shown in Figure 10-11. The benefits of a larger tank are:

- Larger filter surface
- More vertical height to separate the upper collector/distributor from the media
- One of the more important features is less acceleration of backwash water caused by the curvature of the tank as the water flows toward the upper collector/distributor.

There is less chance of losing media with a 10-ft-diameter tank. As a result, it is recommended that 10-ft-diameter tanks be used for horizontal pressure filter applications if practical.

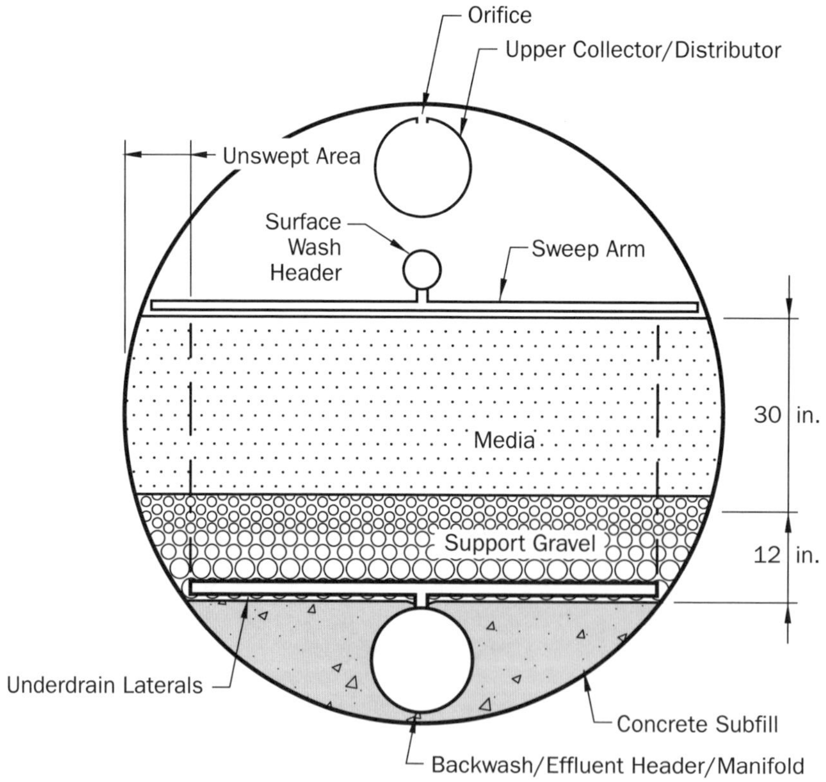

FIGURE 10-12. 10-ft-diameter horizontal pressure filter

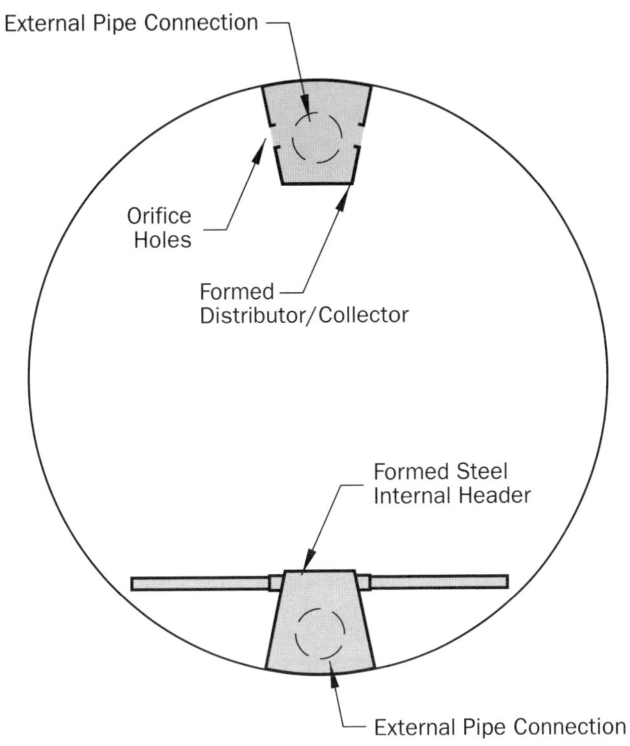

FIGURE 10-13. Horizontal pressure filter, alternate header arrangement

Alternate Header Arrangement

All of the figures used to this point illustrate the use of pipe headers and laterals. An alternative header arrangement is shown in Figure 10-13, in which a formed steel header is used inside the tank instead of a pipe.

An advantage of the formed header is the larger cross-sectional area and lower velocities during backwash. A disadvantage is the difficulty in painting the inside of a formed header after it is welded in place. On the other hand, a PVC pipe header could be used that does not require painting.

Regardless of whether a formed header or a pipe header is used for the effluent/backwash water, concrete fill should be installed.

Another alternative would be to use a formed header for the lower effluent/backwash manifold and a pipe for the upper distributor/collector.

Compartment Design

Horizontal pressure filters have a filter area of up to 400 ft^2 (10-ft diameter × 40-ft long). As a result, the backwash requirements are very high. To reduce the instantaneous backwash flow, some manufacturers subdivide large horizontal pressure filters into compartments that backwash separately, as shown in Figure 10-14. Typically,

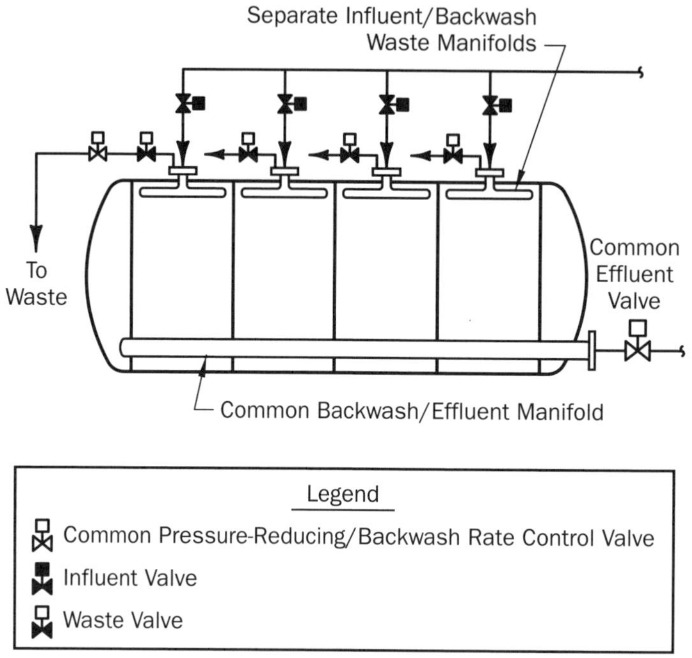

NOTES:
1. The individual influent/backwash waste manifolds may be routed out through one end of the tank instead of out the top as shown.
2. Surface wash piping may be brought out the sidewall of each cell, or out of the end of the tank.
3. Support saddles and other appurtenances omitted for clarity.
4. Each cell should have two manholes, or one manhole and a handhole.

FIGURE 10-14. Compartmented horizontal pressure filter

four compartments are used, so that three compartments can be used to backwash the fourth in a similar manner as shown in Figure 10-6. As a part of the compartmented design, a common effluent/backwash manifold is used with individual inlet and backwash waste connections. The inlet/backwash waste connections can be on top of the tank, as shown in Figure 10-14, or can all be brought out of the end of the tank through the head individually if desired.

To initiate backwash, the common effluent valve is closed, along with the inlet valve to the compartment being backwashed. The backwash waste valve to the compartment being backwashed is opened. The common effluent/backwash manifold collects the treated water from three compartments and routes it to the fourth compartment as backwash and out to waste. A common pressure-reducing/backwash rate control valve reduces pressure to atmospheric for discharge as well as controlling the backwash rate.

The limitations for this system are the same as for multiple vertical pressure filters, as discussed previously, and as shown in Figure 10-6. It is necessary for the three remaining filters to produce an adequate supply of backwash water for the specific media design used and the water temperature, as also discussed previously. Typically, the influent rate is not changed, which means the filtration rate for the three remaining filters is increased to produce the desired flow. Care should be taken to not generate excessive head loss in the filters. Otherwise, the media can become compacted and the formation of mudballs accelerated.

Partition Design

It is not efficient to use formed heads for partitions inside a horizontal pressure filter. As a result, flat compartment walls are generally used, and they are not normally pressure rated. The common effluent/backwash header will equalize the pressure between compartments anyway. Using this method of design, one compartment cannot be taken off-line while the others are operating. The entire filter tank has to be taken off-line if any internal maintenance is required.

Troubleshooting Pressure Filters

Troubleshooting a pressure filter is very similar to the procedure for a gravity filter, as discussed in Chapter 9. The biggest difference is that it is not possible to observe a backwash and the filters cannot be probed or cored during backwash. However, some information can be obtained if the discharge of the backwash wastewater is visible (see Table 10-1).

If the backwash wastewater is still dirty at the end of the backwash cycle, either the time must be extended or the backwash rate must be increased.

Other than not being able to observe a backwash, a troubleshooting evaluation would follow a very similar procedure to that outlined in Chapter 9 for gravity filters. The evaluation would include the items in Table 10-2.

After a backwash, drain the filter and open the hatches for an inspection of the media, as shown in Table 10-3.

When inspecting the inside of a pressure filter, all applicable codes and guidelines must be followed regarding confined-space entry.

Troubleshooting Guide

A troubleshooting guide specific to pressure filters is shown in Figure 10-15. The problems described in that figure are the most common for pressure filters. Unfortunately, they do not often have simple solutions.

In addition to these potential problems, other issues discussed in this handbook such as poor gravel design or equipment failures, as well as many others, also cause failures in pressure filters. For that reason, a complete evaluation may be required, as described in Chapter 9 for gravity filters.

TABLE 10-1: Backwash wastewater

	Yes	No
Does the backwash wastewater clean up in 8 to 10 minutes?	☐	☐
Is the backwash rate too slow or too fast?	☐	☐
Were there still solids being discharged at the end of the allotted backwash time?	☐	☐

TABLE 10-2: Evaluate data and equipment

- Determine the operating conditions.
- Evaluate the raw water conditions.
- Evaluate the pretreatment.
- Evaluate the valves, pumps, and controls.
- Analyze effluent quality.
- Determine backwash frequency.

TABLE 10-3: Inspect the media

- Inspect the media surface for cleanliness and levelness.
- Excavate to the support gravel if necessary.
- Inspect the surface wash sweeps.
- Determine if there has been an underdrain failure.

Summary

Pressure filters are often viewed as an inexpensive method of providing treatment. However, to provide good service over a long period of time, the design recommendations and precautions described in this handbook should be observed. The following suggestions are also given:

1. Provide a small air vent orifice on top of the effluent/backwash header for evacuation of air.
2. Design to the recommended standards for unit flow rate and backwash.
3. Vertical pressure filters:
   - Use a direct media retention underdrain only if it is full diameter.
   - Provide a minimum 72-in. straight-side shell height for larger filters and maintain the recommended freeboard for smaller filters.
4. Horizontal pressure filters:
   - Do not use direct media retention underdrains for any application.
   - Use industry standard backwash rates for the media used.
   - Horizontal pressure filters should be at least 8 ft in diameter, with 10 ft in diameter preferred.

Before troubleshooting a filter, it is often beneficial to determine the operating parameters of the plant. It is recommended that the investigator fill out this form to the extent possible before beginning troubleshooting.

A. Design Flow:	Summer	________	
	Winter	________	
	Average	________	
	Peak	________	
B. Type of Plant:	Wastewater	________	
	Potable	________	
	Slow Sand	________	
	Package Plant	________	
	Industrial	________	
	Commercial	________	
C. Raw Water Source:	Lake	________	
	Impoundment	________	
	Stream/River	________	
	Well	________	
	Domestic Wastewater	________	
	Industrial Waste	________	
D. Chemical Feed:	Coagulants:	Alum	________
		Polyaluminum Chloride (or similar)	________
		Ferric Chloride	________
		Other	________
	Polymers:	Anionic	________
		Cationic	________
		Nonionic	________
	pH Buffers:	Lime	________
		Soda Ash	________
		Sodium Hydroxide	________
	Recarbonation:	CO_2	________
	Disinfection:	Chlorine—Gas	________
		Chlorine—Liquid	________
		Chlorine—Tablet	________
		Chlorine—Powder	________
		UV	________
		Chloramine	________
		Other	________

(Figure continued on next page)

FIGURE 10-15. Pressure filter plant data sheet

		Other:	Powdered Carbon		______
			Potassium Permanganate		______
			Ozone		______
E.	Flocculation:	None	______		
		Hydraulic	______		
		Mechanical:	Turbine		______
			Paddle		______
		Air	______		
		Other	______		
		Detention Time at Design Flow:	______	min	
F.	Sedimentation/Clarification				
	Conventional:	None	______		
		Circular	______	ft^2	
		Rectangular	______	ft^2	
		Square	______	ft^2	
		Settling Tubes	______	ft^2	
		Inclined Plates	______	ft^2	
		Circular	______		
		Rectangular	______		
		Square	______		
		Trough Spacing	______	ft	
		Basin Coverage of Troughs	______		
		Troughs	______	%	
		Ballasted/High Rate	______	gpm/ft^2	
		Other	______		
G.	Package Plant:	Settling Tubes	______		
		Granular Media (Roughing Filter)	______		
		Ballasted High Rate	______		
		Other	______		
H.	Filters:		______		
	1. Filter Sizing/Type:	Number of Filters	______		
		Vertical	______		
		Horizontal	______		
		Horizontal/Compartmental	______		
		Diameter	______	ft	
		Side Shell Length	______	ft	

(Figure continued on next page)

FIGURE 10-15. Pressure filter plant data sheet (continued)

2. Underdrain Type:	Pipe Lateral	________
	Screen Wrapped Lateral	________
	Wedge Wire Lateral	________
	Other	________
3. Manifold Type:	Pipe	________
	Formed	________

I. Valve Data

1. Rate Control Valve Type:	Butterfly	________
	Diaphragm	________
	V-Port	________
	Other	________
2. Actuator Types:	Electric	________
	Pneumatic	________
	Hydraulic	________
3. Actuator Speed:	________ sec	
4. Valve Operation:	Smooth	________
	Jerky	________
	Discontinuous	________
	Other	________
5. Type of Air Valves Used:	Air Release	________
	Combination Air/Vacuum Release	________
6. Year Valves Installed:	________	

J. Media Type:

	ES	UC	Depth, in.
Anthracite	________	________	________
Sand	________	________	________
Garnet Sand	________	________	________
Activated Carbon	________	________	________
Greensand	________	________	________
Other	________	________	________

K. Support Gravel

Layer	Depth, in.
________	________
________	________
________	________
________	________

L. Height from Top of Media to Bottom of Collector/Distributor ________ft

(Figure continued on next page)

FIGURE 10-15. Pressure filter plant data sheet (continued)

M. Control Type:	Constant Rate	__________	
		__________	
	Declining Rate	__________	
		__________	
	Other	__________	
N. Filter Rate:	Normal	__________	gpm/ft^2
	Maximum Demand	__________	gpm/ft^2
	During Backwash	__________	gpm/ft^2
O. Backwash Rate:	High	__________	gpm/ft^2
	Low	__________	gpm/ft^2
P. Backwash Supply:	Vertical Turbine Pump	__________	
	Centrifugal Pump	__________	
	System Pressure	__________	
	Low Pressure Gravity	__________	
	Self-Backwashing/Multiple Filters or Compartments	__________	
Q. Pump Control (if used):	Full Voltage Start	__________	
	Variable Speed	__________	
	Slow Start	__________	
R. Air Scour (if used):	Rate	__________	cfm/ft^2
	Duration	__________	min
Type:	Through Underdrain	__________	
	Air Grid	__________	
	Sequential	__________	
	Concurrent	__________	
S. Surface Wash (if used):	Straight Line	__________	
	S-Shaped	__________	
	Subsurface	__________	
	Rate	__________	gpm/ft^2
	Total Duration	__________	min
	Duration During High-Rate Wash	__________	min
	Number of Sweeps per Tank	__________	
	Diameter	__________	ft
T. Plant Age:	Year Constructed	__________	
	Year Filters Rehabilitated Last	__________	

FIGURE 10-15. Pressure filter plant data sheet (continued)

CHAPTER 11

Filter Maintenance

Introduction

Equipment and materials will last longer and filters will work better when the proper maintenance is performed. In addition, performing the routine maintenance activities will provide early warning of potential problems, so that preventive measures can be taken before failures occur.

NOTE: Follow the manufacturers' recommendations for maintaining mechanical equipment such as pumps, valves, etc.

Recommended Filter Maintenance Schedule

The following task list is very general in nature. The individual chapters in this handbook contain a more complete discussion of each item.

Daily

Record Keeping

- Plant operating data should be recorded in historical file

Optimize Process

- Monitor changes in raw water quality
- Conduct jar tests
- Adjust chemical feed as required
- Monitor clarifier efficiency
- Monitor filter effluent quality

Observe the Backwash Cycle

- Duration (Is the cycle too long or too short?)
- Cleanliness of the filter

Check for the presence of:

- Uncontrolled air
- Hydraulic shock
- Other irregularities

Weekly

- Verify backwash rise rate and media expansion. Adjust backwash flow rate as required.
- Review historical records for consistency and changes
- Hose down sidewalls and residual solids on media surface and backwash. Areas unswept by surface wash arms are a special concern.
- Remove accumulated sludge from the clarifiers as required
- Calibrate chemical pumps
- Calibrate instrumentation (turbidimeters, pH meters, etc.)
- Clean plant and equipment

Monthly

Tube Clarifiers

- Drain and hose-down tubes to remove accumulated solids
- Remove accumulated sludge as required

Inspect Media Surface

- Check for fines
  - — Visual
  - — Sample
- Skim debris as required
- If the surface is flat and smooth, no further action is required
- If there are any irregularities, investigate immediately, per Chapter 9

Periodic Equipment Maintenance per Manufacturers' Instructions

- Grease/change oil on equipment as required
- Change air filter on blowers as required
- Drain water from compressors or blower piping
- Other maintenance—as required

Quarterly

- Check media height—top off as required
- Probe gravel—compare results with historical records
- Inspect media surface—skim as required

Yearly

- Top off to original height and skim

Disinfection Procedure

Filters should be disinfected after installing new media, skimming, topping off, or other significant work. A sample calculation and recommended disinfection procedure follows.

Sample Calculation

- Assume a filter area of 400 ft^2.
- Assume a 4 ft to 5 ft depth of water over the media, as shown in Figure 11-1. The depth of water is dependent on the filter configuration and the amount of media and support gravel used. For media and gravel designs deeper than 42 in., additional water may be necessary.
- Assume a chlorine concentration of 50 ppm as a minimum.
- Assume a chlorine concentration of 5 percent in a liquid form (bleach). If powdered chlorine is used, the calculation should be done in pounds.

$$\text{Chlorine volume} = 400\ \text{ft}^2 \times 5\ \text{ft deep} \times 7.48\ \text{gal/ft}^3 \times 50\ \text{ppm}/(1{,}000{,}000 \times 0.05\ \text{chlorine})$$

$$\text{Chlorine volume} = 15\ \text{gal}$$

Disinfection Steps

1. Fill the filter with water to a level 4 ft to 5 ft above the media, as shown in Figure 11-1.
2. Pour the liquid chlorine into the filter in as many different locations as possible. Stir slightly, if required, to disperse the chlorine in the water. NOTE: If powdered chlorine is used, a respirator should be worn to protect against chlorine gas that will be generated.
3. After the chlorine is dispersed, lower the water level slowly in filter-to-waste mode.
4. Measure the residual chlorine in the filter-to-waste discharge. When a strong level of chlorine is detected, close the filter-to-waste valve and let the filter set for 24 hr.

   It should be noted that the media occupies space in the filter, as discussed in Chapter 5. The void volume of the media will be 50 percent or less. The intent of starting with 4 ft to 5 ft of water above the media is to allow sufficient volume to penetrate the media and also displace water in the underdrain, flume, and filter-to-waste piping while still leaving water over the media.
5. Backwash the filter until clean and return to service.

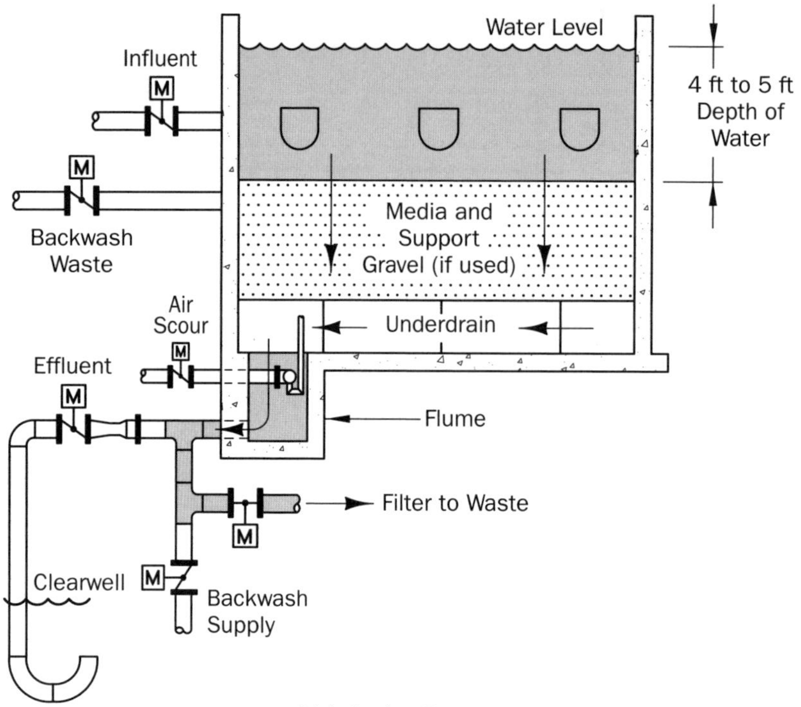

Disinfection Process

1. Fill filter with water to a level 4 ft to 5 ft above the media.
2. Stir required amount of chlorine into filter.
3. Drain filter in filter-to-waste mode until a strong chlorine level is reached.
4. Close filter-to-waste valve and let set for 24 hr.
5. Backwash filter until clean and return to service per Figure 1-2.

FIGURE 11-1. Filter disinfection

Disinfection Assumptions

Disinfection of the filter involves the use of a fairly strong dosage of chlorine. As a result, all of the components of the filter, including valves, piping, air distribution piping (if used), underdrain, caulk, bolts, screws, grout, paint (if used), and any other components in the filter must be compatible with the level of chlorine used.

Historical Records

A database of historical records regarding the operation of the plant should be maintained at all times. Historical data should include the information contained in Figure 11-2 as a minimum.

Having this information available to the operator will help quickly identify the proper chemical feed setting when the raw water quality changes. However, before any changes are made, jar tests should be performed to verify the historical records. In addition, historical data will help in the detection of potential problems. For example,

Date ________			
Raw Water Quality	Turbidity	________	ntu
	pH	________	
	Alkalinity	________	mg/L
	Water Temperature	________	°F (°C)
	Particle Count (optional)	________	
Plant Flow Rate	________	gpm (mgd)	
Chemical Feed Rates	Polymer 1	________	mg/L
	Polymer 2	________	mg/L
	Coagulant	________	mg/L
	Carbon	________	mg/L
	Potassium Permanganate	________	mg/L
	Soda Ash	________	mg/L
	Lime	________	mg/L
	Sodium Hydroxide	________	mg/L
	Chlorine	________	mg/L
	Other	________	mg/L
Clarifier Effluent Turbidity	________	ntu	
Filter Effluent Turbidity	________	ntu	
Run Time Between Backwashes	________	hr	
Particle Count (optional)	________		
Clean Bed Head Loss	________	ft/in.	
Terminal Head Loss	________	ft/in.	
Final Effluent	Turbidity	________	ntu
	Chlorine Residual	________	mg/L
	pH	________	

FIGURE 11-2. Historical records

if the clean bed head loss is increasing over time, the filter media may be fouled. An investigation should be made to determine the cause and any remedial action that should be taken.

As with the recommended maintenance procedures, recording this data takes time and smaller plants may not have the necessary personnel. Many treatment plants, small ones included, are being built with computerized controls and SCADA systems. If this is the case, most of the data can be generated by the controls, with minimal input by the operator.

Pilot Column

A permanent pilot column is highly recommended for all filter plants, especially those producing potable water. A pilot column is a miniature filter, which should ideally have the same support gravel layering and filter media depth as the full-scale plant. It should have the same feed water and be backwashed with the same water as the full-scale plant. If desired, the pilot column may be automated with the same instrumentation and controls as the full-scale filters. The intent is to have the pilot column operate in the same way as the full-scale filters with the same unit operating and backwash rates.

A pilot column may not operate the same as the full-scale plant because of the slick sidewalls and high percentage of perimeter. Sidewalls are always the weakest part of a filter, and a pilot column will always have a higher percentage perimeter than the full-scale plant.

The benefits of a pilot column are that it can provide an early warning of problems, can be used for chemical trials, and is a good learning tool for plant operators.

Early Warning

Many treatment plants have a detention time that is measured in hours. The pilot column can be plumbed to provide a faster response of a filter's effectiveness than the full-scale plant. As a result, the pilot column can be used to provide an early warning of changes needed in the pretreatment. In addition, the sidewalls of a pilot column should be clear to allow observation of the full depth of the media and support gravel (if used). If the plant is having difficulties treating the water, it may be possible to observe mudballs forming in the pilot column, which will be an indication that remedial action is needed in the full-scale plant.

Chemical Trials

Plant operators should continually experiment with different ways that might be used to help the plant operate better. For example, it may be desirable to test a different coagulant or polymer than is currently being used. Instead of putting the main plant at risk, the pilot column could be used for chemical trials. If it works there, then it may be used in the full-scale plant.

Learning Tool

A pilot column is an excellent learning tool for new plant operators. They can run the pilot column manually to determine the effect of their decisions and methods of backwashing. The effect of their operation on support gravel (if used) can be observed, along with intermixing and media classification during backwash. Fines on the surface that need to be removed can also be observed.

If the trainee upsets the gravel bed in a pilot column, it can be replaced relatively easily with little expense. Rebuilding the support layer gravel and filter media load is a good learning mechanism for new operators. It is much less expensive than putting the main plant at risk.

Summary

These maintenance activities will require a considerable amount of time to complete. Smaller plants with limited personnel may have difficulty performing all of this work. It is recommended that as many of the tasks as possible be completed. However, completing this task list should be a requirement at larger plants.

Problems may occur in spite of the best maintenance efforts. Nevertheless, equipment will operate better and last longer if the proper maintenance is performed.

CHAPTER 12

Summary/ Conclusion

Introduction

Whether it is a wastewater treatment plant or a potable water treatment plant, designing, constructing, and operating the facility to produce a quality product involves a total commitment by all employees. Commitment starts with the owner, who must retain a qualified filter design consultant, maintain a staff of qualified operators, and sustain good communications between management and operations.

Design and Construction

Many consultants design and construct water and wastewater treatment plants containing filters. However, as noted in this handbook, there are numerous design details that must be included in a filter plant to ensure proper operation. It is the owner's responsibility to retain a qualified consultant who can produce plant designs that are easy to maintain and operate. Most filter failures can be eliminated through proper design and careful attention to quality control and inspection during construction.

Plant Operators

The owner should maintain a high standard for hiring plant operators. They should be trained in every aspect and piece of equipment of the plant so that they understand its function and purpose.

It is important that the management of any utility delegate responsibility to the operators and make them feel part of a team. The management should listen to comments from the operation staff and involve them in all decision-making processes. In addition to the routine day-to-day work, the operators will feel more like part of a team if they are assigned projects and responsibilities such as conducting pilot tests to improve plant operation.

Operator training is essential for the successful long-term operation of any plant. Better operators are more likely to produce a better product and keep the equipment in better shape, which will result in a less expensive and higher quality product in the long term.

Optimize Operation

When a treatment plant is under construction, it is highly recommended that the future operations staff be involved. It is good for them to know where the underground piping is located and how tanks and other equipment are connected. When the plant is completed and started up, the first goal for the operators is to make sure that all plant processes and equipment are functioning properly.

The second task is to optimize the operation of the plant. Efficient operation of the filters will result in wasting less backwash water and using fewer chemicals. These in turn will result in less downtime for the filter and a higher net production.

Optimizing the plant operation requires the staff to have a high level of experience and dedication to the job. The operations staff is not expected to know how to design a treatment plant, but they should know how to obtain the best performance with the equipment they have. In order to do that, they must have a working knowledge of the issues contained in this handbook.

Maintenance

A filter maintenance checklist is contained in Chapter 11. Heavy mechanical or electrical work can always be subcontracted. However, for most of the tasks listed in Chapter 11, the plant staff should do the work. Maintenance tasks of this sort are a great training tool for the plant operators. It will add to their knowledge of how things work and demonstrate the effect of good maintenance.

Written Procedures

It is important that filters be operated the same way every time, particularly during the backwash cycle. As discussed herein, opening valves or starting pumps out of sequence can easily damage the underdrain, filter media, and support gravel (if used). There are always occasions when it is necessary to run the plant in manual mode. A written set of instructions should be provided that include the sequence of events and the elapsed time for each step. These instructions should be contained in the plant Standard Operating Procedures. Operational problems can easily occur when the operators are not adequately trained and do not have enough information to operate the plant properly.

Troubleshooting Rules

There are several rules that should never be violated while troubleshooting a filter, and they are very similar.

1. Make no assumptions.
2. Take nothing for granted.

3. View everything personally. It is always helpful to have input from the operations staff, management, and other personnel connected with the plant about potential problems. However, the plant personnel may not be as experienced in troubleshooting as the investigator. Therefore, the investigator should see and experience everything firsthand.

Using the tools in this handbook, the operations staff should conduct periodic inspections of the filters and related equipment at least monthly. Some functions may be done more often and some less. When the plant operations personnel have performed their best, a periodic peer review is recommended, even if there are no known problems. The peer review may develop additional information not available to the plant staff.

Repairing filter problems can be very expensive and may take a filter out of service for a considerable period of time. Therefore, the owner and plant operations staff should make every effort to extend the life of the plant.

APPENDIX A

Water Treatment Chemistry and Jar Testing Procedures

Definitions

pH	A log scale used to express the concentration of hydrogen ions ($pH = \log 1/_{H^+}$).
Alkalinity	A measure of the capacity of water to neutralize acid.
Acid	Substances that dissociate to yield hydrogen ions (H^+).
Base	Substances that dissociate to yield hydroxyl ions (OH^-).
Hardness	The amount of dissolved solids that causes scaling and increases the amount of soap required to produce foam, usually calcium or magnesium (expressed as calcium carbonate—$CaCO_3$).

Hardness Classification of Water

Very soft	0–20 mg/L
Soft	20–75 mg/L
Medium	75–150 mg/L
Hard	150–300 mg/L
Very hard	300 mg/L or higher

Giardia and *Cryptosporidium*

Protozoa	Microscopic organisms consisting of a single cell without cell walls. Reproduction is by binary fission.
Giardia	A parasitic flagellate protozoa. Average size: W = 8 µm, L = 12 µm. Average size: 3–7 µm, depending on stage.
Cryptosporidium	Another parasitic protozoa. Average size: 3 to 7 µm, depending on stage.

Treatment of *Giardia* and *Cryptosporidium*

1. Multiple barrier concept
   - Coagulation
   - Sedimentation
   - Filtration
   - Disinfection
2. Filtered effluent below 0.1–0.2 ntu

Coagulation

1. Coagulation occurs when colloids of opposite charge are mixed. The intent is to neutralize the charge of suspended particles and colors.
2. Coagulants used for this purpose are alum, ferric chloride/sulfate, poly-aluminum chloride, etc., which are positively charged.
3. Colloids of concern in water treatment are generally negatively charged.

Optimum Coagulated Dosage

1. Varies with raw water quality
2. Determined by:
   - Jar testing
   - Streaming current monitor (zeta potential)
   - Calibration with jar testing and/or pilot column testing
3. Stability of the colloid is a function of the magnitude of the charge (zeta potential).

Pre/Post Treatment Buffering

Lime	Slaked lime $Ca(OH)_2$
	Unslaked Lime CaO
Soda Ash	Na_2CO_3
Sodium Hydroxide	NaOH

Jar Testing: Equipment Needed

- Jar testing machine (Figure A-1)
- 8–10 1,000-mL beakers (Figure A-2)
- 1-mL pipette
- 5-mL pipette
- 20-mL pipette
- Pipette squeeze bulb

- 100-mL graduated cylinder (or larger)
- Scale for measuring dry powder (if required)
- Distilled water
- pH meter
- Vacuum filter, filter paper, and funnel

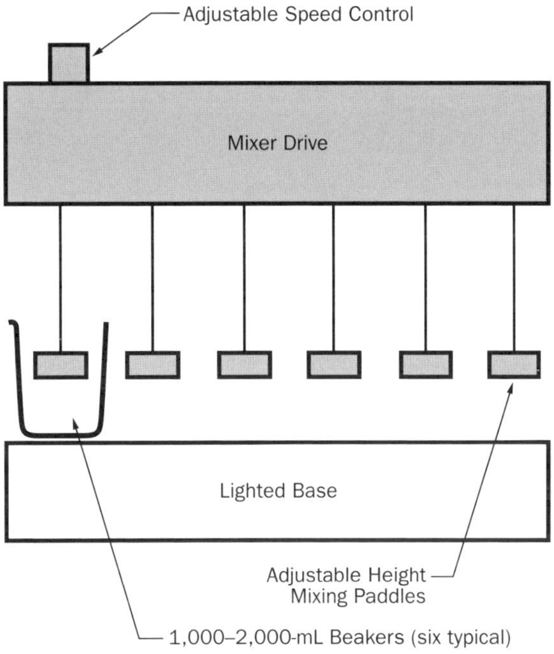

FIGURE A-1. Jar testing machine

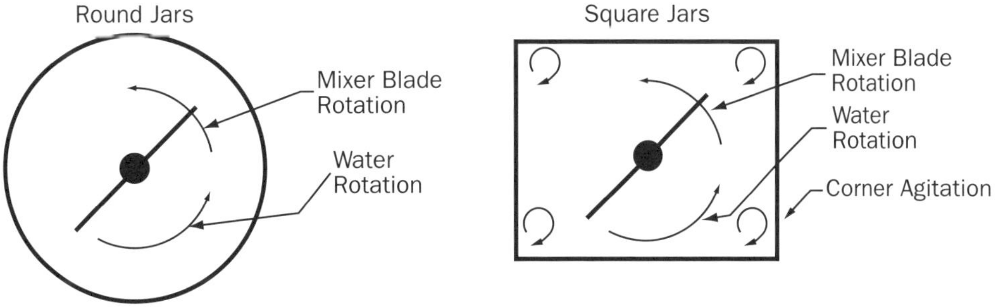

NOTE: Mixing is more efficient with square jars due to corner agitation.

FIGURE A-2. Jar types

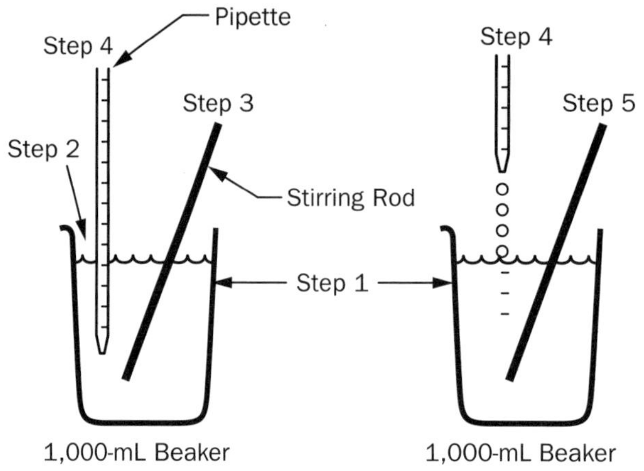

Procedure:

1. Fill both beakers with distilled water.
2. Add 1 g dry chemical or 1 mL of bulk liquid to first beaker.
3. Stir first beaker thoroughly.
4. Take 1 mL of mixed solution and pour into second beaker.
5. Stir second beaker thoroughly.
6. 1 mL of second beaker will now be equivalent to 1 ppm when put into 1,000 mL of raw water.

FIGURE A-3. Chemical test solutions

Jar Testing: Make Up Chemical Test Solutions

- Fill one 1,000-mL beaker with distilled water for each chemical used.
- Weigh out 1 g of dry chemical or 1 mL of bulk liquid. (More can be used. It simply changes the concentrations.) Refer also to the section on Chemical Calculations for a discussion of dry weight versus bulk liquid calculation.
- Put measured chemical into one of the 1,000-mL beakers of distilled water and stir.
- One mL of this solution will be 1 mg/L or 1 part per million (ppm) when put into a 1,000-mL beaker of raw water. (If more than 1 mg or 1 mL is used, the concentration goes up accordingly.)(See Figure A-3.)
- Prepare test solutions for other chemicals used in a similar manner (soda ash, polymer, etc.) It should be noted that the concentration will be different for each.

Jar Testing: Procedure

1. Fill six 1,000-mL beakers with raw water. Make sure to do this quickly before the water becomes warm. Changes in temperature can affect water chemistry.
2. The primary coagulant should be tested first (alum is assumed herein). Use different amounts of alum in each of the six beakers. Use a uniform spread that

_____________________ Water System

Jar Testing Log — Time ________

Test No. ________________ — Date ________

Raw pH ____________________

Raw Alkalinity ____________________

Raw Turbidity ____________________

Jar No.	1	2	3	4	5	6
Alum (mg/L)						
Lime (mg/L)						
Soda Ash (mg/L)						
Potassium Permanganate (mg/L)						
Other (mg/L)						
Polymer (mg/L) Type ________						
Polymer (mg/L) Type ________						
Treated pH						
Results						

NOTES:

Alum ☐ Bulk Liquid Basis ☐ Dry Granular Basis

Polymer A ☐ Bulk Liquid Basis ☐ Dry Powder Basis

Polymer B ☐ Bulk Liquid Basis ☐ Dry Granular Basis

FIGURE A-4. Jar testing log

brackets the anticipated result; i.e., 2, 4, 6, 8, 10, 12 mg/L. (In some cases, it may be desirable to keep one jar as a blank, with no chemical.)

3. Stir rapidly for 30–60 sec, then slowly for 5–15 min, depending on water temperature.
4. Raise the stirring paddles and allow to settle for 10–30 min.
5. Observe and record the results. (Poor/good settling, good/fair color, murky.) The proper alum dosage will produce a visible floc that settles, leaving a clear liquid on top. A typical form is included in Figure A-4. Typical jar testing results are shown in Figure A-5.
6. Measure and record the pH of the liquid on the top of each jar.
7. There is normally a fairly narrow pH range for proper coagulation with alum (Figure A-6). The pH is normally between 6.3 and 6.8 but can be lower. Raw water with a high pH may coagulate slightly higher but should never be more than 7.0. Above that, aluminum begins to dissolve and pass through the filters.

Jar No.	1	2	3	4	5	6
Alum (mg/L)	2	4	6	8	10	12
Results	Poor	Poor	Fair	Fair	Poor	Poor
pH	7.0	6.8	6.6	6.4	6.2	6.0

FIGURE A-5. Typical jar testing results

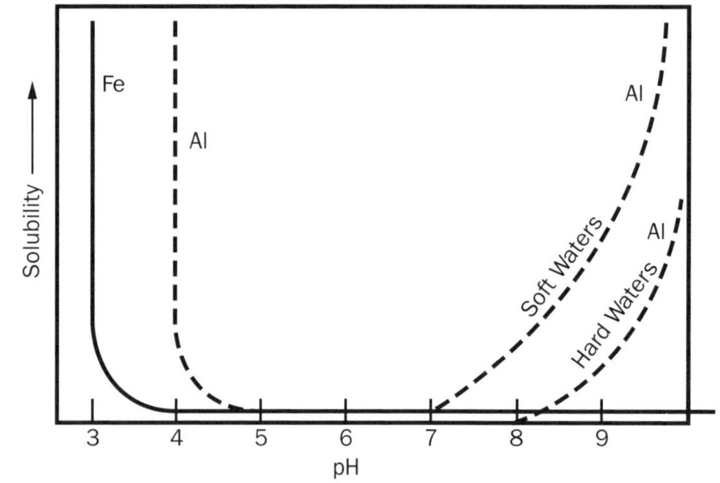

NOTES:
1. 1 mg/L of alum will destroy 0.5 mg/L of alkalinity.
2. 1 mg/L of alum will produce 0.44 mg/L of CO_2.

FIGURE A-6. Solubility curves

8. If the coagulation results from Step 5 are not good, more alum may be required. If the pH in the beakers drops off rapidly from one jar to another, lime or soda ash may be required to buffer the water. Buffering will be especially necessary with low alkalinity water typical of coastal regions or mountain streams.
9. If the pH has dropped off substantially, repeat Steps 1–6 with the same alum dosage and a small amount of lime or soda ash.
10. If the pH range is good and results are not, add more alum and repeat Steps 1–6. If the pH didn't drop off before, it may this time. In this case, repeat the test again with the new alum dosages and more lime or soda ash.

Notes:

1. It may take many jar tests to determine the proper dosages. Keep records of all tests. It is recommended that all jar test results be kept in a binder and sorted by raw pH (6.0, 6.1, 6.2 … 7.0, etc.). A history of your plant will be developed, allowing quicker changes in the chemical dosages when characteristics of the raw water change. Even if historical records are used, the proper feed rates should be verified by jar tests.
2. Never vary more than one chemical at a time in the same jar test. A systematic approach will be faster in the long run.
3. For waters with high raw water pH, it will be necessary to depress the pH by adding more alum, or it may be necessary to switch to a different type of coagulant.
4. For alum coagulation, the coagulation and filtration pH must be in the range of 5–7. Otherwise, soluble aluminum will pass through the plant.

High Quality Raw Water

If the raw water quality is very high, the results from jar tests may not be visible. In this case, a vacuum filter test may be required. The filterability of the water is the ultimate test of the proper chemical feed rates.

1. Decant the clear liquid (supernatant) off the top of the beakers.
2. Run the sample through a vacuum filter or through folded filter paper and a funnel.
3. Measure the turbidity of the filtrate. If it is good, the test is complete. If not, additional jar tests are required.

Polymer Procedures

1. Once the proper alum and lime/soda ash dosages have been determined, various amounts of different polymers can be tried. Use the proper amount of alum and lime/soda ash in each jar and vary the amount of polymer.
2. Polymer dosages should not normally exceed 2 mg/L on a bulk liquid basis. For dry polymer, the dosage should probably not exceed 0.1–0.2 mg/L on a dry weight basis. Verify the maximum allowable dosages with the manufacturer.
3. Caution: Do not overfeed polymers. An excess can glue the filter media together. Figures A-7 and A-8 illustrate examples of excess polymer.
4. Refer also to the manufacturer's recommendations for maximum allowable feed rates. Refer to the section on Chemical Calculations for a discussion of dry weight versus bulk liquid calculations.

FIGURE A-7. Example of excess polymer in a package plant clarifier

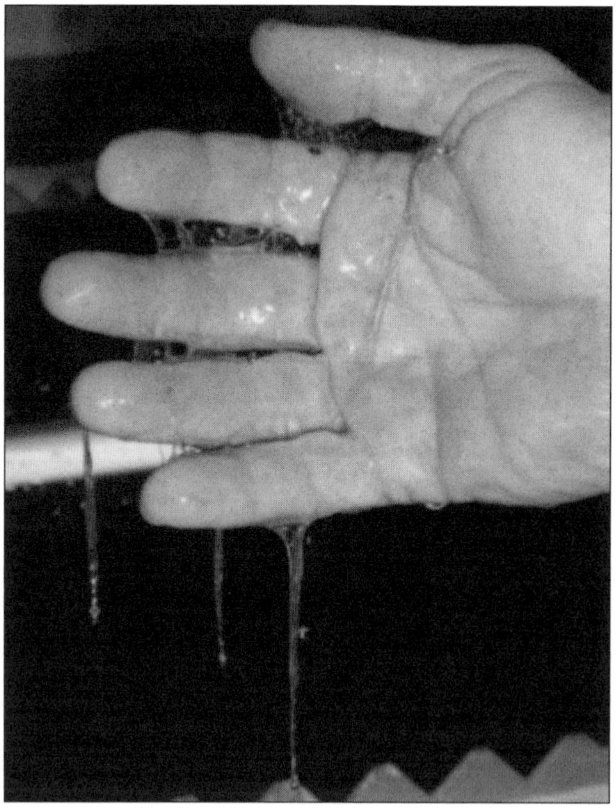

FIGURE A-8. Example of excess polymer in a filter

Chemical Calculations

1. Once the desired chemical feed rates have been determined by jar testing, it is necessary to convert those values to pumping rates.
2. Calculate 1 ppm (mg/L) based on a full day's production for simplicity, even if the plant does not operate that long. There are two methods of making calculations depending on the type of chemical used, either by using dry weight or on a bulk liquid basis.

Bulk Liquid Alum

$$1 \text{ mg/L} = \frac{\text{plant flow (gpm)} \times 1{,}440 \text{ (min/day)}}{1{,}000{,}000} = \underline{\hspace{2cm}} \text{ gpd bulk liquid alum}$$

- The actual concentration of alum may vary with a maximum of approximately 499. However, it should be noted that the actual concentration of the bulk liquid does not matter. If the bulk liquid used for jar tests is the same as the plant uses, the concentration cancels out in the calculations, and it results in simpler computations.

Dry Alum, Powdered Polymer, Lime, or Soda Ash

$$1 \text{ mg/L} = \frac{\text{flow (gpm)} \times 1{,}440 \text{ (min/day)}}{1{,}000{,}000 \times \underline{\hspace{1cm}} \text{ dry chemical concentration (lb/gal of water)}}$$

$$= \underline{\hspace{3cm}} \text{ gpd of mixed chemical}$$

For example, assuming a plant flow of 700 gpm and a soda ash mix concentration of 1/2 lb to 1 gal water (6 percent) would yield the following results:

$$1 \text{ mg/L soda ash} = \frac{700 \text{ gpm} \times 1{,}440 \text{ min/day}}{1{,}000{,}000 \times 0.06 \text{ mix concentration}}$$

$$= \frac{1 \text{ mgd}}{1{,}000{,}000 \times 0.06}$$

1 mg/L soda ash = 16.67 gpd with a 6% mix concentration

Calculate Chemical Flow

- Assume a plant flow of 700 gpm (1 mgd)
- Assume bulk liquid alum
- mg/L from jar test—assume 8 mg/L based on bulk liquid
- 1 mg/L bulk liquid alum = $^{1\text{ mgd}}/_{1{,}000{,}000}$ = 1 gpd bulk liquid alum
- Daily chemical flow = 8 ppm × 1 gpd = 8 gpd

Calculate Alum Pump Setting

- Assume 24-gpd pump
- Pump setting = $\frac{\text{daily chemical flow (gpd)}}{\text{pump capacity (gpd)}} = \frac{8}{24} = 33\%$

NOTE: If the pump has a speed and stroke setting, they must be multiplied together to give the above result.

$$\begin{aligned} \text{Pump setting (\%)} &= \text{speed} \times \text{stroke} \\ &= 57\% \times 57\% \\ &= 33\% \text{ overall} \end{aligned}$$

APPENDIX B

Troubleshooting Checklists

TABLE B-1: Define the problem

	Yes	No
Are the filter runs short because of high effluent turbidity?	☐	☐
Are the filter runs short because of rapid buildup of head loss?	☐	☐
Are the filter runs short because of rapid increase of particle counts?	☐	☐
Is the effluent turbidity not as good as it should be?	☐	☐
Are there visible problems in the filter?	☐	☐
Is the air distribution not as good as it should be (if air scour is used)?	☐	☐
Are there irregularities in the media surface?	☐	☐
Is the filter media not being cleaned properly?	☐	☐

TABLE B-2: Flash mixing

	Yes	No
Has the flash mixing equipment broken down?	☐	☐
Is flash mixing equipment needed?	☐	☐
Is the energy input less than 3 ft of water head?	☐	☐

TABLE B-3: Coagulation/flocculation

	Yes	No
Is the contact time for chemicals short?	☐	☐
Are more chemicals required than indicated by jar tests?	☐	☐
Do jar tests indicate that flocculation is needed?	☐	☐
Is the floc (if visible) entering the clarifier visually different from that obtained in jar tests?	☐	☐

TABLE B-4: Calibrating chemical feed pumps

	Yes	No
Are calibration cylinders required?	☐	☐
Do the pump settings change with time?	☐	☐

TABLE B-5: Jar testing

	Yes	No
Are jar tests being done less than weekly?	☐	☐
Do the jar tests results differ from how the plant works?	☐	☐
Is the raw water pH subject to change?	☐	☐
Is the raw water turbidity subject to rapid changes?	☐	☐
Does the jar test yield different results from the streaming current monitors (if used)?	☐	☐
Is the primary coagulant or polymer less efficient during different seasons of the year?	☐	☐

TABLE B-6: Clarifier design summary

	Yes	No
Are the collection troughs more than 6 ft apart?	☐	☐
Is the overflow rate higher than accepted standards?	☐	☐
Are significant amounts of solids overflowing the clarifier?	☐	☐
Are the filter runs shorter than they should be (significantly less than 24 hr)?	☐	☐
Is the clarifier effluent turbidity greater than 1 ntu?	☐	☐
Does the clarifier operate in plug-flow mode?	☐	☐
Are there collection troughs at only one end?	☐	☐
Are there any clarifier equipment problems or failures?	☐	☐

TABLE B-7: Observe normal clarifier operation

	Yes	No
Is there free fall between the sedimentation basin/clarifier and the filter?	☐	☐
Is there visible floc overflowing from the sedimentation basin/clarifier?	☐	☐
Is the floc over the filter the same size and volume as that in the sedimentation basin/clarifier?	☐	☐
Is the filter effluent quality poor?	☐	☐
Is there visible contamination in the sedimentation basin/clarifier discharge or on the surface of the filter (wood chips, pine needles, leaves, plastic, etc.)?	☐	☐
Is more water coming over the trough weirs in some areas than others?	☐	☐

TABLE B-8: Observe filter backwash

	Yes	No
1. General		
Are clean areas and dirty areas visible on the water surface during backwash?	☐	☐
Are bursts of air or boils visible on the water surface?	☐	☐
Are fine bubbles widespread across the filter during high-rate wash?	☐	☐
Is the apparent backwash rate too high?	☐	☐
Too low?	☐	☐
Is there any visible loss of media?	☐	☐
Are there any apparent problems with the sequence of events or controls?	☐	☐
Was there floating scum or other contaminants that did not wash out of the filter?	☐	☐
Is the amount of media expansion too high?	☐	☐
Too low?	☐	☐
Is any uncontrolled air apparent during the high-rate backwash?	☐	☐
Did the high-rate backwash flow taper off gradually at the end of the cycle?	☐	☐
Was the backwash wastewater still dirty when the cycle was terminated?	☐	☐
Is the distance between the troughs too great?	☐	☐
Is air released at the beginning of low-rate wash (if used)?	☐	☐
Is additional air released at the beginning of high-rate wash?	☐	☐
2. Surface Wash		
Was there any problem with the rotation of the surface wash sweep?	☐	☐
Was any media loss apparent during surface wash?	☐	☐
Were there any problems with the surface wash sequence? (Refer to Chapter 7)	☐	☐
3. Air Scour (if used)		
Did the air come on "hard?"	☐	☐
Did the air come on in one corner or end of the filter and move to the other end?	☐	☐
Was there any jump of water level just before the air came on?	☐	☐
Was there any irregularity of the air pattern?	☐	☐
Did the air turn off before the low-rate water wash started?	☐	☐
Were substantial amounts of air still coming through the media when the water began weiring over the wash trough?	☐	☐
Did the air turn off with the water level too high?	☐	☐
Did the air turn off with the water level too low?	☐	☐
Were there still high amounts of air in the water when the backwash flow ramped up to the high rate?	☐	☐
Was there any boiling or irregularity of air distribution during the concurrent wash step (if used)?	☐	☐

TABLE B-9: Analyze media on flat surface

	Yes	No
Are contaminants present below the surface?	☐	☐
Is the media sticky?	☐	☐
Is the media dirty?	☐	☐
Are there significant amounts of sand fines up in the anthracite?	☐	☐
Is there a significant accumulation of coarse grains of anthracite?	☐	☐
Is there significant intermixing between the anthracite and the sand?	☐	☐
Have significant amounts of anthracite, sand, or garnet sand (if used) been lost?	☐	☐
Are the grains of anthracite worn smooth?	☐	☐

TABLE B-10: Inspect media surface

	Yes	No
Is the surface of the media irregular?	☐	☐
If so, does the surface of the media have large high and low areas?	☐	☐
Are ridges visible?	☐	☐
Are mounds visible?	☐	☐
Are sloped areas visible?	☐	☐
Are craters or "rat holes" visible?	☐	☐
Are cracks visible?	☐	☐
Are there depressions along the walls?	☐	☐
Are there residual solids or contaminants on the media surface?	☐	☐
Does the media need to be skimmed?	☐	☐
Are there clean areas and dirty areas?	☐	☐
Are mudballs present?	☐	☐
If so, are they in concentrated areas?	☐	☐
Are the mudballs large? Record all observations as shown in Chapter 5.	☐	☐
In a filter with surface wash, are the unswept areas still dirty?	☐	☐

TABLE B-11: Limited excavation

	Yes	No
Is there any irregularity in the depth of the various media layers?	☐	☐
Is the interface between the media layers irregular?	☐	☐
Are there mudballs in the media?	☐	☐
Are there mudballs on the surface of the gravel?	☐	☐
Does the surface of the gravel have grains from lower levels?	☐	☐
Is the surface of the gravel irregular?	☐	☐

TABLE B-12: Flume and clearwell

	Yes	No
Is there media in the trap (if provided)?	☐	☐
Is media present in the manifold or flume?	☐	☐
Is there more than a trace?	☐	☐
Is there a large amount (a cubic foot or more)?	☐	☐
Is support gravel present (if used)?	☐	☐

TABLE B-13: Inspection results

	Yes	No
Are there pretreatment problems?	☐	☐
Is there significant contamination in the filter?	☐	☐
Are there any structural failures?	☐	☐
Is the gravel upset?	☐	☐
Is the filter media fouled or upset?	☐	☐
Are the failures slight?	☐	☐
Are the failures significant?	☐	☐
Does the failure appear to be caused by:		
• Hydraulic shock?	☐	☐
• Air scour?	☐	☐
• Uncontrolled air?	☐	☐
• Something else?	☐	☐

TABLE B-14: Hydraulic shock

	Yes	No
Do pressure gauges on the backwash piping fluctuate rapidly?	☐	☐
Are any of the valves or controls malfunctioning?	☐	☐
Is the backwash pump allowed to come on with the valves open?	☐	☐
Is air still discharging from the release valves when backwash water is flowing into the filter?	☐	☐
Does the water level get low enough in the clearwell to allow a vortex to be generated next to the backwash pump?	☐	☐
Are there surges in the pressure gauges on the backwash system?	☐	☐
Do the valves travel from full open to full closed in less than 60 sec?	☐	☐
Do the rate control valves hunt-and-seek excessively?	☐	☐
Is the actuator linkage on the valves worn, broken, or loose?	☐	☐

TABLE B-15: Uncontrolled air

	Yes	No
Is uncontrolled air visible during backwash?	☐	☐
Are the air release valves leaking water?	☐	☐
Do the air release valves allow air in?	☐	☐
Do any of the pipe fittings or appurtenances leak?	☐	☐
Are additional air release valves needed?	☐	☐
Do the controls need to be modified to eliminate air?	☐	☐
If a gravity backwash is used, is a vortex baffle needed?	☐	☐

TABLE B-16: Surface wash

	Yes	No
Are some of the nozzles not working?	☐	☐
Have any of the caps come off the nozzles?	☐	☐
Are there large unswept areas where the media is not being cleaned?	☐	☐
Has it been longer than a year since the bearings were checked?	☐	☐
Are the sweep arms out of level?	☐	☐
Do the sweep arms have trouble rotating?	☐	☐
Is the surface of the media more than 2 in. below the center line of the sweep arms?	☐	☐

TABLE B-17: J-riser tests

	Yes	No
Is the initial J-riser test uneven?	☐	☐
Is the final J-riser test uneven?	☐	☐
Does the air come on hard?	☐	☐
Does the air pattern change over time?	☐	☐
Does the pressure relief valve activate and reset during air scour?	☐	☐

TABLE B-18: Underdrain failures

	Yes	No
Are craters apparent along the sidewalls of false deck underdrains?	☐	☐
Has the clean bed head loss increased over time?	☐	☐
Is the air pattern uneven?	☐	☐
Is there media in the flume or media trap?	☐	☐
Are there grout failures between rows of underdrain blocks?	☐	☐
Is calcification a potential problem?	☐	☐
Are there gaps behind the seal angles?	☐	☐
Are the joints ruptured between underdrain blocks in the same row?	☐	☐
Was the underdrain installed on a very hot or cold day?	☐	☐

APPENDIX C

Historical Records

Date: ______________________

Raw Water Quality			
	Turbidity	________	ntu
	pH	________	
	Alkalinity	________	mg/L
	Water Temperature	________	°F (°C)
	Particle Count (optional)	________	
Plant Flow Rate	________ gpm (mgd)		
Chemical Feed Rates			
	Polymer 1	________	mg/L
	Polymer 2	________	mg/L
	Coagulant	________	mg/L
	Carbon	________	mg/L
	Potassium Permanganate	________	mg/L
	Soda Ash	________	mg/L
	Lime		mg/L
	Sodium Hydroxide	________	mg/L
	Chlorine	________	mg/L
	Other	________	mg/L
	Clarifier Effluent Turbidity	________	ntu
	Filter Effluent Turbidity	________	ntu
	Run Time Between Backwashes	________	hr
	Particle Count (optional)	________	
	Clean Bed Head Loss	________	ft/in.
	Terminal Head Loss	________	ft/in.
Final Effluent			
	Turbidity	________	ntu
	Chlorine Residual	________	mg/L
	pH	________	

Bibliography

Hudson, Herbert E., Jr. 1981. *Water Clarification Processes Practical Design and Evaluation,* Van Nostrand Reinhold Co., New York City, New York

Amburgey, Amirtharajah, Brouckaert, and Spivey, "An Enhanced Backwashing Technique for Improving Filter Ripening," *Journal AWWA*, December 2003.

Index

Note: f. indicates figure; t. indicates table.